IS E.T. HERE?

No Politically But Yes
Scientifically and Theologically

IS E.T. HERE ?

No Politically But Yes
Scientifically and Theologically

6-15-07.
A must to lend.
CHul

Forewords
by

Peter Gersten, J.D.
Executive Director of Citizens Against UFO Secrecy (CAUS) and
trial attorney for Betty Cash et al, in her case against the US
Government — the only lawsuit in history for damage caused by
citizen contact with a UFO

&

Dr. Peter Redpath
Institute of Advanced Philosophical Research, and Professor,
Department of Philosophy, St. John's University

Robert Trundle, Ph.D.

EcceNova Editions
Victoria, BC

For copyright licenses, please contact: Access Copyright, 1 Yonge Street, Suite 1900, Toronto ON M5E 1E5.
www.accesscopyright.ca

Library and Archives Canada Cataloguing in Publication

Trundle, Robert C., 1943-
Is ET here? : no politically but yes scientifically and
theologically / Robert Trundle ; forewords by Peter Gersten &
Peter Redpath.

Includes bibliographical references and index.
ISBN 0-9735341-2-5

1. Unidentified flying objects—Sightings and encounters.
2. Human-alien encounters. 3. Life on other planets. I. Title.

CB156.T78 2005 001.942 C2004-905521-6

Cover image: *Unwanted Interlopers*, Courtesy and © 2004 James Neff

"Robert Trundle has provided all open-minded people with an important service, by so effectively presenting and evaluating crucial evidence related to the UFO phenomenon. With his knowledge of philosophy of science, epistemology, and logic, Trundle is especially well situated to erode the justifications used by many mainstream scientists, journalists, and opinion-shapers when they ignore or attack credible testimony and physical evidence of artificial flying objects not produced by human beings. This well-written book will appeal especially to those interested in a philosophically and technically sophisticated treatment of a phenomenon that defies any easy explanation."

Professor Michael E. Zimmerman, Department of Philosophy, Tulane University, New Orleans. Author of "The Alien Abduction Phenomenon: Forbidden Knowledge of Hidden Events," Philosophy Today (1997)

"Dr. Trundle's book could be titled simply *Is ET Here? Yes Legally* — despite his important explication of the egregious politics. His political analysis strips naked the base motives of hostile witnesses, his philosophy of science eviscerates scientific objections to our testimony, and his theological insights bear on a jury's most profound hopes and fears."

Peter Gersten, J.D., Executive Director of Citizens Against UFO Secrecy (CAUS) and trial attorney for Betty Cash et al, in her case against the US Government — the only lawsuit in history for damage caused by citizen contact with a UFO

"Dr. Trundle's work ... does us a profound service by both debunking the myth surrounding UFOs and seeking to make a study of them more acceptable for scientific, philosophical, and theological considerations. Western intellectuals certainly owe him a fair hearing. For if our scientific past teaches us anything, it is that those who call the Copernicuses of their time 'crazy' often wind up being the ones to whom history more permanently attaches this same epithet."

Dr. Peter Redpath, Institute of Advanced Philosophical Research and Professor, Department of Philosophy, St. John's University

Astronomer Carl Sagan believed that modern physics was not discovered centuries earlier because it was deterred by the superstitions of religion, suggesting that ET's advanced civilization would be due to an unbiased pursuit of science. But science may be related logically to religion in terms of which a religio-scientific belief is shared by aliens and humans. Human pride and an imposition of politics on science are the best explanations for impeded scientific progress and a close-mindedness to one of the most astonishing possibilities in history: that ET is here! While this case is made by the philosophy of science, there is enough science to exclude glib scientific rejoinders and enough philosophy to show that UFOs can violate our scientific theories when their truth is restricted to certain domains. Until there is this defense of ET's possible presence, as evidenced by the very observation that corroborates theories, prevalent confusion will persist on the part of both scientists and the public: On the one hand, the public will think its belief is bad science and many scientists will reinforce that false thought by comparing UFO belief to an irrational belief in religion. On the other, religion and science will be muddled by those who interpret either angels as aliens or aliens as angels. The former interpretation reduces the supernatural to natural phenomena and the latter elevates that phenomena to the supernatural.

Emerging Awareness
of the Problem

"The US Government" claims "that although not all UFO reports can be explained, there is no evidence that Earth has been visited by aliens. Most scientists and leading journalists agree with this.... However, these same scientists believe that there must be many advanced civilizations on planets orbiting the billions of stars they estimate to exist in the universe. The gap between these two positions is generally explained by the assumed inability of even the most advanced society to travel the enormous distances separating the Earth from even the nearest stars. Yet there are thousands of sightings of novel, high-performance craft in our skies, reported by highly skilled and experienced observers. There are also hundreds of other reports of... humanoid beings in their vicinity."
Rockefeller Report. Directed by Dr. Peter Sturrock, Prof. Emeritus of Applied Physics and Emeritus Director – Space Science & Astrophysics, Stanford University.

Meanwhile, scientists "are laboring along progressively more torturous trails... in convoluted attempts to accommodate [anomalous phenomena such as UFOs and strange creatures]." We must redefine "sacred scientific tenets, such as rigid replicability" and permit foreign concepts such as "teleological causality." In regard to science, the scientific community is more concerned with "its politics" and "administration than with advancement of its basic understanding."
Dr. Robert Jahn, Prof. Aerospace Science, Dean Emeritus of the Mech. & Aerospace Engin. Dept., Princeton University, "20th & 21st Century Science," JSE 15 (2001).

Courageously pursuing understanding and truth is as necessary for the fulfillment of human nature as it is for survival. But since the 1960s, there has been a *cowardly* deference to a political correctness in the university. "The university is politicized, the politicizers say. But they do not recoil, appalled at their conclusion that every scholar deep down is a politician... No, they embrace it. They furthermore say, 'It's necessary to replace the politics we've had up to now with our politics, or, rather, my politics.' This is a claim of tyranny, somewhat disguised by the demand... to democratize everything. " We "must begin to oppose things we professors have allowed to happen.... We mustn't let things get by we know are wrong; we must start to raise a little hell."
Dr. Harvey Mansfield, Frank G. Thomson Professor of Government, Harvard University. Remarks delivered to the Salvatori Fellows on 26 June 1991 in Washington, D.C.

Contents

Scientific Smugness? • Examples of Complacency • The Need for a Proper Modesty • *On the One Hand* • *On the Other hand* • The Roots of Extraterrestrials as Myth • Epistemological Totalitarianism • A Peculiar Backdrop for Aliens as Gods • *The One Extreme* • *The Other Extreme* • The Gods Must be Crazy • Creation and Metaphysics Re-evaluated • A Supreme Being of Aliens and Humans • Metaphysics of ET?

The Claims of Adamski • *The Credibility of his Claims* • *Interpreting the Claims* • *The Human Form Reconsidered* • Subsequent Claims About the Anatomy • *A First-Hand Report?* • *Descriptions by Abductees* • *Are There Many Alien Species?* • Reports of Extraterrestrial Technology • *Descriptions in Some Case Studies* • *Alleged Inner Features of Extraterrestrial Spacecraft* • *Reports of Additional Technologies* • *A Top Secret/MAJIC Report, April 1954* • *Reasons the Technology "Cannot Exist"* • *Behind the Story of Why the Technology "Cannot Exist"*

8 Since the Missions: Controversy Over Religion and Science 223
US Air Force Academy: Physics 370 • *Unidentified Flying Objects* • *33.1 Descriptions* • *33.2 Operational Domains* • *33.3 Some Theories as to the Nature of the UFO...* • *33.4 Human Fear and Hostility* • *33.5 Attempts at Scientific Approaches* • Does ET Share Our Religio-Scientific Belief? • A Final Sign of Hope? • Politics, God and Science

Foreword

by Peter A. Gersten

FROM 1971 TO 1994 I was a criminal defense trial attorney in New York City. During the day I was in court arguing to a jury that they shouldn't believe four eyewitnesses who testified that it was my client who shot the victim. But that was easy compared to what I was involved with after business hours. I represented the public-activist group Citizens Against UFO Secrecy (CAUS) which sought to pry loose UFO documents from government agencies under the Freedom of Information Act (FOIA). Trying to convince federal court judges that the CIA, FBI, NSA, Army, Navy, Air Force, Department of Defense, and FAA (to name only a few) possessed UFO documents was much more difficult than getting a murderer acquitted. Certainly, it is always easier to prove that someone is lying than to prove that someone is

Engaged in practicing law in New York City for over 20 years as a prosecutor and criminal defense trial attorney, and still licensed with over 30 years of legal experience, Peter Gersten, J.D., is the Executive Director of Citizens Against UFO Secrecy (CAUS). The latter is a nonprofit organization which holds that there is overwhelming legal evidence that Earth is being visited by one or more species of nonhuman intelligence and that citizens have a right to know. His efforts have led to the following: ∗ His being Chief Counsel for plaintiffs in the case of Betty Cash *et al* V. the US Government — the only lawsuit in history for damages caused by a UFO ∗ Release by the CIA of over 900 pages of UFO documents ∗ Appeal to the US Supreme Court, enjoining the National Security Agency to release 135 UFO documents ∗ A case pending against the Department of the Army over information on alien bodies and autopsies ∗ A case pending against the Defense Department over information on triangular UFOs that are seen continually in our skies.

telling the truth. And with the subject of UFOs the truth is completely irrelevant.

To give you some idea of how ridiculed the subject of UFOs is consider the following: Only one credible witness is legally sufficient for a conviction in a murder case and sometimes mere circumstantial evidence is sufficient if it is convincing beyond a reasonable doubt. But when it comes to eyewitness reports of strange craft in our skies hundreds, if not thousands, of witnesses are inadequate to establish the existence of UFOs. Whereas during the day Jane Doe could testify in court about seeing a crime and be believed, if on her way home she reports seeing an unidentified flying object she will be derided for seeing nonexistent things.

⋀

There is overwhelming legally admissible evidence that our species is interacting with another form of intelligence and UFOs represent the primary manifestation of that contact.

⋁

During my 30-year law practice I have participated in hundreds of criminal hearings and trials. I am familiar with the evidence allowable in a court of law. Presently, there is overwhelming admissible legal evidence that our species is interacting with an alien form of intelligence and UFOs represent the primary mode of that contact. I can prove this in a court of law beyond a reasonable doubt. Indeed, Dr. Trundle's book could be titled simply *Is ET Here? Yes Legally* — despite his important explication of the egregious politics. His political analysis strips naked the base motives of hostile witnesses, his philosophy of science eviscerates scientific objections to our testimony, and his theological insights bear on a jury's most profound hopes and fears. Eyewitness testimony, photos, experts, government documents, and even scientific reports are all admissible, if relevant, in a court of law. Not surprisingly, there is both qualitative and quantitative evidence to prove the existence of UFOs. The book you are about to read discusses this evidence in all of its intriguing manifestations.

So why, with all the stunning proof, have geometric shaped craft, orbs, spheres, saucers, rods, fireballs, and other UFOs been formally ignored for over 50 years? Why after all this time do their identity, origin, and purpose remain unknown to the public? The answer to that question can be found more in the attitudes and beliefs of our species than in the surreptitious nature of the alien objects.

Though the US government has possessed UFO documents for over 50 years and is presently still withholding hundreds of UFO documents because of national security concerns, it continues to deny that it has any interest in the subject. In this age of terrorism and in light of the attack on 9-11, the government's position on strange aerial objects appears, to say the least, to be misleading. But government is synonymous with secrecy and thus we can not expect the truth in matters of 'national security'.*

So where is the public expected to look for explanations and answers? The obvious place is within our scientific community because some UFOs, with their unconventional performance characteristics, most likely represent another form of intelligence, possibly even extraterrestrial. And while science should be a constant exploration of the normal and paranormal, the latter is mostly neglected for fear of the same scorn and ridicule as that suffered by UFO witnesses. Thus science's responsibility to explain what is unexplainable, but which is in plain sight, is overwhelmed by its need to stay within certain 'acceptable parameters'. But then again how does a scientist understand, let alone explain, an aerial craft that was created by a super-advanced intelligence? Possibly by first admitting that the object exists.

Has the status quo been institutionalized to the point that any evidence of extraterrestrial craft in our skies is so threatening that they must be suppressed at all costs? If not, then Dr. Trundle's excellent analysis and interpretation will hopefully inspire a revitalized UFO trek for the truth.

---

* *Author's note:* The national security issue brings to mind the now much admired give-'em-hell President Harry Truman. He established the CIA in the same month and year as the Roswell UFO crash in July 1947. Decades later he said that "Secrecy and a free democratic government don't mix." Did this express the mellowed naïveté of an old man or hindsight of a wiser one? *Cf.* Merle Miller, *Plain Speaking* (Berkley Publishing, 1973), p. 392.

Foreword

by Peter A. Redpath

I N THE 13TH CENTURY, St. Thomas Aquinas wisely cautioned Christians not to claim to have natural knowledge of either the existence or nature of *beings* that exceed the power of our natural reason to grasp. For example, he held that Christians should not pretend to know, by reason and experience, that God is a Trinity or that the universe was created in time as opposed to having a perpetual existence — then, the received view of science.[1]

In both instances, his reasoning was similar. Since we cannot positively know God's essence, we cannot properly define Him. Our inability to define Him positively prevents us from making indisputable claims of knowing He is a Trinity. Knowledge of what something *is* is a necessary element of evidence. And thus evidence cannot be adduced for knowing by natural reason whether God is, or is not, a Trinity of persons since He cannot be properly defined.

The legion of honors conferred upon Peter Redpath, Ph.D., Professor of Philosophy at St. John's University, range from his election to the Board of Trustees of the Institute for Advanced Philosophical Research to being a panelist for the Carnegie Council on Ethics & International Affairs as well as being invited by Assist. Secretary of State Elliot Abrams to participate in the International Conference on Religious Liberty at the U.S. State Department, Washington D.C. Professor Redpath has authored countless books and articles, is a member of the Board of Editors of the Dutch publishing house Editions Rodopi, Editorial Advisor to *Contemporary Philosophy*, past Vice President of the American Maritain Association, and contributor to a host of proceedings. These include the International Congress on Patristic, Medieval, & Renaissance Studies at Villanova University and the International Association of NeoLatin Studies at St. Andrews University in Scotland.

Dr. Robert Trundle knows a good deal about the intellectual ending of this historical tale and helps modern science avoid making the same mistakes that religious believers made in the Middle Ages.

⋎

Because definitions abstract from individual instances and consider universals, they abstract from the here and now. In demonstrations, we predicate the definitions abstractly and universally of spatiotemporal conditions as being everywhere and always true of their subjects. Without possessing a proper definition nothing can be irrefutably known or demonstrated since a thing's definition is one of the principle's of incontrovertible knowledge. And thus we cannot know or demonstrate either the eternity or noneternity of the *created* universe since, even if we can establish that it has no intrinsic necessity to exist through its own makeup, this lack does not imply that "human beings, the heavens, or a stone did not always exist" through the support of an outside power that has the necessity to exist by the intrinsic ability of its essence.[2]

Thomas chided religious believers who try to demonstrate the Trinity's existence since they derogate their faith in two ways. First, they diminish its supernatural status by lowering the ontological condition of God to human understanding. Our intellects can recognize only beings whose natures are related directly or indirectly to the sensible and material. This is what St. Paul means in *Hebrews* 1:11 when he says that faith is of the unseen. Thus we can admit of the unseen supernatural God only by grace through a supernatural elevation of our mind. Second, when believers attempt to prove as irrefutable what are in principle refutable, when they seek to demonstrate as non-mysterious the mysterious, they can only do so by arguments that lack either soundness or cogency. And this provides disbelievers with ammunition to ridicule believers.[3]

Sadly, many believers during and after the Renaissance and Enlightenment lacked Aquinas's wisdom. He always sought to avoid crediting too much or too little to our powers in this world. He discerned that we are prone to exaggerate and underestimate, and that error lies largely in these two tendencies. Lacking his prudence, some Renaissance humanists, following Francesco Petrarch, were more concerned with elevating the *political* status of poetry and rhetoric in the growing Academy than they were with *truth*. They violated the known sensorial limits of natural objects in a sophistic grab for power. In so doing, they appealed to William of Ockham's nominalism, thereby boosting superstition, subverting science, and blurring the natural and supernatural orders of being. Galileo Galilei properly rebuked them for violating the 'rights' of physical nature, the mind, and mathematical science as well as for restricting the imagination and reasoning of intellectuals to see farther than their predecessors.

Unhappily, Galileo exaggerated the power of mathematics by limiting reality to phenomena responsive to it. He claimed hyperbolically, about his book *Nature*, that it was written in a language of math and that only mathematicians could read it. Happily, Dr. Robert Trundle knows a good deal about the intellectual ending of this historical tale and helps modern science avoid making the same mistakes that believers made in the Middle Ages and the sophists in Galileo's time. Trundle knows that philosophy never established a metaphysical basis for modern science. A dangerous situation, indeed.

Readers unfamiliar with metaphysics might wonder why I make such a critical observation. Simple: Unlike physics or mathematics that address *only some things* for establishing truths based upon principles derived from definitions of their subject, metaphysics involves a study of the subject's truth conditions and the underlying principles of *everything* insofar as this can be done through classification. Consequently, the classifying scope of metaphysics can assess the nature and limits of physics but not *vice versa*. That is, what is true about everything in general is true about some things but what is true about some things, as things in physics, is not necessarily true about everything in general.

Trundle also understands that political correctness, like other forms of fideism or naïve fundamentalism, precludes reasonable studies from an examination in the Academy.

Before starting to do physics, metaphysics indicates that we must know some things that are generally true about everything. If we did not, we could not reason. For example, we know that effects are unequal to causes; that causes cannot be absent when there are events; that order presupposes privation; that negating a judgment is not identical to positing it; that necessity is different from contingency; that extremes are not the same thing as midpoints; that not all opposites are contradictories; and that contradictory and contrary opposites cannot both be true simultaneously. These sorts of truths are denied at our peril.

Precisely, this is what modern scientists began to do when, following the

Father of Modern Philosophy René Descartes, they tried to base modern physics on principles of doubts and dreams instead of on the sensible apprehension of physical things. At the beginning of his *Metafusikh (Metaphysics)*, Aristotle wisely shows us how all our experiential, artistic, and scientific thinking presupposes our prior acceptance of metaphysical truths.[4] None of us can reason about either universe or our place in it unless we have some idea of who we are, where we are, how we got here, and where we are headed.

This is so true that the renowned priest-physicist Stanley L. Jaki has been easily able to trace the rise of modern physics in the West to the metaphysical conviction that the universe is created and mono-directional in time, not existentially and temporally cyclic.[5] Aristotle's *Metaphysics* intimates the need to establish a divine dependency for scientific knowledge when he argues for finite explanations in science. No science in principle may study an actually infinite number of beings through an actually infinite number of principles. Were it to attempt to do so, it could never establish the truth of anything and would lack any rightful claim to the name 'science'. Any science must study a finite class of things in light of a finite number of principles that essentially and necessarily relate to a subject. Thus to accept a subject of science is to accept a reality of limits, or finitude, about what is studied and the principles that are used to study it.[6] Something creationists easily do.

Much to the detriment of today's higher learning, and for many past generations to boot, Kant and his intellectual epigonies were able to pull off an intellectual sleight of hand without a peep.

No wonder then that Aristotle sought to limit physical and metaphysical explanations by appeal to an Unmoved Mover, just as Plato did by a world of Forms and the Good. No wonder too that Ptolemy's *Almagest* presupposes Aristotle's metaphysics — as does Claudius Galen's *On the Natural Faculties*, that Galileo was a believing Catholic, and that Sir Isaac Newton prided himself more as a theologian than physicist and structured his *Philosophiae Naturalis Principia Mathematica (Mathematical Principles of Natural Philosophy)* as an elaborate argument for the existence of God. Even modern and contemporary evolutionists presuppose, metaphysically, evidence of a final causality in their assumptions about ever increasing orders of perfection in being, a providential order intrinsic to the being of every finite thing.

Trundle recognizes that the sophists of our Enlightenment tried to go beyond Descartes by impossible dreams of a boundless system of empirical science. Physics was thereby confused with metaphysics. As he shows, this explains the rise of murderous twentieth-century projects: a pseudo-scientific Marxism, an equally irrational national socialism of Nazism, and now myopic ideologies of terrorism and a decline of higher education in the fascistic hands of politically correct utopians who suppress dissent for being 'anti-science'. Trundle realizes that political correctness, as other forms of fideism or naïve fundamentalism, bans reasonable studies in the Academy. By reducing metaphysics to physics, truths about everything in general are reduced to truths about some things in particular. Kant was a grand master at using this trick in the eighteenth and early nineteenth centuries. Shrinking scientific reason to an anthropocentric Newtonian world fit perfectly with his reactionary restriction of knowledge to particular sensations *a posteriori*. Neat trick, if you can pull it off.

Much to the detriment of today's higher learning, and for many past generations to boot, Kant and his intellectual epigonies were able to pull off this intellectual sleight of hand without a peep. By so doing, they imperiously excluded from a domain of rightful philosophic and scientific examination many things upon which we are entitled to reflect. The time has come to jettison from serious consideration a *backward progressiveness* of Enlightenment sophists.

Does ET actually exist? I have no idea. I do know that for centuries, from ancient pagan to Medieval and Renaissance Europe, philosophers and theologians allowed for intelligent beings who exist in and between planets. This speculation was part of the metaphysico-astronomical considerations of the circular universe of Aristotle and Ptolemy. I know also that due to discoveries in modern physics, those bold conjectures are now excluded. Discoveries in physics made previous metaphysical truths seem unreasonable. Analogously, today's more expansive physics of the universe prevents us from straightjacketing metaphysics within Kantian dreams that limit *genuine* scientific enlightenment. The time has come for the Academy to scrap the arbitrary limits that the Enlightenment sophists have placed upon reason.

⅄

Today we know that nothing is metaphysically or physically impossible about the existence of extraterrestrial beings, even their traveling here! Religious believers cannot demonstrate the nonexistence of ET by appeal to mysteries of faith.

⅄

True, Einsteinian physics precludes travel faster than light. Interstellar travel requires speeds approaching or exceeding it. But Trundle rightly questions slavish acceptance of Einstein as the last word, especially in light of recent advances. Physics has discovered forces below the level of our ordinary sense experience: a strong force to explain fusion, fission and bonding; an electromagnetic force; a weak force in radioactive decay; and gravity. We are aware of these in virtue of a mathematical, negative, and analogous reasoning that resembles some of the best work in Medieval negative theology, metaphysics, and mysticism. Thus Trundle is right to stress a possible spiritual cause of the Shroud of Turin and ET as a possible cause of atypical phenomena. And if phenomena can exceed the speed of light, then, given a history scientific progress, interstellar travel in short amounts of time is theoretically viable.

If modern physics reveals anything, it is that the universe is far more mysterious than Descartes, Hume, Kant and their 'enlightened disciples' ever dreamed. Attempts to remove physics from metaphysical experience is impossible. Worse, are attempts to reduce metaphysics to physics. Metaphysics is a theoretical study about what must be universally true. Its principles cross-fertilize with other sciences. For this reason metaphysics began in the West as a conflation of mythical poetry and philosophical physics. The more we discover about our universe, the more we tend at one time to reduce, at another to expand, the metaphysical domain. This understanding is a rule of historical philosophical experience.

At one time the heavens were the subject of speculative thought. This is no longer so. Today, we know that nothing is metaphysically or physically impossible about the existence of extraterrestrial beings, even their traveling here! Religious believers cannot demonstrate ET's nonexistence by appeal to mysteries of faith. Discoveries in physics increasingly suggest that interstellar space travel is a determinate possibility. Such being the case, as strange as Dr. Trundle's work may appear to some intellectuals, he does us a profound service by both debunking the myths about UFOs and seeking to make their study more tenable. Scientists, philosophers, and theologians certainly owe him a fair hearing. For if our scientific past teaches us anything, it is that those who call the Copernicuses of their time 'crazy' often wind up being the ones to whom history more permanently attaches this same epithet.

Notes

[1] St. Thomas Aquinas, *Summa Theologiae*, 1, 32, 1 and 1, 46, 2 respondeo.

[2] *Ibid.*

[3] *Ibid.*

[4] Aristotle, *Metaphysics*, 98^{a1}–98^{a25}.

[5] Stanley L. Jaki, *Science & Creation: From Eternal Cycles to an Oscillating Universe* (Edinburgh: Scottish Academic Press, 1986). See also Paul Haffner, *Creation & Scientific Creativity: A Study in the Thought of S.L. Jaki* (Front Royal, VA: Christendom Press, 1991).

[6] Aristotle, *Metaphysics*, 993^{b1}–1012^{b30}.

Former Army Air Force Lt. Walter Haut (left) and author Dr. Robert Trundle (right) in the summer 1997. Lieutenant Haut was the Public Information Officer at Roswell Army Air Base who reported — the first official report in history — that a UFO had been recovered by the military on 8 July 1947. Haut, a personable and articulate man, told Robert Trundle during a two-hour interview that he and intelligence officer Maj. Jesse Marcel were having drinks in a lounge in Roswell in the summer of 1975, talking about the UFO crash, when Marcel became very animated in insisting that the debris he found was that from a UFO. Haut told the author that he had no doubt that the crash occurred and that he had been given direct orders by Col. William Blanchard to report the crash to the local news media.

Preface

ODDLY, MY FIRST UFO book, *UFOs: Politics, God & Science,* was the first philosophy book ever written on a likelihood that ET is here. This new work embellishes the old one with illustrations as well as augments it with fresh facts and analyses. While the analyses address some of the most controversial people in the field, from Robert Lazar to George Adamski, new points are raised for consideration without a naïve acceptance of their claims.

In virtue of his alleged contact with one species of ET, for example, Adamski claimed that the human form is virtually universal. While this universality was derided by scientists who assumed vast genetic differences, it is consistent with a cosmological principle accepted universally by the scientific community. And intriguingly, this principle is echoed by purported intelligence documents that state, in regard to a crashed UFO: "The laws of physics and genetics may have a genesis in a higher structured order than once previously thought." That this thought would be questioned does itself reveal a remarkable degree of thoughtfulness about an import of the crash, adding credibility to the document's veracity.

And besides the fact that attacks on Lazar's credentials amount to an *ad hominem* when they are used to dismiss his account of how allegedly recovered UFOs bypass the speed of light, the accepted light speed is now debatable. In Princeton, New Jersey, NEC physicists made the startling announcement that they conducted a duplicated experiment in which a light pulse behaves as if it takes a negative time to traverse the distance along its direction of travel. Their work stunned the scientific community into questioning interpretations of our most cherished laws of physics, if not challenged the physics itself. Reports on their research invariably state that scientists are starting to accept feasible "interstellar space travel."

In point of fact, travel to Earth by ET cannot in principle be dismissed by science. Taken with thousands of reliable reports of novel high-performance craft and nearby humanoid beings, that fact renders illogical the dismissal.[1] That is, the dismissal is at odds with both the reports and well established principles in the philosophy of science such as a *pessimistic induction*: In virtue of the history of science in which every major theory was superseded by another that limited its truth to a domain, as Einstein's theory restricted Newton's to phenomena that did not approach the speed of light, it is *inductively* reasonable to hold that either more advanced ETs can visit us now or we can visit them in the future due to a *pessimism* about the total truth of Einstein's physics. Thus although this previously unquestioned physics excluded the travel because of the accepted light speed, it was probable that limits such as that speed would be surpassed.

The surpassed light speed, among other recent scientific revolutions, would pose perplexing questions. But the latter do not obviate the fact that we can show philosophically that human psychology, biology and physics need not be entirely false to permit the truth of more advanced theories that could afford the fantastic. But this fact is tempered by the sobering point that revitalized proofs of a First Cause indicate that theology is related logically to science with the import that a more mature alien science and culture may evoke ET's belief in Nature's God. Do we and ET share a theologico-scientific belief? Certainly, our most cherished religious beliefs need not be endangered. Harvard Medical School's John E. Mack, M.D., states that the French COMETA (Committee for In-depth Studies) challenges "the notion that accepting the reality of UFO phenomenon would be a threat to religion." COMETA's elite scientists, government leaders, and military officials ask: "Would religions founded on the existence of a God... be shattered by such apparitions? Nothing is less certain." The Earth's great religions do not deny "other inhabited worlds."[2]

Notes

[1] This support of ET's presence on earth is made by Stanford University Professor Emeritus of Physics Peter Sturrock in the *Rockefeller Report* (1997).

[2] Letter to Dr. Robert Trundle from John E. Mack, M.D., on August 29, 2000.

About the Author

ROBERT TRUNDLE received his Ph.D. at the University of Colorado, was Outstanding Junior Professor for Teaching and Scholarship at his university, was awarded early tenure and full professorship, is a member of the New York Academy of Sciences, the Scientific Research Society of Sigma Xi, the Federation of American Scientists, the American Philosophical Association, and the Phi Kappa Phi Honor Society. He was nominated for *Marquis Who's Who in American Education*, *Marquis Who's Who in the Southwest*, *Who's Who Among American Teachers*, *Directory of American Scholars* (Gale Group), *Contemporary Authors*, and *Dictionary of International Biography* (Cambridge). He has been an invited referee for the journals *Philosophy of Science* (Philosophy of Science Association), the Université Laval's *Laval Théologique et Philosophique*, and *Dialogue: Canadian Philosophical Review* (official journal of the Canadian Philosophical Association). Among various lectures, he was invited by physicist Dr. Maria Falbo-Kenkel to address the Scientific Research Society on how theology is related logically to natural science. His research includes over 40 articles for the following: *Bulletin Ind. Inst. of the History of Medicine* submitted with Michael Vossmeyer, M.D., of the CC Hospital Medical Center, *Method & Science: Study of the Empirical Foundations of Science* in The Netherlands, *Res Publica* at Leuven University's Political Science Institute, *Studies in Conflict & Terrorism* at RAND Corporation, *Aquinas* at the Pontificia Università Lateranense (Vatican City), *Logique et Analyse* at the Belgium Research Center in Logic, *The National Forum* of the Phi Kappa Phi Honor Society, *Philosophy in Science* edited by William Stoeger, S.J., Ph.D. in Astrophysics at Cambridge University (Kraków Papal Academy of Theology, Vatican Astronomical Observatory & the University of Arizona), *Thought: A Review of Culture* at Fordham University, *Sensus Communis: Research on Alethic Logic* in Italy, *Journal of Business Ethics* at the University of North British Columbia, *Review Journal of Philosophy & Social Science* at Arizona State University, *Idealistic Studies* at Clark University, *The Modern Schoolman* at St. Louis University, *Epistemologia* at Genoa University, *Dialogos* at the University of Puerto Rico, and invited articles such as one in *New Horizons in the Philosophy of Science* (1994), edited by Dr. David Lamb at the Medical School of Birmingham University, England. Professor Trundle's books include *Illustrated News of the Unbelievable* (2005) with Maj. George Filer, U.S. Air Force Intelligence (Ret), *Camus' Answer: No to the Western Pharisees* (2002), *UFOs: Politics, God & Science* (2001), *From Physics to Politics* (1999), *Medieval Modal Logic and Science* (1998), *Ancient Greek Philosophy* (1994), and *Beyond Absurdity* (1986) with R. Puligandla, Ph.D. Philosophy, M.S. Physics.

1

ET and Terrestrial Politics

I S ET HERE? I am a philosophy professor who sometimes chats with my colleagues about 'weird things'. In the course of discussing them, one faculty member said she wished super-advanced ETs were visiting earth in UFOs. Their presence would offer a hope for the human condition that rivaled religion. A student, by contrast, said she did not believe in the visits because an alien intelligence would conflict with Christianity. She believed that we alone are made in God's image and that this image centers on a human mind and body which uniquely resemble Jesus.

However, even traditional views of the Bible may support a consistency of Judeo-Christianity with extraterrestrial UFOs. And a messianic view of aliens that rivals religion may *not* be consistent with their intentions. An important point is that if it is respectable for philosophers to discuss religion and God's existence, then it seems at least equally fitting to consider the existence of alien UFOs and their implications.

What would be the implications of UFOs for religion and science? While the deep-space probe *Pioneer 10* was launched with earthbound instructions for aliens almost as many years ago as the start of NASA's radiotelescopic *Search for Extra-Terrestrial Intelligence (SETI)*, any philosophic attention to UFOs seems to have as much credibility among scholars as Santa Claus. The analogy to Santa Claus is an apt one since that name was allegedly the US astronauts' code word for UFOs and since this icon is largely relegated to myth, along with religion, by a politics of the academic community.

On the one hand, academia is rooted in modern intellectual revolutions far more open to dramatic possibilities. After publishing his wave theory of light in *Traite de la luminere*, physicist Christiaan Huygens inferred in 1690 that there are intelligences on other planets from the Copernican revolution in astronomy. Our planet is not the central heavenly body and is nothing special.[1]

A

H.G. Wells suggested that contact with advanced extraterrestrials might change the end of Human history as it was conceived by Karl Marx. Lenin looked at him thoughtfully and finally admitted: "You are right."

Y

Later, Immanuel Kant modeled his philosophic revolution on the Copernican theory that the earth moves around the sun. He held our mind also 'moves' in the sense that it is active in the acquisition of knowledge. That is, knowledge involves interpretation due to a mental activity. And notwithstanding a failure to exploit the point, there was a suggestion not only that this activity may be peculiar to a psychobiological nature of the human species but that other species in other solar systems might have different interpretations that spawn more advanced sciences.

On the other hand, the scientific and philosophical revolutions spurred rabid political responses. Despite a rise of democracy during the industrial revolution, these revolutions embodied anthropocentric worldviews at odds with extraterrestrial issues. Whereas Marxian science posed a dialectical determinism for liberating the world's workers, fascism exalted a mystical Superman *(Übermensch)* for exploiting workers and defending nationalism. Though fascism embraced a human-centered new world order and radicalized elements of conservatism, Marxism promised a utopian end of human history and became an idealistic extreme of liberalism.

However, although both extremes committed unprecedented atrocities and politicized science—as Einstein's theory was derided as 'Jewish physics' by the Nazis and 'Bourgeois physics' by the Soviets,[2] the Soviets became our ally in World War II after its pact with Hitler was broken. Thus its mass murders were never exposed by war-crime trials such as one of the fascists at Nuremberg. The lack of a Nuremberg for the Left led many liberal academics to delude themselves about a Marxian 'scientific rationality'.[3] And in being as reactionary as religious belief, a belief in UFOs would be equally inimical to a progressive politics.

Tellingly, the renowned Oxford university philosopher Anthony Kenny notes that, with the dissolution of the Soviet Empire, a survival of Marxist ideology depends on "its devotees in universities of the West."[4] Though a conservative status quo had hindered UFO studies, we may presume that a typical Western professor would ignore ET's visit even if it was the most astonishing event in human history. "I started out," admits one professor, "conforming to all the expected academic dogmas of liberalism. I was a

ИОСИФ ВИССАРИОНОВИЧ
СТАЛИН

While inconsistent with the anthropocentric ideology of Marxsm, Stalin took seriously the report of a UFO crash in New Mexico by his spies at Los Alamos. His interest contrasted to a studious disinterest of the US academic community.

Marxist... feminist" who studied the "'Christian right' for the same reason I studied serial killers.", Indeed, it is an understatement to say that many academics in the ivory tower stress a politicized *terrestrial* progress.

A progressive politics of academic liberals collides head-on with traditional religion and any other 'unearthly' concern. In echoing liberal journalist Wendy Kaminer's *Sleeping with Extraterrestrials*, which exorcises our silly fascination with "aliens... and virgin birth of Jesus Christ," we may reasonably suppose that the liberal professor would augment the exorcism by exposing more sinister motives behind a 'fascination' of her less anthropocentric and more otherworldly colleagues: intentions that are "economic, psychological or political."[6]

Ironically, a politicization of religion and the UFO issue, in open societies, contrasts to an earlier candor of Lenin when he spoke to H.G. Wells. Wells suggested that contact with aliens by a development of human technology might change the end of history as conceived by Marx. "The Marxist conception itself," he said to Lenin, "would then become meaningless." Lenin admitted "You are right... All human conceptions are on a scale of our planet. They are based on the pretension that the technical potential... will never exceed the *'terrestrial limit'*."[7]

⋀

"Although dreams of interstellar voyages fueled by antigravity are... a half century away, the idea is now in the scientific mainstream," admit NASA scientists.

⋎

Again, the foregoing is not a mere polemic against liberal academics. A conservatism of the 1950s was equally fatal to any inquiry into UFOs. The point is that a liberal to left outlook in academia since the mid-1960s, when UFO sightings began to challenge our scientific theories, has largely shaped its intellectual *Weltanschauung* and focused that worldview on practical agenda. The agenda of liberationist educators at all levels, notes Carol Iannone, Editor of *Academic Questions*, was meant "to drain the curriculum of academic content.... If you ever looked at recent furors over condom distribution and 'self-esteem' and wondered how on earth the schools had gotten involved in such things to begin with, this is the reason."[8] And though many new fields in higher education such as business ethics or ethnic studies contribute valuable insights, they belie antipathy towards the 'starry heavens', as Kant might have put it, as well as towards theory.

Other academics are quick to provide a theoretical apologetics for both an ideological obsession with practical progress and a studious disregard of ET. In the evident confusion of physical possibilities with ones excluded logically in *How to Think About Weird Things*, professors Theodore Schick and Lewis Vaughn express alarm in a flier: "Do your students believe in *weird* things... UFOs?" By rejecting physical possibilities that conflict with our science, some professors promote an uncritical belief of many scientists.

Many scientists suppose that it is not physically possible for interstellar UFOs to be here, since although it is logically possible, their presence would conflict with our physics. One theory of physics holds that the speed of light is not exceeded. Thus aside from either time travel from the future or extra-dimensional visitors from parallel universes, which are even more bizarre, aliens from another solar system would have to travel many tens of thousands of years from even the nearest star system. The problem alone of traveling to even this close star system is a major source of skepticism.

Skepticism is expressed by the SETI Institute whose elite scientists included the late Cornell astronomer Carl Sagan. UFOs are rejected out of hand since the distance to the nearest star "is over 4 light years. That's about 24 trillion miles away.... This poses a daunting engineering problem even for a more advanced civilization."[9] But equally qualified physicists at a prestigious US military academy caution their future pilots about assuming that our knowledge is complete. For example, consider the laws 1) all actions must have opposite and equal reactions, 2) all particles attract all others with a force proportional to a product of the masses and inversely as the square of the distance, 3) energy, mass, and momentum are conserved, 4) no material body can have a speed as great as c, the speed of light in free space, and 5) the maximum energy E that can be obtained from a body at rest is $E = mc^2$ where m is the rest mass of the body.

These physicists note that, while laws 1 and 3 seem fairly safe, only 3 is valid now from a relativistic perspective. And that perspective alone warrants laws 4 and 5. Yet relativistic physics completely revised our concepts after 1915. For a much longer period, Newton's physics functioned effectively and reined supreme! Further, they add, it needs to be said that general relativity, despite its exaltation, has not yet been fully verified. Consequently, civilian

4

"Scientists are more closed-minded on the subject of UFOs than the general public." ~ Dr. Bernard Haisch, Director, California Institute for Physics & Astrophysics, Editor – *Astrophysical Journal,* and staff scientist at Lockheed Palo Alto Research Lab.

Philosopher Neal Grossman, at the University of Illinois, says "it has taken me twenty years to gain the courage to be able to reply simply and honestly to questions pertaining to my interests. The taboo against having any interest in the paranormal except for... debunking it has persisted in academia." The "punishment for violating this taboo is to be marginalized by colleagues. My fear... has now left me."

scientists find themselves in an odd position of having to certify the impossibility of alien controlled UFOs by such laws:

> ... yet three of the laws are recent in concept and may not even be valid. Also, law number 2 has not been tested under conditions of large relative speeds or accelerations. We should not deny the *possibility* of alien control of UFOs on the basis of preconceived notions not established as... relevant to UFOs.[10]

Since those scientists who *have studied* UFOs disagree, the agreement of a majority of other scientists is irrelevant. And insofar as the relevant experts lack expertise in philosophy, even their voice is limited. One philosophical principle renders probable that future scientific theories will permit what is today deemed impossible. Even today, the possibilities include space voyages that are interstellar. "Although dreams of interstellar voyages fueled by antigravity are a half century away, the idea... is now in the scientific mainstream," admit scientists at NASA.[11] Does NASA deny a reverse possibility for extraterrestrials?

Evidently, the possibility that aliens could come here cannot be made public or there is a denial that aliens could be fifty years more advanced. In regard to ET's presence made public, secrecy may be due to a fear of panic, security concerns about crashed UFO technology falling into the wrong hands, and a government's understandable refusal to inform citizens that it cannot protect them from a superior science of aliens whose intentions remain unclear. In regard to how advanced aliens might be, let us hope for the sake of even a scintilla of humility that no one denies that aliens could be fifty years in advance — if not in advance by millions of years. Since their presence *is not* excluded by our science and *is* reported by eminent witnesses, ignoring them is not merely at odds with a life of the mind: It is insensitive to what may pose the gravest threat or most beneficial contact in human history.

Some recent history, herein, illuminates these provocative alternatives. But it would be a sham to address them if seemingly fantastic reports were not vividly detailed. Accordingly, the serious reader is being cautioned now about these detailed reports which, the author holds, *cannot* be rejected out of hand. Even if there is only an iota of likelihood that nonhumans are visiting earth, no society could regard that possibility with indifference. Indifference and a psychological state of denial are rife, however, among those who should be the most epistemologically liberal. Predictably, cavalier rejections will come from academics to whom, sadly, an ancient saying applies: *Damnant quod no intelligunt.* Meaning "they condemn what they do not understand," many scholars will pedantically profess that these reports are fit only for a pop culture of pseudo-science. For example, a professor with one publisher said of my work: "This reviewer is somewhat confused as to why a psycho-clinician

Image of an interstellar spacecraft, displayed by NASA, with a "negative energy" induction ring that is inspired by theories on space being warped with the energy for hyperfast travel. NASA has a Breakthrough Propulsion Physics Project for 1) propulsion systems that have no conventional propellant mass, 2) propulsion for maximum physically possible speeds, and (3) a breakthrough energy production for power.

has been asked to evaluate a manuscript about extraterrestrial intelligence written by a philosopher who dabbles in science fiction." And with a sarcasm that would amuse those who have endured an academic toadyism that panders to scholastic insecurities, this psychologist adds: "Perhaps, I thought, what is really being asked for is a diagnostic surmise as to why a serious scholar should immerse himself in pseudo-science. Frankly, were it not the case that Dr. Trundle has published two earlier books [with that particular publisher], I would have been tempted to regard the assignment as a bit of a prank."[12]

But although claims about aliens will disconcert many academics, if not the religious community, the scholars have themselves often acquiesced shamefully to a political correctness that has subverted the university's traditional mission of pursuing truth. So, despite both an unpopularity of my attempt to uncover some truth and a risk that this work may seem patronizing, I am not inclined to worry about offending academic sensibilities in my endeavor to be substantive. This book is not subordinated to either an academic politics or egocentric psychoclinicians and gives new life to well established principles of philosophy. Stunningly, philosophical principles are applied to religio-scientific beliefs that, notwithstanding most science-fiction writers, aliens and humans may actually share. The shared beliefs would displace a prosaic view that aliens who perform virtually miraculous feats are the new substitutes for angels or God! And the principles include a novel mode of reasoning which, without paradox, should make us confident as well as skeptical about today's science.

Given a scientific *progress* over the ages, the unqualified truth of any theory is reasonably doubted. Today, there should be doubt about one that excludes UFOs or even bizarre reports of their occupants. Finally, in mainstream science itself there are in principle inconsistent but equally good accounts of all phenomena in terms of which there may be extraterrestrial hypotheses. These apply to the most intriguing prospect we have ever faced. On a more pedestrian note, when we must finally face an official news release of ET's presence, this book will be amply rewarded by all the professors who are heard muttering "Well, I'll be damned!"[13]

Notes

[1] See "Life on other planets a possibility," Holland 1690, *Millennium Year by Year* (London: DK Publishing, 1999), p. 310.

[2] See R. Trundle, *From Physics to Politics* (NJ: Transaction Publishers/Rutgers, The State University, 1999), pp. 128-133.

[3] See S. Courtois, Ed., "Introduction," Tr. by M. Kramer, *The Black Book of Communism: Crimes, Terror, Repression* (Harvard University Press, 2000).

[4] Anthony Kenny, Ed., *The Oxford History of Western Philosophy* (Oxford: Oxford University Press, 1994), p. 368.

[5] Interview of Professor Kerry Jacoby, Ph.D., Perdue University, *Our Sunday Visitor* 87 (1999) 6. Jacoby now rejects her former feminist views. Sadly, however, for a scholarly assessment that supports their being typical of higher education in America, which has become a gold standard for the international academic community, see Christina Sommers' "Argumentum Ad Feminam," *Journal of Social Philosophy* 22 (1991) 5-19, as well as Daphne Patai and Noretta Koertege's *Professing Feminism: Education and Indoctrination in Women's Studies* (Lexington, 2003). Themselves former professors of Women's Studies with expertise in literature and philosophy of science, Patai and Koertge note that a 'social constructionism' of mainstream academic feminism ignores biological differences of men and women (*Biodenial*), that it is the engine behind political correctness, and that it incoherently holds 'truth' to be relative to whoever has political power — in the past dead white males who constructed concepts such as 'gender' and 'truth' to enhance their power to dominate. Paradoxically, the feminist idea that 'truth' is relative to whoever dominates — domination

now being sought by feminists — does itself have its roots in the dead white male Karl Marx: "it is only since Marx that this relativism has gained the intellectual substance to win them broad support [142]." Thus as Marx said philosophy should not just interpret the world but change it, changing it is the aim of politically correct feminists who reject objections *a priori* as patriarchal social constructions. These include traditional male-dominated sciences and 'otherworldly' concerns irrelevant to their agenda.

[6] See N. Vincent's "Unbelievable," *Nat. Rev.* Vol. 51, No. 22, 1999, p. 54, for the quote on Kaminer and *Our Sunday Visitor*, p. 6, for Jacoby.

[7] Alexandrov, V., *L'ours et la Belaine*, Stock, Paris, quoted in Jacques Vallée's *Anatomy of a Phenomenon* (London: Neville Spearman, 1966), p. 114. From Timothy Good 's *Above Top Secret* (NY: William Morrow, 1988), pp. 248, 249, emphasis.

[8] Carol Iannone, "They'll Never Learn," *Nat. Rev.* Vol. 52, N. 17 (2000) 57. American students, being *less* educated, are *more* receptive to indoctrination and are now last among industrialized nations in math and science. Besides threatening their economic survival and requiring foreign students to do the research in US universities (see "Security Impedes University Research: [Foreign] Students' Visas Denied" in the *Cincinnati Enquirer* on 7 Sept. 2003), the concern at hand is that dumbing down students cripples their appreciation of science and how scientific progress is spurred by 'weird' anomalies. The most pronounced anomalies include UFOs, say eminent physicists from Princeton's Robert Jahn to Stanford's Peter Sturrock. Nevertheless, serious attention to UFOs remains politically incorrect since its 'otherworldliness' lessens attention to political agenda for changing *this* world. Accordingly, ET is banished to inane cartoon parodies on TV for children.

[9] The SETI Institute Website, 6 March 1999.

[10] See Maj. D. G. Carpenter, Maj. L. A. King, Dr. P. D. Lowman et al, per the Department of Physics, Physics Class 370, "Chapter 33: Unidentified Flying Objects," *Introductory Space Science*, Vol. II (Colorado Springs, CO: US Air Force Academy, 1968), p. 465.

[11] Robert Boyd, "Universe may be under power of a 'dark force'," *Knight Ridder News Service*, 25 June 1999, p. D4.

[12] These remarks were sent from Transaction Publishers – Rutgers, The State University, in March 2000, with whom I published *From Physics to Politics* (1999, 2nd Edition 2001). The Editorial Director, to his credit, admits that the reviewer was "clearly not sympathetic" and that "Given the wide interest and enormous amount of writings in the area of unidentified flying objects, it is important to reach out beyond polemics...."

[13] *Cf.* Nick Pope, Secretariat (Air Staff), British Ministry of Defense, *Open Skies, Closed Minds* (NY: Dell Publishing Group, 1998), p. 236.

2

Science on Angels and Aliens

THERE HAVE BEEN two main sorts of scientific input into the UFO controversy. Sadly, in being influenced by the politics of higher education, both have been overwhelmingly negative. Before addressing this negativism and the extremes of either elevating extraterrestrials to angels or lowering angels to extraterrestrials, if not consigning belief in both to illusion, there is attention to a complacency of most scientists amid the dispute.

Scientific Smugness?

The Catholic priest Stanley L. Jaki, Ph.D. Physics and Ph.D. Theology, Distinguished University Professor at Seton Hall University at Princeton, NJ, notes in regard to the 20th High Energy Physics Conference at the University of Wisconsin in 1980 and the 25th Conference in 1990 that a majority of the physicists were looking for things to do when physics was over since further major discoveries were discounted.[1] He ties the attitude to world-renowned scientists who include Planck, Rutherford, Schrödinger and Bohr who, as great as they were, were "prone to believe... that they had formulated the last word in physics, or were within striking distance of it."[2]

It is one thing for the apparent complacency to be shaken by recent developments in quantum cosmology, chaos theory, and familiar phenomena in contexts that "cannot possibly exist" in terms of well established theories. It is another thing for these theories to be challenged by bizarre UFO phenomena that reveal point blank their fallibility. Since the problem of fallibility was not even remotely entertained in the above developments, it is not difficult to illustrate the complacency in academia from Hawaii to Oxford.

Computer-enhanced image of the Shroud of Turin. Though most scientists believe that UFOs are natural phenomena, despite the fact that they remain a mystery going into the 21st century, religious phenomena are equally anomalous. One anomaly is this Shroud that shows, what some believe is, the body of Christ. STURP scientists concluded that no paints, pigments, dyes or stains caused the image. The only plausible explanation that they have been able to deduce is that some form of unknown radiation was flashed *from the body* with enough intensity to chemically alter the fiber 0.2 millimeters in length. In a lecture on the Shroud in the early 1980s by scientists at the US Air Force Academy who investigated it, the author heard several of them state that they had been converted to religious belief.

Examples of Complacency

Oxford University physicist David Deutsch affirms that there can be time travel by Closed Time-like Curves (CTCs) that are perfectly compatible with general physics. A complacency is evident insofar as he supposes that ET exists in the future with a far more advanced science but would not visit today's earthlings because we are at a primitive stage of technological fire. Professor Deutsch not only seems to assume that our fire-stage science is the cutting edge paradigm in the present universe, since otherwise visits could be made today by ET, but disregards reports of a probable alien presence. The presence is given credence in the United Kingdom by notable officials who include Nick Pope, the former Secretariat (Air Staff) British MoD, and Admiral of the Fleet Lord-Hill Norton GCB, Chief of Defense Staff (Ret). Indeed, Professor Deutsch says that his theory explains "why we have not yet encountered any extraterrestrials!"[3]

Moreover, today's chaos theories seem to challenge a Newtonian-Einsteinian cosmology that is deterministic. But while determinism alone is accepted by Deutsch, University of Hawaii astrophysicist Victor Stenger argues that the indeterministic chaos of an inflationary Big-Bang Theory is rooted in accepted theories. These include Einstein's general relativity for relating a protonic-size black hole's energy to a mass of our universe ($E = mc^2$) and Heisenberg's Uncertainty Principle for explicating the hole's fluctuation at Planck time, that violate "no principle of physics."[4] And although the impossible rate of spin of a pulsar at Supernova 1987A led a physicist to admit of "incontrovertible evidence of something that cannot exist,"[5] this major anomaly did not lead any astronomer in *The Astrophysical Journal* to even momentarily express any fundamental inadequacies with our current understanding of the cosmos.[6]

Given surprises which the cosmos has had in store for scientists over the ages, philosophers and scientists should wonder about the rationality of such a complacency. What is its import for an open-minded scientific curiosity which sparks scientific revolutions? A belief there will be new revolutions, if not quantum leaps in progress by possible contact with an alien intelligence, is based on the reasoning of a Pessimistic Induction which is well known to philosophers of science. The history of science renders inductively reasonable the notion that any major theory, at any historical time, will be replaced by a new theory's revolution. Scientific progress features a historical development of revolutions wherein things deemed impossible by current theory, which in the twentieth century includes UFOs exceeding the speed of light, yield to new theories which allow for the possibility.

The Need for a Proper Modesty

Modesty, based on a history of scientific revolutions, is also supported by logical problems with a complacent attitude. Despite Schick and Vaughn who ask "Do your students believe in *weird* things... UFOs?,"[7] philosophers of science should not be as dismayed as scientists by UFOs that execute maneuvers inconsistent with current science. Most of these philosophers are familiar with the Underdetermination-of-Theory-by-Data (UTD) Thesis which specifies that any data or phenomena are subject to logically inconsistent but empirically equivalent theories. Simply stated, the Thesis brings to mind two detectives with opposing theories which equally explain the forensic data of witnesses.

Dr. Hynek, Professor of Astronomy at Northwestern University, noted that the mere mention of UFOs in academia, in any open-minded way, would incur "the equivalent of scientific tar and feathers."

By analogy, UFO data afford inconsistent interpretations. These range from social-science theories which specify that UFO observers underwent mass delusion and, paradoxically, that delusion was not undergone to competing theories of physics that equally explicate a dynamics and kinematics of UFOs as natural phenomena and, inconsistently, as phenomenal machines which really exist. It would miss the point to object that the existence is impossible since an interstellar origin means the speed of light was exceeded. The point is that besides a rational pessimism about knowing *in toto* a truth of our science, scientific theories do not in principle have the last word on what is real; much less on certifying that reality is exhaustively explicable by a theory.

This brings us back to negative responses on UFOs by academics. On the one hand, some professors have apparently kept their eyes open, on behalf of the government, to snitch on colleagues who are pursuing the question. On the other, the question is utterly ridiculed by other professors. The latter have often tolerated an unparalleled political correctness in the name of 'science'. At the same time they have derided their UFO-minded colleagues in terms of what one eminent scientist calls an "epistemological totalitarianism."[8]

On the One Hand...

The late Lt. Col. Philip Corso (Ret.), former acting Chief of the US Army's Foreign Technology Division at the Pentagon, notes that select professors were part of the Advanced Research Projects Agency (ARPA). Formed in response to the Soviet Union's launch of Sputnik in 1957, the ARPA is said to have begun using academics not only for research but for networking among their peers to report on those who did not disclose to the government various sorts of forbidden inquiries. The latter included technology on UFOs and aliens, called "extraterrestrial biological entities" (EBEs) and on theories about either their existence or their intentions. "In other words, in addition to being a conduit for research," said Corso, "ARPA was another intelligence-gathering agency."[9]

Many of Corso's reports on academia are not new. What makes them intriguing is his authoritative background. Admittedly, the background would be scorned by many baby-boomer professors. They would be those born from 1946 to 1964, notes one cultural critic, who are "a certain type of liberal." Though having exalted mind-altering drugs in the era of Woodstock, he adds wryly that they now "pay fealty to their youthful ideals by demanding no-smoking signs at the Barnes & Noble."[10] But in addition to their preoccupation with political agenda in *this* world, these countercultural boomers are also products of an anti-war movement who viscerally hate the military and skew its history.[11] These points bear on their predictable lack of responsiveness to Corso's disclosure. Having been an intelligence officer on Gen. MacArthur's staff in Korea and member of Eisenhower's National Security Council, he won nineteen decorations and ribbons for meritorious service over many decades. Finally, after his retirement, he served Senators James Eastland and Strom Thurmond as a staff specialist for national security.

With a caveat that anyone collaborating with Corso on his book could have altered his reports, his long held posts, patriotism, and heroic military

The Department of the Army's DA Form 66 for Lt. Col. Corso that, while blacking out personal/ classified information, confirms his military record and position to know classified information on UFOs.

achievements strengthen the case for their general reliability. Besides an integrity necessary for them, it would be irrational for him to discredit his own reputation and career by going on record with a phony story about a 'Whitewater' cover-up of cosmic proportions. A spirit of official trust was not violated if, as he says, national security was no longer at stake — at least to the degree of warranting a cover-up — in virtue of alien threats having been neutralized by recent events. These, as noted later in more detail, include the Strategic Defense Initiative.

Thus it is significant that, in addition to his forestland knowledge of academic ARPA members, Corso suggests that they promoted the idea that an actual UFO crash in 1947 at Roswell was merely a myth. So effective was the promotion by the professors that a technology development consultant, Dr. Paul Fredericks, reportedly reacted to a government request to analyze the crashed UFO debris by asserting "Funny, but I always thought Roswell was a kind of legend."[12] The term 'legend', in alternatively denoting mythology, is related to another face of the academic community.

On the Other Hand...

At the 93rd American Anthropological Association Meeting in December 1994, Dr. Benson Saler presented "Explanation, Conspiracy & Extraterrestrials: The Roswell Incident." The Incident of a crashed UFO was said to be a mere myth.[13] Dr. James Hopgood sought my response. After noting that Saler committed a fallacy of *suppressed evidence* by ignoring facts against his case, I was challenged as missing the point. The point was my apparent misunderstanding of what 'mythology' means in social science. In closing this section with my continued respect and admiration for Dr.

Hopgood, a copy of my follow-up letter to him is quoted. It goes to the heart of one obfuscation in the academic community:

December 6, 1994

Dear Jim,

Thank you for the article on the Roswell UFO crash. Let me say that I didn't miss your point about scientific investigations of mythology — that they don't question evidence for cultural stories. This accords with a standard social-science view that "the importance of a myth is not its truth... inaccuracy is irrelevant" (per *Sociology*, 1990, T. and S. Dunn, for instance).

My point was that the author not only ignores facts on Roswell but argues that "UFO aficionados" convert evidence against it "into support." Thus he cannot say that accuracy is irrelevant if he argues against it. And hence he is not examining myth in a neutral textbook way despite his verification reasoning wherein something counts against hypotheses or theories.

Interestingly, theories have been politicized by an anti-science Marxian agenda in the university. The historians of physics Sir Karl Popper and Imre Lakatos used a counting-against principle on Marxism to show that *it* was pseudo-scientific since nothing was permitted to count against it. But they never reduced it to mere myth! Should they have? [This question bore sarcastically on an academic New Left that, having led to a political correctness since the late 1960s, did not bode well for scientific candor in the universities.] My question goes back to differences between mythology, ideology, and science as well as to a double-edged sword of evidence. What evidence would count against the author's *own* disbelief in the Roswell UFO? Might his own disbelief proceed *pari passu*, by his *own* reasoning, with a fact disregarding anthropocentric myth in its own way?

— Bob Trundle

Hopefully, this letter captures some equivocation on the word 'myth' that characterizes many academic responses to the UFO issue. When there is an attempt to *appear* impartial, the word is said to mean something other than untrue stories. Precisely, however, such stories may be intended at professional meetings or in scholarly works. The works are reminiscent of a 'trick memo' to which Professor Allen Hynek refers, noted below, to appease a bias of professors who are loathe to acknowledge their own myth and ideology. And the latter are evidently rife in most universities.

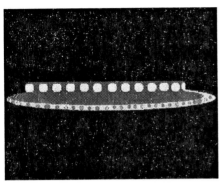

Child's drawing of a UFO sent to NASA, titled "E.T." Tuska wrote the agency, stating: "My picture is *real!!!!!!!!* My mother and my grandfather sow one flying soucee here*!!!!* I live in Rio, Brazil. Thank you *!!!!!!!!!!!!!*"

The Roots of Extraterrestrials as Myth

The forgotten roots of extraterrestrials as myth began in the late 1960s. The *Condon UFO Report* from the University of Colorado, named after Dr. Edward Condon, Chair of the Physics Department, was commissioned by the US Air Force to assess whether or not UFOs were alien visitors in response to growing public alarm over reported sightings. Academic interest faded by the mid-1970s because of Condon's ambiguous conclusions, although the conclusions were at odds with the overall findings of his own investigators that supported a reality of UFOs and that sparked unprecedented attention to the phenomenon by a few mainstream scientists. Also, professional meetings led to heated disagreement and there was UFO testimony before an apathetic Congress with disappointing results.

Physicist James McDonald's suicide in an Arizona desert, after he was both ridiculed by a Congressman for supporting UFO research and abandoned by his wife, seemed to cap off an era of serious interest. If there was any doubt about an end of the era, it seemed to evaporate when Northwestern University astronomer Dr. Allen Hynek died in 1986 after a long bout with cancer and more than a decade since he had been outspoken in defense of continued academic research into the phenomena.

There is an irony of political ideology parading as science: UFOs draw such heavy fire for being unscientific when these unscientific ideologies are outlandishly contrary to any objective research.

Hynek anticipated the intellectual shenanigans of many academics in "Science is Not What Scientists Always Do," a chapter in his pioneering book *The UFO Experience: A Scientific Inquiry* (1972). He noted that the mere mention of UFOs in academia in any open-minded way would be "the equivalent of scientific tar and feathers."[14] And he referred dismayingly to a

15

university Dean's memo that the "trick" would be for professors to make their research *appear* objective. That is, the trick was to feign objectivity for predetermined conclusions that observers of UFOs were deluded, hysterical, or confused mythology with reality.

Though there was some academic literature in the 1980s, most of it rehashed skeptical arguments that were uninformed and questionable. One of the most notable exceptions was the work of NASA Scientific Researcher and Professor Richard Haines. Then, a renewal of fresh inquiry spawned in the mid to late 1980s. As scholarship applied to practical endeavors such as business ethics and women's studies influence popular thought, thought-provoking literature at this level influences scholars as well. There was a flurry of publicized UFO sightings and abductee reports, some of which appeared as televised documentaries, and well written books by nonacademics. The books ranged from Budd Hopkins' *Intruders* (1987) and Timothy Good's *Above Top Secret* (1988) to the *UFO Crash at Roswell* (1991) by Kevin Randle and Donald Schmitt.

Clearly, this UFO literature for mass audiences revitalized the interest of scholars. Haines continued contributions such as *Advanced Aerial Devices Reported During the Korean War* (1990). But there were also Temple University Professor David Jacobs' *Secret Life: Firsthand Documented Accounts of UFO Abductions* (1992), with a foreword by Harvard psychiatrist Dr. John E. Mack, and Mack's own *Abduction* (1994). These books inspired other scholars to apply their expertise and resulted in the interviews of Jacobs and Mack on popular television programs.

Appearing on *WCET-PBS's* "Charlie Rose Show" in 1994, Mack noted the interest in UFOs by Tulane University philosopher Michael Zimmerman. Professor Zimmerman enhanced the scholarship in 1997 with his article "The Alien Abduction Phenomenon" in the peer-reviewed journal *Philosophy Today*. This kind of scholarly exposure stimulates an increased credibility of an already fascinating subject that further fosters mass-media coverage. A two-hour TV special "Larry King Live: The UFO Cover-Up" was presented on *CNN* in October 1994 with millions of viewers who saw interviews of several investigators. One was a physicist who worked outside a Los-Alamos government research area in Nevada where UFOs are reported. But a problem of the off-and-on academic interest in UFOs, as presaged by Professor Hynek, is that the vast majority of academics are convinced that serious research is either silly or biased in favor of UFOs.

In point of fact this community has suffered the prejudice. In avoiding even the appearance of biased scholarship, Jacobs and Mack made painfully clear that they did not begin their pioneering inquiries by either supporting or

Pulitzer Prize-winning psychiatrist John E. Mack, M.D., of Harvard University's Medical School, bravely resisted peer pressure to cease study of UFO phenomena and to deny the truth of his patients' testimony about their alien abductors. Dr. Mack questions: "What if the alien encounter phenomenon was subtle in the sense that it may manifest in the physical world but derives from a source which by its very nature could not provide the kind of hard evidence that would satisfy skeptics for whom reality is limited to the material? What if we were to acknowledge that the phenomenon is beyond our present framework of knowledge?"

debunking UFO abductees. Many of the witnesses came to be believed, after careful study, as telling the truth as they understood it. These reassurances did not soften the disposition of their academic critics. Their intransigent attitude raised the troubling specter of an epistemological totalitarianism.

Epistemological Totalitarianism

A report on Dr. Mack in *The Chronicle of Higher Education* in July 1994 was as deplorable as the 'trick memo' of a university Dean to deceive the public, noted by Professor Hynek. The *Chronical* revealed a larger intellectual dishonesty of scholarly responses to UFOs. In response to feedback that his study evoked dismay from almost all his colleagues, Mack noted that there is an epistemological totalitarianism among scholars on taboo topics.[15] Epistemology is an area of philosophy that involves the study of 'truth'.

⋀

Questions ensue of whether professors who glibly debunk UFOs, without any research, are either victims of scholarly self-deception or deliberately foster confusion to satisfy their own insecurity and prejudice.

⋁

Taken with 'totalitarianism', Mack suggests that there is an ideological policing intended to make faculty and students avoid subjects which are politically incorrect even if what is 'incorrect' is empirically true. This conformity collides head-on with the university's traditional mission of pursuing truth even if it threatens vested academic interests and insecurities. To make matters worse, *The Associated Press* reported 5 May 1995 that the Harvard Medical Committee was looking into whether "Mack's alien abduction research meets the school's standards for scholarship." His attorney noted that this sort of intimidation violated "the tenure system, which gives jobs for life so professors can feel free to pursue radical research."

The point was lost on one of Dr. Mack's colleagues who, with no research to support his misgivings, shamelessly quipped that Mack is "a brilliant fellow who occasionally loses it, and this time he's lost it big time." Clearly, there is plausibility to Mack's assertion that there is an epistemological totalitarianism in higher education. It was said of this pedagogical despotism in *Method & Science*, a scientific journal in The Netherlands, that "scholarly inquiries are too often... sanitized in order to avoid insensitivity to mainstream academic views (that include political ideologies parading as science, *e.g.* radical feminism and liberal to Marxist analyses in the social sciences)."[16]

A lecturer in black studies at the University of Massachusetts, sociologist Clinton Jean, adds to this dismaying phenomenon. He admits that a feigned *exactness* of the obligatory liberal to Marxian views made him extremely uneasy at both Brandeis and Columbia Universities, "at least as... social science described it."[17] The irony of political ideologies parading as science is that UFO studies draw such heavy fire for being unscientific when there is the patent hypocrisy of these ideologies that are so contrary to unbiased scientific research. The University of Pittsburgh's Nicholas Rescher, Editor of the *American Philosophical Quarterly* and former official of the American Philosophical Association, wrote the troubling editorial "Where Wise Men Fear to Tread":

> It is a cheerless... fact that trained philosophers [today] are just as captivated by the political fashions and frenzies of their place and time as other people... But since they presumably realize the pitfalls involved, it is some what more distressing that philosophers are not epistemologically more cautious about voicing their opinions.[18]

Now this is not as scathing as a quip of one cultural critic whose wit rivals that of the legendary Will Rogers. He would rather be governed by those listed randomly in a telephone directory than by Harvard's faculty. And philanthropists, he opines, could greatly benefit humanity by paying the most influential present-day philosophers to not think! But Rescher's statement is not flattering either. Dismayingly, he quotes famous philosophers who defend not only terrorism but a political correctness which does not bode well for any area of unpopular research. Questions ensue of whether professors who glibly debunk UFOs, with no research, are either victims of scholarly self-deception or deliberately generate confusion to satisfy their own prejudice.

Certainly, there are cogent arguments against some UFO cases by academics, as opposed to broadside dismissals of all the phenomena. But the

CE-5 by NASA Senior Research Scientist Dr. Richard Haines, (Ret.). "CE-5" refers to human behavior which evokes a response from aliens or UFOs. He rejects a reduction of aliens, who have a physical nature, to either myth or angels. In a letter to me about my book, Dr. Haines stated, "You clearly captured the currently held world view of science." As a product of the Humanistic movement, science "set out (rather secretly) to discredit man's faith in God and what is seen." Science "erected laboratory alters where its priests (scientists) would worship, wearing white robes (lab coats) and O_2 masks and filters, reading from good books (Handbooks of Chemistry and Physics), and indoctrinate younger priests into the [esoteric] 'mysteries' etc."

facts outlined reflect common responses. And they are worsened by demands that there be material evidence — not that it is lacking — in order for there to be rational belief. Finally, complacency and an epistemological totalitarianism are exacerbated by a prejudice against theology in secular universities. In viewing a theological metaphysics as being properly superseded by the physics of our enlightened era, these universities have encouraged the few UFO researchers there are to treat scientifically superior aliens as virtual high-tech gods. This treatment will only incur more skepticism among those with a liberal education who might otherwise take seriously the research.

A Peculiar Backdrop for Aliens as Gods

In regard to aliens having a superior science, a formidable challenge to traditional theology will come from a scientific worldview *(Weltanschauung)*. This modern worldview is often called "scientism." In other words, science alone is held to determine what is real and true, and it will make evil vanish as well. Having philosophical seeds in the Enlightenment in which Baron d'Holbach's *System de la Nature* (1770) held that man is unhappy because he does not know Nature, modern metaphysics was typically confused with physics.

That is, the 'enlightened belief' that our cosmos is like a mechanical clock composed merely of material atoms, a mastery of which will resolve all earthly moral problems, was not a theory of physics. It was a convenient assumption for secular scientists. Influencing later interpretations of such things as the Virgin Mary to be either delusional or even UFO phenomena, this view of science has its origin in modern secular philosophy. The latter was inspired by physics and contemptuous of medieval metaphysics.

In sum, a metaphysics of theology yielded to modern empiricism,

Marxism, pragmatism, and logical positivism. They hid their own metaphysics under a rubric of 'verifiable sense experience.' Bearing poignantly on the UFO issue, this new idea of experience was more restrictive than that afforded by a common sense evidence of Nature. By the same token, a supposed 'neutral territory of reason' replaced a natural reason rooted in Nature's God. Finally, a philosophical method of empirical verification, which was not itself verifiable, nurtured the view that science was *the* knowledge-yielding enterprise for truth about reality. And 'reality' was skewed for later generations.

⋏

Modern culture has a superior physics. But its deterministic assumption is inferior to a metaphysics of traditional theology for understanding the nature and limits of science.

⋎

Generations of scholars with this view have now educated generations of students who have influenced others without a higher education. Though the view has not been entirely accepted by the general public, it has become a respected public norm of reasonableness that finds expression in our modern 'therapeutic culture'. Bearing on the promise of a vastly superior alien science that will ameliorate our social ills, many churches now accept a liberal notion of 'sin' as a social pathology rectifiable by science. "The more I thought about what was at issue," noted science and religion editor William Tammeus at a national conference of Christians, "the more I saw evidence of... our therapeutic culture."[19]

> A recently retired Los Alamos physicist and I were leading a seminar... about science and religion. Our group... was taking note of the astonishing places physics has taken all of us... [And given both the success of physics and the absolute lack of any progress in religion for making human beings feel good about themselves since the time of the *Fall* in Genesis,]... we are encouraged to imagine that the right therapy, self-help books, counseling and meditation techniques will allow us to reach some kind of personal nirvana... sad to say, many religious communities have bought into this world view so completely that they may as well call themselves not a First Church but First Therapeutic Support Group.[20]

Tammeus adds that "assumptions behind our therapeutic culture are so widespread... that many of us no longer question them."[21] The assumptions beg for exposure in light of the question at hand: How much more might our ills be remedied by an advanced science of aliens? Many academics answer with untenable extremes. One extreme pridefully rejects any UFOs since they are impossible by our science. Paradoxically, the other appeals to this science for viewing advanced aliens as either angelic beings or divine creators. Both extremes beg for some scrutiny.

The One Extreme

Rejecting UFOs since they violate science proceeds *pari passu* with rejecting religion for the same reason. Here, religion is challenged by a deterministic and materialistic metaphysics that is presupposed by physics. Thus it is timely to compare physics to a traditional theology that is more liberal in what it permits to be real. While *science excludes* extraphenomenal realities such as free will, 'good' and God, *theology includes* them as well as a reality of phenomena. Without dismissing phenomena that are studied scientifically from a naturalistic perspective, the limitations of this perspective are captured uncannily by St. Paul in 1 Corinthians:

> The natural man does not accept what is taught by the Spirit of God. For him, that is absurdity... because it must be appraised in a spiritual way. The spiritual man, on the other hand, can appraise everything... [2:14].

One spiritual man who illustrates this point is Georgetown University Professor Patrick Heelan of the Roman Catholic Society of Jesus (S.J.). Having a Ph.D. in Physics from the Louvain and a Ph.D. in Philosophy from St. Louis University, he gives a relevant critique of physicist J. Maxwell's famous formula.[22] In being specified surreptitiously as

$$\emptyset s \ are \ real = df \ \emptyset s \ exist \ as \ phenomena,$$

the formula was used metaphysically — not scientifically — to exclude both spiritual entities and phenomena inexplicable by known laws. A problem with the laws will become obvious for UFO reports. Suffice it to say at this point that the reports are deemed to be impossible by most academics. Many of them are overly influenced by a deterministic and materialistic metaphysics of physics which improperly excludes God as well as moral praise and blame because the latter presuppose free will.

The exclusion of free will and moral behavior by religious skeptics may be related to the defamatory remarks about their colleagues who pursue UFO research and even to some complicity in government cover-ups. But theologians who are scientists such as Dr. Heelan, with a more liberal view of reality (*ontology*), might view their skepticism in terms of a nefarious pride. Pride is the primary sin. Despite a knee-jerk conviction that religious belief is

Painted by the talented artist Monarca Lynn Merrifield, the caption says: "When Elijah had apparently finished his life on earth, we read that 'the Lord was about to take Elijah up to heaven by a whirlwind.'... One gains the impression that it was customary for Elijah to disappear in some sort of UFO, for Elijah's fellow prophets ordered a search for Elijah." Her theological advisor is Dr. Barry Downing, who wrote *The Bible and Flying Saucers* (1998), for which she illustrated the cover.

irrational, their pride would be that which is specious since it is rooted in a dogmatic faith in the exhaustive scope of today's scientific knowledge.

Pride in this knowledge has resulted in an association of religious illusion with the belief in a superior alien technology. Thus besides a studious disregard of the technology leading to UFO reports being relegated to sensational tabloid newspapers, California's *Orange County Register* (Feb. 17, 1994) connects a disbelief in alien abductee reports by "most psychiatrists" and the late atheistic astronomer Carl Sagan to those reports having thrived in a "Bible belt town." A disparagement of this town because of its openness to the Bible and UFOs is as silly as the other extreme.

The Other Extreme

The other extreme uses a dubious philosophy of science to adhere to current science for linking UFOs with a miraculous technology o f divine alien *creators*. The religious community can expect some of those who replace religion with science to be in awe of a superior extraterrestrial technology. With a faraway look and tone that suggested despair over the human condition and a hope for salvation, a university colleague sighed, during a conversation with her about UFOs, "If only I could believe *They* are here." A super-advanced science of extraterrestrial cultures, which affords their UFO visits, may augment the belief that science alone explains our evolution and offers hope for our survival.

Need one rely on personal encounters? Notwithstanding his admirable courage amidst the perfidious behavior of his colleagues, Harvard psychiatrist

John Mack might have added inadvertently to a supernatural hope. Certainly, he has no naive faith in science and there is no intention to disparage his remarkable career — as if it could be defamed! But while we are beholden to the enormous scholarship and prestige he brings to UFO research, he may have given some voice to the messianic hope when he says that an alien intelligence might not be "indifferent to the fate of the earth, regarding its life forms... as one of its better or more advanced creations."[23] That this creator might be understood as divine seems clear when Dr. Mack adds that if our techno-destructive acquisitiveness were diagnosed by ET, "we really do not know how the *divinity* might experience itself and its creation...."[24]

A

Evidence of God would exclude a belief in the divine status of aliens as well as suggest a unique bond between aliens and persons.

Y

What lies behind the idea of such a naturalistic divinity? Even if we were genetically tampered with by an alien species, subtle figurative responses by either Dr. Mack or others could be construed more literally. An alien intelligence who is advanced enough to have interstellar spacecraft and whose behavior is a numinous mystery is precocious enough to be the actual 'Lord' or 'Creator,' of our traditional religions. This divine extraterrestrial intelligence could now have second thoughts about the "further evolution of [our] consciousness or whatever the *anima mundi* has in store."[25]

Before examining the topics of creation and evolution as they bear on UFOs, it needs to be noted that an untarnished truth of fantastic alien scenarios does not need to be accepted to accept serious concerns for theology. Theologians can rationally take at face value some relatively modest scenarios for the purpose of considering the most radical challenges to Judeo-Christianity in history. If the challenges are responded to forthrightly, this fact would evidence doctrinal security, broaden hermeneutics, and allay religious cults as well as quasi-spiritual responses that are already being formulated by some influential UFO advocates.

As insightful as the following are, for example, various spiritual responses are evident. There is Whitley Strieber's movie produced *Communion* and *Transformation* in which an extra-dimensional alien nature is said to be one which is advanced spiritually beyond our comprehension. And Dr. Jacques Vallée, a former investigator for the US Defense Department's computer projects, compares a paranormal domain of alien UFOs to both mysticism and

Many well educated persons who do not believe in God also compare UFO belief to an irrational belief in the spiritual, whether good or evil. *Above*, the evil is fought by the Church of Ireland's William Lendrum who says he routinely confronts the Devil. "In our modern scientific world," states journalist Gail Walker, "it is commonplace to offer rational explanations for strange occurrences and more rare for people to believe they are Satan's work." The work of either Satan or God is related to the supernatural. And although a reality of them and their work is not discountable on that account, the supernatural begs for a distinction from naturalistic or scientific interpretations of the UFO phenomena.

religious miracles in *Confrontations*. Even if one denies that there are enough credible reports to assess the nature of aliens, Bud Hopkins' *Intruders* is noteworthy for reminding us that "One simply cannot reconcile the idea of kindly... all-powerful [alien] 'Space Brothers' — a science fiction cliché now dear to Spiritualist Cults — with the ethically complicated reality of these unsettling UFO accounts."[26]

The Gods Must be Crazy

A response to neo-enlightened views of science does not take much imagination. This is the case even if the response has been largely ignored in scholarly literature. Besides the fact that scientists are not doing physics but rather metaphysics, there is a question of how our extraterrestrial creators were *themselves* created and how *they* evolved. The question of evolution is dealt with before metaphysics.

In the metaphysical theology of traditional Judeo-Christianity, evolution is complemented by a Mosaic understanding that man is created by God.[27] In supposing that this understanding is false on a scientific assumption that evolution is a purely accidental process and our creators were extraterrestrials, there are several intriguing points. First, a divinity of our traditional Creator is related to our creation out of nothing (*ex nihilo*) because, though everything else is causally dependent, He is not dependent on anything. This accords with the Mosaic revelation "I AM THAT I AM." And since scientific inquiry presupposes that everything is caused, it is contrary to science itself to hold that our 'creators' made us out of nothing and that they are divine in this sense.[28]

Second, the creation of humans by aliens — keeping in mind that we are aliens to them — raises a specter of the creation of aliens by other aliens and they by others and so on. There is no more reason to grant one species a divinity than the others. For the creations of aliens would involve biogenetic processes of science. And science *per se* specifies a naturalistic interpretation of a blind evolution of the first aliens. Not either they or an alien science can evidence the supernatural in terms of a such an interpretation.

Λ

UFO researchers have not merely tended to reject traditional religion. They have ignored religio-scientific modes of reasoning that may be *common* to extraterrestrial and terrestrial beings.

Y

Finally, the materialistic and deterministic assumptions of modern science exclude our reasoning from what *is* the case about our psychobiological nature to what *ought* to be the case it. Called a *naturalistic fallacy*, the fallacy excluded reasoning from our nature to how it ought to be fulfilled. The fallacy was defended by the eighteenth-century radical empiricist David Hume. He rejected theology and exalted science. If his criticism is correct, however, what aliens say ought to be the case has no logical weight. Indeed, concluding we ought to cease our techno-destructive behavior because our alien 'creators' may punish us would itself amount to a *naturalistic fallacy* as well as a fallacy of *appealing to force*. But Hume is correct only if there is no Creator who created a cosmos as it ought to be. For if the cosmos is as it ought to be in virtue of a Creator, then we can reason from scientific descriptions of our physical nature to how it ought to be fulfilled morally or spiritually. And thus the spiritual does not depend on alien benefactors regardless of Hume.

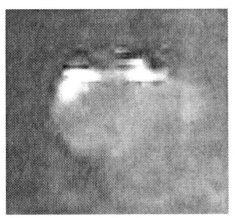

Do UFOs that are seen from Ohio to Baku indicate that ET is a benefactor, if not our creator? According to information provided to *Olaylar news* by Fuad Gasimov, Chairman of the Cosmic Seismological Department of Azerbaijan National Aero Cosmic Agency, UFOs may symbolize alarm. "If we analyze the processes going on in the world [*e.g.* Iraq], we would observe that a probability of war becomes high." In the genre in which aliens are our saviors or benefactors, he added that UFOs may manifest themselves as a sort of ominous omen that warns against our wars.

Creation and Metaphysics Re-evaluated

Understood traditionally as "things after physics" *(ta meta ta phusika)*, metaphysics studies Reality and what renders coherent truth about

Has an extraterrestrial species impregnated women for hybrid babies and, having visited earth for millennia, virtually created us and monitored our progress?

it. Modern philosophy transposed medieval efficient causality into a causal principle necessary for intelligible scientific inquiries. But the inquiries permitted only a material nature of events that are causally determined. One difficulty is that contradictory truth-claims would be equally truth because equally caused. And this problem is worsened by materialism. For thought would itself be a mere phenomenon. And since 'truth' is not ascribed to phenomena but rather to thoughts or statements about them, even scientific 'truth' collapses as a coherent notion.

A more liberal medieval notion of efficient causality included voluntary *and* natural (material) causes. Insofar as we have bodies subject to deterministic laws, we are natural causes among others. And inasmuch as we have free will, we are voluntary first causes, analogous to a First Cause or Creator, who are not entirely determined. When we lift something, both voluntary and natural causes render intelligible the truth 'He lifted his arm,' as opposed to 'His arm is lifted,' since we are limitedly *free from* a determined spatiotemporal realm. Thus although modern society has a superior physics, its metaphysical determinism and materialism are inferior in scope and coherence to a medieval metaphysics of theology.

Before turning to a creationist metaphysics based on efficient causality that bears on UFOs, we note that a metaphysical relation of God to intelligent beings was rooted in classical notions such as "Perfect" that was said to be embodied by God: To talk of some things being more perfect than others is to admit of the Perfect, per se, who renders intelligible truth about comparable things in virtue of being an ideal standard. Though all intelligent beings are unlike God in not being perfect, they are like Him in seeking moral, aesthetic and rational perfection. A quest of perfection finds expression in scientific theories in terms of which in *Genesis* we are sanctioned to both subdue and be stewards of Nature. How are these points connected to UFO reports?

A Supreme Being of Aliens and Humans

Many reports refer to an extraterrestrial religion. Thus it is odd that this possible religion is neglected by interpreting either religious visions as UFO phenomena or this phenomena as religious data. Without supposing the truth of various reports, there is an intriguing account of Sid Padrich. Evoking the attention of the FBI and US Air Force Intelligence which questioned him, he alleged that aliens in a flying disc asked him if he would like to pay respects to a "Supreme Deity":

> "We have one, but we call it God [said Padrich]. Are we talking about the same thing?" He [the alien] replied, *"There is only one"*... So I knelt and did my usual prayer... Until that night I had never felt the presence of the Supreme Being....[29]

This Being's incarnation in Jesus may complicate a spiritual status of aliens. But the Church's refusal to accept them into full sacramental rights does *not* imply a refusal to recognize God's love for them as a part of His creation. Hence a central part of our creation, analogous to the Trinity, includes our mind, body and will of which, said St. Augustine, we are incontrovertibly aware. Taken with our ability to love unconditionally, a similar potential of aliens admits of their creation in God's image without denying a unique image of humans.

When posing unparalleled challenges to our culture, religion and science, extraordinary phenomena demand exceptional explanations.

There are, however, polar alternatives of theology. A theological chauvinism is avoided by admitting, for example, that aliens may not need salvation if they did not suffer a 'Fall'. Or salvation may be excluded for aliens who, say, in psychologically resembling insects by lacking individuality, lack a sufficient analogical relationship to God. The first alternative was noted by Cambridge theologian C. S. Lewis: We may "find a race which was, like us, rational but, unlike us, innocent." And "If we were wise, we should fall at their feet."[30] The feet could turn out to be those of a race whose typical member has the innocence of a child but the mind of a thousand men. A second alternative was noted by Col. Corso. Whether or not his claim of top-secret

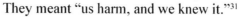

information is true, he gives voice to the view that EBEs are not friendly visitors looking for a polite way to say "'Hello, we mean you no harm'." They meant "us harm, and we knew it."[31]

So impressive is Professor William Lane Craig in proving the existence of Nature's God that he has been interviewed repeatedly in the news. One of his proofs of a First Cause is "The Ultimate Question of Origins: God and the Beginning of the Universe" in *Astrophysics and Space Science* (1999). His publication in this prestigious journal and others reveal that scientists are starting to take seriously what they had relegated to a mere speculative metaphysics. Having a Ph.D. from the University of Birmingham (England) and D. Theol. from the University of Munich, Craig argues that the absolute origin of the universe in the Big Bang singularity contradicts the naturalistic assumption that the universe always existed. Quantum gravity models, such as the Hartle-Hawking and Vilenkin models, must themselves appeal to a metaphysically dubious and physically unintelligible device of "imaginary time" to avoid the universe's beginning. Contingency implied by an absolute beginning *ex nihilo* points to a transcendent Cause beyond space-time.

These points are ignored by many UFO authors who are either ignorant of theology or antagonistic to it. Theology involves a metaphysics of creation that until recently was relegated to confused seminarians by most Anglo-American philosophers in the positivist tradition. "Astonishingly," says Oxford philosopher Anthony Kenny, even the "argument for the existence of God, regarded in the 1950s as the most exploded weapon in the philosophical armory, has been... redeployed to effect on the contemporary battlefield of philosophical theology."[32] In viewing theology as a jumble of senseless metaphysical sentences, modern philosophy has fostered naïve assumptions that continue to *dumb down* serious attempts to study UFOs, not to mention considerations of whether or not they are piloted by ET.

A Metaphysics of ET?

The late Ludwig Wittgenstein of Cambridge University, of whom most philosophers are said to be in 'awe', pioneered both logical positivism and analytic philosophy in the twentieth century. While holding that there are only senseless answers to the question 'Why is there something rather than *nothing*?', he was astonished by the 'mystery' of there not being *nothing* rather than a world.[33] Bearing on ET who may have related a vastly more advanced science to the theology of a First Cause in virtue of a superior mind or older culture, or both, the answer may be that there never was nothing but always *something*. Traditionally, this *thing* is understood as both God and the Creator of the world.

Surely, if one reasons that it is physically

28

impossible for there to be a world when there is no God and rejects God, one would be astonished by the world's existence. It *is* possible that aliens are burdened intellectually, as are many modern philosophers, by a bias against proofs of a Creator. But in terms of new developments in modal logic — a logic confused with a conjectural metaphysics, the compelling case for a Creator seems clear. And it seems equally clear that reasoning which is sound for one intelligent species would be sound for another in any world where scientific inquiry presupposes a causal principle and this principle implies a First Cause who is understood as Nature's God.

A

However ET relates God to science, a *possibility* of ET's science relies on an *impossibility* that its world either caused itself or came from nothing.

V

Reasoning to this God is not countered by the possibility of different 'worlds' with varying physical laws. For the laws would be yielded by scientific inquiries which, in presupposing causal regularities, ultimately presuppose that the worlds have a causal Creator.[34] There may not be a decisive rejoinder to an objection that there can be many Creators. But it conflicts with a perennial Socratic reasoning in which many virtues beget one Virtue as many gods do one God. Viable evidence of God as noted by St. Thomas, as incongruous as it is with a modern antireligious bias influenced mostly by David Hume, throws into question a divine status of ET as well as expresses a unique bond between ET and humans.

Rooted in Thomas' reasoning and based on our sense experience, the proof of God logically connects natural theology to science, if not to religious faith. Since the proof is not merely about faith, there has been an explosion of interest, from William Lane Craig's "The Ultimate Question of Origins" in *Astrophysics & Space Science* (1999) to Robert Trundle's "A First Cause & Causal Principle" in *Philosophy in Science X* (2003). The proof's ability to relate natural theology to science is important if a science of aliens is stressed. And it has been stressed to the virtual exclusion of everything else. Whatever aliens call the reasoning to God, the coherence of their scientific inquiry presupposes a physical impossibility that phenomena come from nowhere and out of nothing. The impossibility is not a logical one, say the universe both did and did not come from nothing. Rather the impossibility is a physical one whose denial violates both a thermodynamics of physics and what it makes sense to say in ordinary language.

Affirming the existence of Nature's God is illustrated by *Many Worlds: The New Universe, Extraterrestrial Life and Theological Implications* (2000) whose contributors include eminent scientists such as Christian de Duve, Bernd-Olaf Kueppers, Arthur Peacocke, and the Editor Steven Dick of the US Naval Observatory. Professor Dick proposes a cosmological theology: "a new world view that theologies ignore at their peril and must eventually accommodate if they are to remain in touch with the real world." A bio- and intelligence-friendly universe has generated abundant possibilities beyond Earth. "As we learn more about our place in the universe and as we physically move away from our home planet our cosmic consciousness will only increase. With due respect for present religious traditions whose history stretches back four millennia, the natural God of cosmic evolution and the biological universe... may be the God of the next millennium." The irony of this statement is that traditional theology is consistent with the proposed God of the next millennium.

In missing the point, one might insist nevertheless that something being uncaused is conceivable. The point is that the lack of a cause, though logically possible, would not be deemed physically possible. A child who told her mother that a cookie came out of nowhere would evoke the mother's humorous reply "That's impossible!' In addition, the lack of a cause would exclude an intelligible scientific inquiry even if physical laws differ in different worlds. A more formal note about such worlds is perused later.

Precisely, to say that it is impossible for terrestrial and extraterrestrial beings to come into existence out of nothing, because it violates a causal dependence of all things, is to say *all* things are dependent. This point holds for all worlds subject to science, even ones of aliens since they would need to suppose that everything is causally dependent for an intelligibility of their science. How could any science predict behavior of physical systems if the systems' past and future states are not related by causal regularities? Even if the regularities compose an infinite series, the series would still be dependent and could not exist if there is no non-dependent Cause.

⋏

This work is written with a caveat that it is a modest tribute to these courageous scientists and begs for further philosophical inquiry of my more able but pendulous colleagues.

⋎

A quiet soft-spoken man who was renowned for his integrity, Dr. James McDonald was a professor at both the Institute of Atmospheric Physics and the Physics Department at the University of Arizona. Professor McDonald stated that UFOs "are probably extraterrestrial vehicles" on a "reconnaissance operation" — the best hypothesis "for the fascinating array of UFO phenomena." After being ridiculed in Congress for this testimony, marginalized by colleagues, and abandoned by his wife during the liberation counterculture of the 1960s, he committed suicide at the cañon Del Oro in the Arizona desert in 1971.

Dr. Allen Hynek with a photo of the famous Adamski Saucer. Chair of the Astronomy Department at Northwestern University and a US Air Force consultant, he pioneered the classification of UFOs and coined the words 'Close Encounters', adopted in the film *Close Encounters of the Third Kind* in which he appeared. Having worked with military intelligence, he said that "CIA and NSA protestations of... lack of interest in UFOs are nothing short of prevarication." No one can deny that "intelligence agencies of our government were long cognizant of UFOs."

The causal dependence of all things on a non-dependent Cause renders coherent inquiries of astronomy into the cause and structure of the cosmos. Again, if the cosmos either caused itself or had merely caused causes *ad infinitum*, not to mention if it came from nothing, it would be both unlike all the other phenomena studied by science and not be subject to scientific inquiry. Indeed, the excluded inquiry would turn the cosmos into the very *mystery* that those who deny a First Cause seek to avoid. Moreover, this Cause is a necessary condition for a *verisimilitude* in which theories are understood historically to be more and more true in virtue of phenomena comprising causes and effects that permit increasingly successful predictions. And in regard to the subject at hand, a greater predictive and manipulative power of theories is necessary for even speaking intelligibly of extraterrestrial sciences that may be more advanced than ours.

To say science and theology are related logically is to say that while it is *logically* possible there is a world when there is no causal Creator, it is *physically* impossible — as surely as the impossibility, thermodynamically in physics, for matter to either create itself or disappear in the sense of going clean out of existence.[35] That is, necessarily if there is no Creator, there is no world. Since there is a world, there is the Creator. Having noted this ignored reasoning of modal logic, this introduction is not the place to more formally express it and its profound import for belief systems of scientifically advanced aliens.[36] But it is sufficient to cast doubt on virtually all studies which have either elevated aliens to divine creators or denied a religio-scientific belief that we may have in common.

This common belief, one may object, is suspect when it is based on an arcane proof. But besides its being presented lucidly and supported by scholarship, many scientists submit hypotheses on both the cosmos and UFOs that are equally arcane and extraordinary. Extraordinary phenomena, when posing unrivaled challenges to science and religion, demand exceptional responses. The present time is no time to be timid. Did not Sir Karl Popper propose bold conjectures for all quantum leaps of progress?

In order to make this work interesting to scholars and laypersons as well as to those who are both familiar or unfamiliar with the UFO controversy, the following chapters range from reports of alien anatomy to technology and military projects, and are peppered also with philosophic commentary. The commentary is made in honor of Dr. James McDonald, who suffered so personally from the social-political upheavals of the 1960s, and Dr. Allen Hynek whose *The UFO Experience: A Scientific Inquiry* suggested that an inquiry of science be complemented by philosophy.[37] Thus a philosophical inquiry into the UFO experience is written. It is written with a caveat that it is a modest tribute to these courageous scientists and begs for further philosophical inquiry of my more able but pendulous colleagues.

Notes

[1] See Dr. Stanley Jaki, Princeton Research Center, "The Last Word in Physics," *Philosophy in Science* V (1993) 9-32. In this same volume, see my article "Applied Logic: An Aristotelian Organon for Critical Thinking" where I quote Dr. David Martin of the London School of Economics. He notes other effects of closed academic minds on popular culture: "When my wife and a colleague talked at length to a hundred young people not long ago... [they] had a vague sense that some scientist or other (was it perhaps Darwin?) had certified that 'the Bible wasn't true'. Their view of Scripture was entirely *precritical*, and their notion of scientific method *nonexistent*. But they had a reverence for scientific works." This naïveté was parodied by a humorous UFO movie. Though aliens who visited earth had an advanced technology, they were as ignorant of the theory behind it as humans who know how to use lights but not how lights work!

[2] *Ibid.*, p. 20.

[3] See Oxford University Professor of Physics David Deutsch and Professor of Philosophy Michael Lockwood, "The Quantum Mechanics of Time Travel," *Scientific American* 270 (1994) 68-74. An objection that time travel *is* consistent with our present knowledge in physics would miss the point. Precisely, the point is that the travel is possible by a principle of pessimistic induction even if it is inconsistent with that knowledge.

[4] See physics professor Victor Stenger's "The Face of Chaos" in *Free Inquiry* 13 (Win 1992-

93) 13-14. This journal is antagonistic to revealed religious belief as well as belief in Nature's God. But viable interpretations of chaos theory at the cosmological level can be exploited for revitalizing proofs of God as a First Cause. Though Stenger disputes the Cause's connection to new Cobe-Satellite data on the Big Bang Theory, he does not address the Theory in terms of modal logic. And even apart from this logic, a case for an Aristotelian-Thomistic teleology of a self-organizing universe is made in R. Trundle, "Quantum Fluctuation, Self-Organizing Biological Systems, & Human Freedom," *Idealistic Studies* 24 (1994) 269-281.

[5] From G. Chui, "Pulsar Vanishes: Astronomers Baffled," *Knight News Service*, 13 July 1989.

[6] See Professors P. Ubertini, A. Bazzanu, R. Sood, T. Sumner, and G. Frye, "Hard X-Ray Spectrum of Supernova 1987A on Day 407," *The Astrophysical Journal* 337 (1989) L19. Also, subsequently, Dr. Ray Villard of the Space Telescope Science Institute (STScI) stated, "Images taken by Hubble show a dim object in the position of the suspected source of the celestial light show." And Dr. Chris Burrows of the STScI/European Space Agency concluded, "This is an unprecedented and bizarre object. We have never seen anything behave like this before."

[7] Dr. T. Schick and Dr. L. Vaughn, *How to Think About Weird Things* (Mountain View, CA: Mayfield Publishing, 1995), in a flier on this book.

[8] It is refreshing that none of my colleagues have openly disparaged UFO studies. Some of the professors in my department who most notably supported me are Dr. Jim Hopgood, Dr. Robert Lilly, and Dr. Prince Brown.

[9] Lt. Col. Philip Corso (Ret.), *The Day After Roswell* (NY: Pocket Books, 1997), p. 237. Consider Karl Pflock's criticism of Corso's book in February 1999 (www.jse.com/bookreviews). The *ad hominem* criticisms are ignored in order to address less flagrantly irrational ones. He takes some of Corso's errors, *e.g.* the Discoverer satellites were a *NASA* project, to trump the unlikeliness Corso would risk dishonoring his military career by lying. But it seems silly to conclude "there is simply no good reason to take Corso's tale seriously" because there were errors of this mundane sort among all the other data. This holds especially in light of his advanced age and possible faulty memory of details. Besides the fact that some details may turn out to be true in virtue of military intelligence to which Corso was privy, a glance at historical data shows how entangled many projects were. *Facts on File Yearbook 1959* reports that the Discoverer "capsule was thought to have landed by parachute in the North Polar region" and that a research program to send satellites over that "region was announced Apr. 20 by the Natl. Aeronautics & Space Admin. [NASA] in Washington."

[10] Jonah Goldberg, "X Marked the Spot," *National Review*, Vol. LII, July 2000, p. 27.

[11] To suggest that there may be pathological motives which might explain skewed teaching on recent politico-military history, by many professors, does not commit an *ad hominem* fallacy if the suggestion is made after cogently arguing that the teaching *is* skewed. The suggestion rationally addresses the question of why it occurs. For a case that it has, see R. Trundle's "Has Global Ethnic Conflict Superseded Cold War Ideology?," *Studies in Conflict & Terrorism*, RAND Corp, Washington D., V. 19, No. 1, 1996, pp. 93-107. For a more detailed analysis of the ideology, see Trundle's "Cold-War Ideology: An Apologetics for Global Conflict," *Res Publica*, Leuven University, Pol. Science Inst., V. 37, 1996, pp. 61-84.

[12] Corso, *The Day After Roswell*, p. 132.

[13] Professor B. Saler, "Explanation, Conspiracy, and Extraterrestrials: The Roswell Incident," *93rd Annual Meeting of the Am. Anthropological Association*, 4 Dec 94, at Atlanta, Georgia.

[14] Professor A. Hynek, *The UFO Experience: A Scientific Inquiry* (Chicago: Henry Regnery Co., 1972), p. 210.

[15] Dr. P. Monoaghan, "Encounters With Aliens," *The Chronicle of Higher Education*, 6 July 1994, pp. A10, A16.

[16] R. Trundle, "Extraterrestrial Intelligence & UFOs: Challenges to Physics, Metaphysics, and Theology?," *Method & Science* 27 (1994) p. 73. Support of political ideology, when UFO research is derided hypocritically as unscientific, is referenced by social-science professors.

[17] *Cf.* Sociology professor Clinton Jean's *Behind the Eurocentric Veil* (MA: University of Massachusetts Press, 1992, p. xix. Also, see anthropology professor J. Peacock's "MultiCulturalism in the USA," *AnArchaey Notes II, Issue I* (Oct. 1994), p. 4, given to me by anthropologist Dr. Sharlotte Neely, and sociologist F. Lynch's "Willy Loman, Angry White Guy," *Special to the Los Angeles Times*, 22 Mar. 1995, p. A10. C.

[18] See Professor Nicholas Rescher's editorial essay "Where Wise Men Fear to Tread," *American Philosophical Quarterly* 27 (July 1990), p. 259.

[19] William Tammeus, "Nobody's a Sinner Anymore: Psychobabble *Vs.* Religion," *The Cincinnati Enquirer*, 13 Aug 00, G2. This criticism of the therapeutic culture in churches is not peculiar to any denomination. Given the naive exaltation of science as an elixir for the human condition, however, one needs to add that those which are more orthodox are not the ones whose members are necessarily more ignorant!

[20] *Ibid.*

[21] *Ibid.*

[22] Professor Patrick Heelan, S.J., Ph.D. Physics, Ph.D. Philosophy, "Quantum Mechanics & Objectivity," *The Problem of Scientific Realism*, Ed. by E. MacKinnon (NY: Appleton-Century Crofts, 1972), p. 267, fn. 6.

[23] Dr. David Jacobs, Foreword by John E. Mack, M.D., Professor of Psychiatry, Harvard Medical School, *Secret Life: Firsthand Documented Accounts of UFO Abductions*, NY: Simon & Schuster, 1992), p. 12.

[24] *Ibid.*, p. 13, emphasis added.

[25] *Ibid.* p. 13.

[26] B. Hopkins, *Intruders: The Incredible Visitations at Copely Woods* (NY: Random House, 1987), p. 297.

[27] Both the traditional word 'mankind' and a male gender for God are defended by women in scholarly works. They include Miriam Moran's *What You Should Know About Women's Lib: A Christian Approach to Problems of Today* (CT: Keats Publishing, 1974) and Helen Hitchcock's *The Politics of Prayer: Feminist Language and the Worship of God* (San Francisco: Ignatius Press, 1992). Political ideology, *e.g.* 'correctness,' not only affects the UFO issue but marketplace. One student told me that she was advised by a liberal bookseller, at odds with her job to sell books, to *not* buy the book by Hitchcock! Indeed, Spence Publishing

reports that a major book wholesaler received seven-hundred orders for attorney Carolyn Graglia's *Domestic Tranquility: A Brief Against Feminism* (1998). But the wholesaler refused to order her book, due to feminist sentiment, until the vice president of a bookstore chain complained to officials of the wholesaler (www.spencepublishing.com, 26 July 99). This illustrates how ideology, spreading from the university to publishing to business, may especially suppress a free flow of ideas when they combine two or more politically incorrect views. Both criticism of feminism and scholarly support for UFOs fall into this category at this time. In the future, who knows what will be politically incorrect even if true. In any case, traditional differences of males and females, which conflict with a gender feminism, are related to extraterrestrials in the next chapter.

[28] In classical and relativistic and quantum physics alike, future states of physical systems are related to present states of the systems by causal regularities. The regularities are expressed by a causal principle 'All events have either exactly or inexactly measurable causes' wherein even a supposed indeterminism of quantum physics is not obviated because its laws are deterministic of probabilities.

[29] See T. Good, *Above Top Secret: The Worldwide UFO Cover-Up*, Foreword by the Former Chief of Defense Staff, Lord Hill-Norton, G. C. B. (NY: William Morrow, 1988), p. 297.

[30] Professor C. S. Lewis, *Christian Reflections* (Grand Rapids, Mich: William B. Eerdmans Publishing Co., 1971), pp. 174, 175.

[31] Corso, *The Day After Roswell*, p. 122. EBEs refers to 'Extraterrestrial Biological Entities.'

[32] A. Kenny, ed., *The Oxford History of Western Philosophy* (Oxford: Oxford University Press, 1994), p. 368. Kenny prefaces a new-found respect for medieval proofs of God by stating: A "fragmentation of the analytical monolith has led to a revived renaissance of medieval studies: medieval philosophy, once a... nursemaid to theology taught only in seminaries, is now taught... as a significant element of the philosophical heritage."

[33] 'World' means universe. See Norman Malcolm's *Ludwig Wittgenstein: A Memoir* (NY: Oxford University Press, 1984), p. 59. Wittgenstein was not a disbeliever but expressed the greatest respect, among all thinkers, for the theologians Soren Kierkegaard and St. Augustine — virtually "awe" in the case of Augustine to the dismay of many of his followers. By contrast, Martin Heidegger was a disbeliever and rejected the proof of a Creator. He asked paradoxically "Why are there... [existents] rather than nothing?" According to Heidegger, this was the most profound and troubling question that a human being could ask. An answer to the question may be that there never was *nothing* but always *something*. When one construes the *thing* as a First Cause, wherein it is not physically possible there is a world when there is not the Cause, a paradox about there being something rather than nothing seems more evident if the Cause is denied. See Heidegger's *Introduction to Metaphysics* (NY: Doubleday Publishers, 1961), p. 3.

[34] For varying possible laws of physics in different worlds see J. Nolt's *Logics* (CA: Wadsworth., 1997), pp. 351-356.

[35] This proof was presented by invitation of physics professor Maria Falbo-Kenkel to the Scientific Research Society on 8 Oct 97 at the Natural Sciences Center of NKU.

[36] For example, see R. Trundle's "A First Cause & Causal Principle" in *Philosophy in Science*

X (2003), Ed. by William Stoeger, S.J., Ph.D. Astrophysics, Cambridge University (Kraków: Papal Academy of Theology, the Vatican Astronomical Observatory & University of Arizona, 2003), pp. 1-17. When a complimentary copy of this article was sent to Indiana University Professor Emerita Noretta Koertge, Chief Editor of *Philosophy of Science*, she requested that I referee a submitted article for that journal. Also see my article "Thomas' 2nd Way: A Defense by Modal Scientific Reasoning," *Logique et Analyse* 146 (1994), at the Belgium Center for Research in Logic. An expansion on my proof was also accepted as a book by two scholarly publishers. One sent an expert reader's evaluation stating, "The text is well researched and effectively worked through, with a well-extended focus and depth." And Dean of the Catholic University School of Philosophy Jude Dougherty, though not representing a publisher, stated "I believe you have produced an important work which deserves publication." Without suggesting that the proof does not need development, the foregoing is noted because it bears dramatically on a religio-scientific cosmology that bears on terrestrial and extraterrestrial beings for a coherence of science in any high-tech scientific culture.

[37] See Hynek, *The UFO Experience: A Scientific Inquiry*, p. xi. Hynek acknowledged a unique role for philosophers on UFO investigations when he stated, "philosophers sense the limitations of the present more quickly than do scientists, absorbed as the latter are in their immediate problems." Unhappily, no academic philosopher, since the time he wrote his seminal work until my 1994 article in *Method & Science*, took up the task with a publication in a professional journal to support of UFOs. This point is not made to exalt the author but rather to question why an inquiry into extraterrestrial UFOs, which lies *within* a context of scientific studies, would be less credible than studies of the existence of God which are often held to lie *outside* a domain of science. In our culture, science is typically construed as a paradigm knowledge-yielding enterprise despite knotty epistemological problems addressed herein. If the problem of ignoring the issue of alien UFOs was due to a fear of ridicule from colleagues — a difficulty faced bravely by Harvard psychiatrist John Mack, this fact would bode ill for the *raison d'être* of academic tenure, much more for the daring that is necessary for revolutions in either philosophy or other scholarly areas.

3

Alien Anatomy and Technology

HAVING NOTED THAT a novel case is being made for extraterrestrial visits, it needs to be said that this prospect will seem ridiculous to those whose only news is the mainstream media. Its coverage of Orson Welles' 1938 broadcast of *The War of the Worlds* caused public panic. And the media believes that it too hastily disclosed a US military announcement, which was later retracted, that an alien UFO was retrieved in 1947 at Roswell, New Mexico. There is the old saying 'fool me once, shame on you; fool me twice, shame on me.' But shame may be on the American press. In contrast to many of its European counterparts, it has virtually conspired with a government repression.

The press release "UFO Theorists Gain Support Abroad but Repression at Home" in the *Boston Globe* (May 21, 2000) reports that "High-level officials — including retired generals from the French Institute of Higher Studies for National Defense... concluded that "numerous manifestations observed by reliable witnesses could be the work of craft of extraterritorial origin and that, in fact, the best explanation is 'the extraterrestrial hypothesis'." And this hypothesis was backed by the Defense Department of Mexico when film of UFOs, taken by the Mexican Air Force, was made available to the public on May 11, 2004. By contrast, "The Joint [US] Army Navy Air Force Publication 146 threatens to prosecute anyone under its jurisdiction... pilots, civilian agencies, merchant marine captains, and even some fishing vessels — for disclosing reports of sightings relevant to the US."[1]

Nevertheless, there is vital news below the mainstream US media that is as real as an ignored world below our oceans. The oceanic world includes reports of alien anatomy and technology that were brooked until recently only in terms of secondhand reports. Before an import of the reports for politics and religion is investigated, we focus on the descriptions. Even if their truth is dubious, and some are grotesque, they deserve some brief reiteration. For the descriptions are infused into the literature, they cannot be dismissed *a priori* by our conventional science, and they would bear on one of the most dramatic events in human history. Intriguingly, there seems to be new firsthand testimony that could confirm many less commanding claims. Let us begin with the claims of one of the most disputable figures.

The Claims of Adamski

The late George Adamski is one of the most vilified and honored men in UFO lore. He was in his 60s when he claimed to have filmed UFOs, talked to aliens, and received rare information from them. His first UFO sighting in 1946 was followed by the fantastic report in 1952 that he had made contact with aliens at Desert Center, California. The film he took of saucers there and

at Silver Spring, Maryland, became classics after which there followed a stream of fake photos by others. Nonetheless, S. Darbishire and A. Myers filmed similar objects in 1954 at Lancashire, England. And the same sort of craft was officially reported by policemen Tony Dodd and Alan Dale in 1978 at Yorkshire.

The Adamski saucer resembles a metal lamp shade with a rim of portholes over a flared slope and light bulbs below. Some scientists tried to discredit his film at Silver Spring by ridicule. NASA's Goddard Space Flight Center sought to relegate them to mere artifacts. But the film was

In the golden UFO days, George Adamski piqued our imagination.

also assessed independently by optical physicists who confirmed its likely authenticity.[2] This was partly due to a geometry of the images and distortions, of no significance to Adamski, in which the craft appeared blurred out of shape. Skewed shapes of UFOs would be hard to fake and were later related by many scientists to gravity distortions caused by electromagnetic processes of UFO propulsion systems. The latter were explained in NASA scientist Alan Holt's *Field Resonance Propulsion Concept*, a book praised by physicist Dr. Hal Puthoff, Director, Institute for Advanced Studies at Austin.

The Credibility of His Claims

Adamski's false claims about his background were taken to bear on those about UFOs.

He spoke of himself as a professor in his contact with an alien at Desert

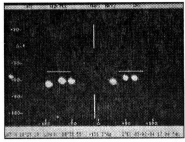

Almost 60 years after Adamski was ridiculed for reporting a UFO, this infrared UFO film — taken from a military jet — was released by the Defense Department of Mexico in 2004. Pilot Maj. M. Castanon said "I believe they could feel we were pursuing them." Another pilot stated, "We are not alone! This is so weird."

Center and alleged professional affiliations with the Mt. Palomar Observatory when only helping out in the Palomar Gardens restaurant. He also claimed that aliens said all intelligent beings have human forms, and he seemed even more suspect when he asserted that alien spacecraft took him to meet aliens on Venus and Saturn. Astronomers have long held, given their relative uninhabitableness — Mars less so than other planets in our solar system, that the probability is *nil* that intelligent life-forms could evolve and live there.

In all fairness to Adamski, an exaggerated résumé is not a fault limited to a few persons who feel insecure when they find themselves in the limelight. Also, the American and Soviet space programs of the 1960s have not entirely excluded aliens evolving in our solar system. But even if this system excluded the evolution, it is not absurd to consider that aliens could have deceived Adamski about his location — given that Star-Trek speeds to other solar systems are now entertained by physicists, or that he might have mistakenly assumed he was in our solar system. While his claims in *Flying Saucers Have Landed* were bizarre by standards of 1953, reports in later decades were equally amazing and new scientific theories make his story now seem less fantastic.

Interpreting the Claims

There is the report in 1990 of a US Army General who said he was informed that aliens had populated our solar system.[3] In terms of our knowledge of this system, his claim and that of Adamski might both be deemed false. But Adamski had little education in contrast to the General whose rank also elicited respect. My point is *not* that the General's report should be accepted because of his elite status. Status, in fact, makes little difference in what aliens allegedly say. The point *is* that ridicule or a presumed falsity of what an abductee says does not render false a report. Given our own space probes, a possibility of intelligent species in our solar system is more subtle than little green men on Mars.

A supposed howler of Martians traveling about our solar system may belie the howler's arrogance. Lately, scientists have employed well regarded theories that permit the visits by exceeding or bypassing the speed of light. The theories include those of astrophysicist Miguel Alcubierre at the University of Wales and Russian Nobel Prize winner Dr. Andrei Sakharov at the Physical Institute's Academy of Sciences. In "Warp drive: hyper-fast travel within general relativity" (*Class.Quant.Grav* 11 (1994) L73-77), Alcubierre states that in a "framework of general relativity and without the introduction of

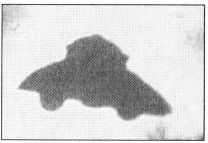

Author's computer-enhanced photo of an Adamski-type Saucer, location and time unknown, with spheroidal structures below. Canadian government scientist W.B. Smith stated that he and other state scientists learned that the 'balls' were not landing gear, as thought, in virtue of a contact with alien *people*! Called the "boys from topside," these *people* allegedly informed the scientists of novel technologies that were easily tested.

wormholes, it is possible to modify a space-time" to allow "a spaceship to travel with an arbitrarily large speed." By a local expansion of space-time behind a spaceship and an opposite contraction at its front, "motion faster than the speed of light as seen by observers outside the disturbed region is possible. The resulting distortion is reminiscent of the 'warp drive' of science fiction." And Sakharov notes the fantastic possibility that regions of space billions of light years apart are linked by wormholes or parallel entrances. Time crossing space may be so short, as in the film *K-Pax*, "we would appear in the new place quite unexpectedly... someone would suddenly appear next to us."[4] Other possibilities are all the more viable by more advanced sciences of aliens with a comparable intelligence in older civilizations or in civilizations comparably old where ET is more intelligent.

Λ

Astronaut Story Musgrave, M.D., stated that "It's almost a statistical certainty there are beings millions of years more advanced than we. It's one chance in a billion I'd succeed in making contact."

v

Persons who claim they were taken by ET to planets in our solar system, such as Adamski, may assume that other systems would take much longer to reach; perhaps 1,000,000 years or more? But if aliens can exceed the accepted speed of light despite special relativity and various paradoxes — a feat we might duplicate in the near future, then they may be able to travel millions of light years in less time than it takes us to ignite NASA's rockets. This travel is conceivable and every indication is, *via* the history of science, that alleged physical impossibilities will be possible in terms of future theories.

In addition to the theories already noted, NASA has begun to develop interstellar travel by new energies for a propulsion with no propellant mass to attain unparalleled transit speeds. Attention is to the coupling of gravity and electromagnetism, quantum vacuum, hyperfast travel, and super-luminal quantum effects.[5] Given there need be no causal paradox of Einstein's physics in holding that after breakfast we can travel two million light years to a galaxy and be back in time for lunch, according to Alcubierre, should there not be more humility about some of George Adamski's claims?

The Human Form Reconsidered

Despite skeptics who doubt ET would look human, Adamski said it was revealed to him that the human form is universal. Admittedly, he may have been misinformed by the alleged aliens. Or, aliens who have our bodily form could be those who, piloting UFO probes, are genetically engineered for space travel by the more typical species who do not have human forms. But research in the journal of *Nature* by the Neuroscience Prize winner Dr. David Perrett, at St. Andrews University, stated that he and his colleagues established by empirical investigation that *one* universal idea of human beauty actually transcends *many* different cultures.[6] This intercultural universal is remindful of Plato's theory of Ideas and may afford reasoning to interstellar universals that include a universal form of humans. And although Plato's metaphysics was not scientific by even ancient Greek standards, his metaphysics may be fortified in modern physics by various principles. In terms of a well established principle of homogeneity, for example, it would be the case *prima facie* that there are relatively uniform biophysical forms of intelligence throughout the universe.

A rancher reported four-foot tall green creatures with no noses or hair, egg-shaped hands that resembled hooves of cattle, and strides that suggested an organic robotism.

With no more of a hasty generalization than concluding that all phenomena of a given sort are subject to a scientific law when it holds in one part of the universe, we can infer a universal human form for all regions of a certain sort in the cosmos when it holds in this cosmic region. Without excluding other forms of intelligent life, this understanding is supported not only by phenomena being homogeneous and isotropic (Cosmological Principle) but by the Big Bang Theory. Given an evolution of *psycho-biophysical* systems from

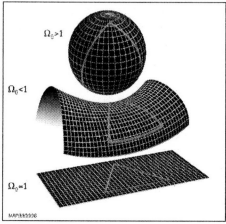

$\Omega_0 > 1$

$\Omega_0 < 1$

$\Omega_0 \approx 1$

MAP990006

Given that matter in the universe is homogeneous and isotropic (Cosmological Principle), says NASA, the distortions of space-time — due to gravitational effects — has 1 of 3 forms *(above)*. One would have evolved in terms of the Big Bang Theory whereby the universe was created 10 to 20 billion years ago by a cosmic explosion. In 1927 the priest Georges Lemaître proposed that the universe began with the explosion of a primeval atom, after analyzing the red shift in distant nebulas and a model of the universe based on Einstein's relativity. Later, Hubble found experimental evidence to support Lemaître's theory, finding that galaxies in every direction are moving away from us at speeds proportional to their distances. Interestingly, although the Big Bang Theory is inextricably related to the theory of evolution, NASA says that this Theory will probably "never be proved."

biophysical systems and they from *physical* ones with the same origin of a protonic-size black hole, intelligent life forms of a similar sort can be reasonably expected in the universe. We may expect, that is, many forms of extraterrestrial intelligence to have the same or similar morphologies as we do on Earth.

This morphological hypothesis seems as tenable as the theories of either a holographic universe or alien intervention. Princeton physicist David Bohm and Cambridge biochemist R. Sheldrake hold that a holographic universe interacts *deterministically* in virtue of a tiny element, like a hologram, that embodies all information of the universe.[7] The transmission of a 'morphological resonance' occurs by an evolutionary blueprint of the universe: The universe is rife with life and life on planets, if it evolves, evolves by the same pattern. A difficulty with the pattern, despite its support of a universal human form, is that it supposes an odd epistemic realism in which a causally determined reality is the truth-condition for truth. Thus the realism renders incoherent our truth-claims. The claims that p and $\sim p$ would be equally true because equally determined. And ironically, the deterministic dilemma holds for both an affirmation and denial of the holographic theory itself.

What about a theory that ET intervened in our evolution? While this spin on evolution cannot be dismissed *a priori*, it is both different from saying that ET is our creator and is more conjectural than a simple theory of evolution that includes a principle of homogeneity and Big Bang Theory. Still, a theory of alien intervention is not necessarily inconsistent with an acceptance of the Big Bang theory and Biblical Creation. In regard to Creation, for instance, Berlin Technical University physicist Dr. Wilhelm Schaaffs defends a

A researcher at the US government's Sandia National Laboratories explores new on-line communication methods that allow her to be "two places at once," using collaborative virtual environments *via* the Internet that in many ways are "as real as face-to-face interactions." She creates a reality "in virtual space by inserting organizational culture and intercultural communication cues into the situation." The cues "can be as simple as a mouse click on an avatar — a graphical representation of a human being in a computer-generated world...."

consistency with the Big Bang by noting *inter alia* that virtually all scientists had relegated a Biblical beginning of the world to pure myth until an expanding universe implied its beginning "in a very small place." And in agreement with a Scriptural notion that the sun was preceded by light, light is now construed as a primary electrical and atomic effect, and formations of fixed stars are often interpreted as "secondary macrocosmic effects."[8]

Finally, aliens who have technologies advanced enough for interstellar travel, by propulsion systems we do not understand, are advanced enough to have a genetic engineering to produce living automatons that *may* or *may not* look human. Policeman Gabriel Valdez saw footprints, near cattle mutilations, which were triangular-spaced 4 inch-diameter pods that led to landing marks of a craft. Another report by a Texas rancher involved four-foot tall green creatures with egg-shaped hands, like cattle hooves, whose behavior suggests an organic robotism. In short, aliens who appear inhuman may be consistent with Adamski's claim of a universal human form. Other forms of aliens might be engineered biogenetically by a 'master race.' Though they do not confirm his claims, these examples indicate an important point. Attempts to defame Adamski's character in order to discredit his claims commit an *ad hominem* fallacy. And this fallacy, which also reveals a less than generous spirit, is as irrational as disregarding interpretations of the claims of this personable man who passed away on April 23, 1965.

Dr. Sarbacher notes that aliens may withstand their craft's tremendous acceleration in virtue of being built like insects wherein the inertial forces would be quite low.

Subsequent Claims About the Anatomy

Since UFOs have been ignored by the scientific community, it is a foregone conclusion that claims about alien anatomy will be equally disregarded. Before considering it, a point needs to be made. *Something that at first seems irrational may, by relatively modest steps of reasoning, turn out to be eminently reasonable. Given that extraterrestrials exist and that some are thousands of years more advanced than humans, we may assume they have technologies that can achieve what we deem impossible. This includes interstellar travel in short times that are as unbelievable to us as our flying to the moon was to scientists in previous centuries. And even supposing that some aliens are only a century more advanced, it is still reasonable to assume that they could have already visited earth.*

That many scientists believe in alien intelligence is shown by the radiotelescopic *Search For Extra-Terrestrial Intelligence (SETI)*. And that there are well educated persons in the US space program who accept the visitation is clear by Astronaut Story Musgrave who holds six degrees including that of MD. In an interview with *USA Weekend* on 26 November 1993, he stated: "It's almost a statistical certainty there are beings out there millions of years more advanced than we.... It's one chance in a billion I'd succeed in making contact."

Contact need not follow from a reality of UFOs since a desire by aliens to communicate with humans is not implied by their visits. But if the visits are reasonable, then it is reasonable also to suppose that the craft may sometimes crash and that the crashes would yield living as well as dead aliens. Viewed this way, a specter of alien anatomy is not the same as unscientific 'weird things' such as angels, witches and demons. That some academics associate both fantasy and the supernatural with extraterrestrial beings is evident, nevertheless, in the popular text book *How to Think About Weird Things*. Here, Professors Theodore Schick and Lewis Vaughn compare a belief in demons to an unscientific belief that "extraterrestrial beings have visited earth."[9]

Having noted the patent silliness of making such a comparison, we note that an alleged anatomy of extraterrestrial beings ranges from ambiguous to detailed descriptions based on firsthand witnesses and those with whom they shared their secrets, with various degrees of creditability. Some reports address only one or two species and others a half dozen or more. One report stems from the *UFO Crash at Roswell*, coauthored by former US Air Force reserve Capt. and psychologist Dr. Kevin Randle.[10] His discussion centers on autopsies by military pathologist Jesse Johnson, M.D., after a reported crash of an object in 1947 at Roswell.

A rare photo of Professor Robert Sarbacher – President and Chair of the Washington Institute of Technology. Dr. Sarbacher was a consultant for the Canadian Government's Research & Development Board (RDB). He confirmed that Dr. Robert Oppenheimer, as Dr. Vannevar Bush (the science advisor to the US President on the Manhatten Project), was involved in secret research on "flying saucers." Sarbacher said that "very light materials" had been analyzed after the crash of a saucer. He thought it regrettable that governments continue to wrap UFOs in mystery. A letter by Sarbacher revealed that reports on the crashed UFO indicated unknown gear ratios, anomalous instruments, and people who controlled the machines who were of an extraordinarily light weight — sufficiently light to minimize the terrific effects of the acceleration and deceleration caused by their propulsion system. These extraterrestrial people, he stated, may have been biologically similar to insects so that they could withstand the G-forces.

A female's more passive psychic nature was rooted biologically in her less aggressive sex drive and embryonic matter passively receiving a 'rational' seminal form from a male.

In piecing together interviews of witnesses who either knew Dr. Johnson or observed the bodies after autopsies, a summary was compiled. Reportedly, four retrieved bodies measured 3 1/2 to 4 1/2 feet tall and weighed about forty pounds. The entities had large almond-shaped eyes, without discernible pupils, which made the heads appear oriental. The heads had small openings on the sides without ears, were large in proportion to the bodies, and had indistinct noses and small mouths. Evidently, the mouths did not function for either communication or eating because they were similar to 'wrinkle-like' folds only two inches deep.

Finally, they had thin hairless bodies with small bone mass. This brings to mind a reply of biophysicist and government consultant Dr. Robert Sarbacher to alleged government descriptions of aliens. Alien pilots may have been able to endure "tremendous deceleration and acceleration... with their machinery" since they "were constructed like certain insects... wherein... the inertial [G] forces would be quite low"[11] — though this explanation ignores a technology by which G forces might be neutralized. Also, it was stated that the creatures had a pinkish-gray skin which was tough and leathery and, under magnification, had a mesh-like structure. Reportedly, the internal examination specified that the bodies contained a colorless liquid without red blood cells, no apparent gastrointestinal tract, no reproductive organs, and not either a lower intestinal canal or a familiar rectal area.

OPERATION ANIMAL MUTILATION

REPORT OF THE
DISTRICT ATTORNEY
FIRST JUDICIAL DISTRICT
STATE OF NEW MEXICO
BY
KENNETH M. ROMMEL, JR.
PROJECT DIRECTOR
JUNE 1980

PREPARED FOR THE
CRIMINAL JUSTICE DEPARTMENT
UNDER GRANT # 79-DF-AX-0034

DISTRICT ATTORNEY

K.M. Rommel was a counterintelligence FBI agent for 28 years. Then, he accepted a bizarre assignment analogous to a real-life Fox Mulder of *X-Files*. Rommel became the Director of "Operation Animal Mutilation," an investigation by the District Attorney's Office of New Mexico's First Judicial District — funded by the Law Enforcement Assistance Administration of the Justice Department.

A First Hand Report?

A claimed firsthand report on alien anatomy comes from Lt. Col. Philip Corso who was certifiably in a position to know. Though his report is less succinct, it is more important in virtue of his official capacity. Recalling his job as intelligence officer at Ft. Riley, Kansas, he states that five deuce-and-a-halfs and low-boy trailers from Fort Bliss stopped over in route to the Air Materiel Command at Wright Field in Ohio. After the cargo was unloaded in a warehouse on 6 July 1947, he said that his position and friendship with a sentry led to a shocking sight. He pried off the lid of a coffin with an interior glass container. There, floating in a blue liquid was a four-foot humanoid figure, like a child, with "bizarre-looking four-fingered hands... thin legs and feet, and an oversized... head that looked like it was floating over a balloon gondola for a chin."[12]

Corso tried to forget the horrible container with its army intelligence document that said the body came from a crashed UFO in Roswell and was routed for a morgue's pathology lab at Walter Reed Army Hospital by way of Wright Field's AMC. Ironically, as head of Army R&D years later, with oversight of material from a crashed saucer, he obtained a photo of an autopsied alien like the one he saw at Ft. Riley. Showing a four-foot body, a report said that "organs, bones, and skin composition are different from ours" since the "heart and lungs are bigger." And the "bones are thinner but seem stronger as if the atoms are aligned... for a greater tensile strength."[13] Evidently, the creature was adept to long-distance space travel because its huge heart and lung capacities permitted a slow metabolism. Coupled with its fibrous, thinner and flexible skeletal frame suited to extreme G-forces, some members of his group believed that the entity was genetically *bioengineered* for such travel.

A medical examiner's report on the alien's body chemistry bore on the craft and cattle mutilations. In regard to the mutilations, there were advanced lasers that, with no tissue trauma, surgically removed organs and blood. A fluid serving as alien blood, while composed of known elements, had new aggregates of organic compounds that regulated bodily functions, similar to our glandular secretions. That is, there was a combination of blood and lymphatic systems wherein nutrients and waste were exchanged in the alien's epidermal tissue. Thus by the tissue's novel dynamics of bodily exchange, hygienic utensils were replaced by the creature's skin. Was its skin an entry point for a cow's blood? If the blood was *not* ingested orally, does that explain the craft's lack of culinary facilities?

Descriptions by Abductees?

The above reports are augmented by Temple University Professor David Jacobs' *Secret Life: Firsthand Documented Accounts of UFO Abductions*, with a foreword by Harvard psychiatrist John E. Mack.[14] In compiling research on persons who claimed to have been abducted by aliens, both Jacobs and Mack summarize what many other reports have in common. 'Small' aliens are between 2 and 4 1/2 feet tall and are reported more often than relatively 'tall' ones. And whereas the tall ones are from 2 to 6 inches taller, which makes the tallest about 5 feet, both species seem to have comparatively frail humanoid bodies with head, arms, two legs, feet, hands, and fingers. More detailed features, of course, could vary significantly from humans, as noted shortly.

Skin colors of both kinds of creatures are described as being dark to light gray, tan to tannish gray, or white to pale. And although colored lights inside UFOs could alter the tone, their skin usually appears tinged with gray. Those who claim to have touched them say the skin feels rubbery or plastic-like for the small creatures and rough for the taller ones, though a leatheryness of the small entities was noted in the autopsy report. Typically, there is not either hair or dermatological imperfections such as warts or moles.

Genetics expert Dr. Leon Visse said he complied with a US Government request, under high security, to analyze an extraordinary alien fluid at Wright-Patterson AFB in Ohio.

Claimed details of the anatomy have led to descriptions of both clothing and demeanor. The attire does not tend to express personality. Does a lack of *ego* reflect a psychology with biological roots in an insect-like physiology? At times there is insignia of serpents, jagged lines, or other shapes. In the corroborated Socorro case of 1964, a well regarded Police Chief stated that an egg-shaped object had a crescent-shaped insignia with a vertical arrow

A small part of Wright Paterson Air Force Base in Ohio. It was here that genetics expert Dr. Leon Visse said he responded to a US government request, under high security, to analyze blood from two blond 7-foot tall extraterrestrials who had been autopsied after a UFO crash.

and line beneath, though it was not on clothing.[15] Aside from the occasional insignia, a lack of individual style in the clothing is manifested by demeanor. Generally, the aliens discussed by Jacobs lack distinctive facial features and expressions which suggest sadness, happiness, or anger. Their nebulous morphology seems to be complemented by an indifferent and humorless, but efficient, behavior.

As with the various autopsic scenarios, the aliens are claimed to have heads large in proportion to bodies with bulbous craniums over the eyes, small mouthlike slits, large eyes with no pupils, corneas, or irises, and mere bumps for noses. The fact that many abductees do not see aliens with functional mouths, taken with secondhand reports at Roswell of the aliens' mesh-like skin, has led some investigators to speculate about the livestock mutilations even before Col. Corso's firsthand report. Given that the carcasses had no blood, they believed that alien nourishment came from blood absorbed through the creatures' skin.

Finally, while abductees do not generally report appearances of genital areas or bulges in clothing that indicate gender differences, they sometimes distinguish males from females by either telepathic communication or physical comportment. The abductees often report that females reveal a subtle emotionality and seem more gentle and nurturing than the males. The descriptions of aliens raise a specter of universal gender traits and Aristotle's psychobiological distinction of male from female humans in his *Generation of Animals* (729[b]-730[a]). Intriguingly, this study also has import for a physical and psychological nature common to humans and aliens. A female's more passive psychic nature was rooted biologically in both her less aggressive sex drive and her embryonic matter that passively receives the male's 'rational seminal form'.

Aristotle is attacked by politically-correct gender feminists as a sexist 'dead white male'. But their *ad hominem* muddles the fact that their position untenably ignores Nature entirely in favor of nurture (*biodenial*) — gender being a mere social construct whose acceptance is conditioned by culture.[16] In fact, Aristotle's study finds novel expression in abductee reports. Given that they reflect what abductees observed, the issue is related to a serious

Moscow Aviation Institute's Dr. Feliks Zigel revealed startling facts in 1981 after a study of UFOs in the Soviet Union: They were measured at speeds up to 100,000 km/h -28 km/s, are noiseless in virtue of creating a pneumatic vacuum, appear and disappear at will, and can stop electrical power stations, radio sources, and engines. *All of the foregoing effects are caused by an intelligence higher than that of man.* The Institute possesses 50,000 reports of precise observations and 7 cases in which extraterrestrial machines landed between June 1977 and September 1979. The report says that most of these facts should not be published (in *samizdat*), to prevent a panic of the public.

consideration of anthropomorphic traits. Usually scorned by contemporary scientists, the traits bear on both conventional and unconventional phenomena that may include alien beings.

Are There Many Alien Species?

Beyond the small greys, reports abound of other species. In *Classified: Secret AUSTEO (Australian Eyes Only)*, laboratory chemist William Chalker noted in 2003 that personnel at a military research facility learned of alien 'males' who were identical to humans externally but had no other internal organs, save for smaller lungs. Moreover, Police Chief Jeff Greenhaw reported to *NBC-TV News* in 1973 that after responding to calls about a UFO, he saw a metallic, six-foot creature standing; it then ran away at over 40 mph. And Jean-Charles Fumoux stated in "Preuves Scientifiques" (*OVNI* 1981) that genetics expert Dr. Leon Visse reported seeing autopsied blond aliens over seven feet tall at Wright Patterson AFB, after first examining a peculiar fluid that turned out to be their blood.

Many other worldwide reports by abductees and military personnel describe similar beings, as well as other types. Robert Dean, retired US Army Command Sergeant Major, stated that he had seen top-secret reports of at least four alien races when he was assigned to NATO's command center, the Supreme Headquarters Allied Powers, Europe (SHAPE). While one race was allegedly indistinguishable in appearance from human beings and were the most friendly, the second was reportedly the same except for gray pasty skin. A third race was said to be the little gray species, and a fourth reptilian with vertical pupils.[17] Dean held that all of the species operate their craft openly to demonstrate their capabilities — the visits leading to contacts with humans. In this regard, the much esteemed Soviet scientist Dr. Felix Zigel named three classes of extraterrestrials. A Professor of Cosmology at the Moscow Aviation Institute, these beings include *spacemen, humanoids,* and *aliens*:[18]

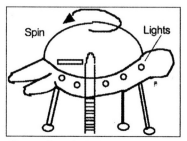

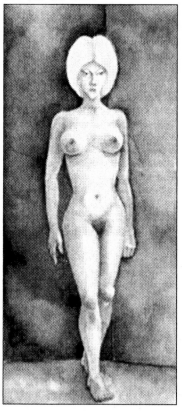

Above: Antonio Villas Boas saw an egg-UFO, as he plowed a field in his tractor, that made its headlights fail. **Below:** Taken aboard, he allegedly mated with a humanoid female who, when done, pointed to her belly and to the stars as if to indicate the future birth of a *star child*.

Spacemen: Wear bulky space suits with head gear, are as tall as 3 meters (9+ feet), have extraordinary strength, and can run as fast as a automobile in third gear as if they are from gravity heavy planets. Spacemen tend to fit reports made by Greehaw and possibly by Fumoux.

Humanoids: Similar to human beings in appearance, these 'people' could have mated with men and women who they abducted, and might even have infiltrated into our societies. The woman who was encountered in a spacecraft by Brazilian Villa Boas could have been one of these entities.

Aliens: Spacemen and humanoids are usually viewed as aliens. But Dr. Zigel restricts aliens to the small and tall gray species. Though tall only relative to each other, the aliens average a meter in height and resemble humans in basic bodily features: heads that are large in proportion to their bodies, long arms, four fingers and toes, and indistinct noses and ears. Further, as expanded on later, physicist Stanton Friedman argues for the authenticity of a classified 1994 government document *Majestic-12 Group Special Operations Manual: Extraterrestrial Entities & Technology, Recovery & Disposal.* Marked "TOP SECRET/MAJIC Eyes Only," it referred to EBEs of types I and II. There is no definite knowledge of their origin, "but it seems certain that they did not evolve on earth... [and] may have originated on two different planets."[19]

Type I EBEs: could be mistaken for oriental persons. While the rounded cranium is larger, their body parts are proportional to ours. These EBEs are bipedal, up to 5 feet 4 inches tall, and weigh 80 to 100 pounds. Their pebbled skin is thick with a pale chalky-yellow color, and almond-shaped wide-set eyes have large pupils and a pale gray cast instead of white. They have small ears high on the head, thin long noses, and almost lipless wide mouths. They have no facial hair, little body hair except at the groin and armpits, and their bodies are thin with well developed muscles. Their hands are small with four long digits, there is no humanlike webbing between fingers, and opposable outside digits may function like thumbs. Finally, their legs are bowed with long splayed feet.

Dr. Pierre Guerin of the French Institute of Astrophysics states that "a hyper-sophisticated nonhuman technological activity... could not be regarded with indifference."

Robert Lazar, an EK&G employee subcontracted to work at a top secret installation at Groom Lake, NV, appeared on *Las Vegas* and *Japanese TV* to claim that retrieved UFOs were being reverse engineered.

Type II EBEs: also have a humanoid appearance. But they are bipedal, 3 feet 5 inches to 4 feet 2 inches tall, and weigh 25 to 50 pounds. The head is proportionally larger than that of either humans or Type I EBEs and has a slight peak extending over the crown with a large elongated cranium. Over the eyes there is no brow ridge. The eyes appear black with no noticeable white, and are also larger than the Type I with a slant that wraps about the sides of the skull. The nose is located well above a lipless mouth that looks like two small slits. There are no external ears. A fine-celled smooth skin is bluish-gray and darker on the back. There is no hair on the body or face, and the entities do not seem to be mammalian. The arms are long relative to the legs, and the hands consist of three long tapered fingers and thumbs, the latter being almost as long as the fingers. Small and narrow feet issue in four toes joined by a membrane.

However, even the general classifications are eluded by other reports, from aliens who look like robotic devices to ones reminiscent of the dwarfs, fairies, and leprechauns of ancient legends. Their beards, bad tempers, mischievous behavior, and large pointed ears will tempt those of the scientific community, in perhaps committing a hasty generalization, to conclude that all UFO literature is mere mythology in the most pejorative sense of that term.

Reports of Extraterrestrial Technology

Dr. Pierre Guerin of the French Institute of Astrophysics states that "the presence of a hyper-sophisticated nonhuman technological activity... could not be regarded with indifference."[20] References to alien technology suggest its enigmatic nature and vast superiority to our own, which took tens of thousands of years to develop. Given even its remote reality, an alien science could not be entertained aloofly by societies that view science as an ideal knowledge-yielding enterprise.

The former Chief of the US Army's Foreign Technology Division, Col. Philip Corso, says that although the propulsion system of a UFO recovered in 1947 has not been satisfactorily duplicated, analyses of other artifacts led to revolutionary developments of integrated circuit chips, fiber optics, lasers and supertenacity fibers.[21] Though his claims are authoritative, they follow a long stream of intriguing reports.

First, there is a long documented history of reports, based on military and civilian witnesses, of external features and mechanical behavior of apparent extraterrestrial craft. The latter are typically luminescent in the evening, make no noise, and have no visible nuts or bolts or joints when viewed close up. The machines can execute acute turns at velocities as high as 22,000 mph, and appear to accelerate almost instantaneously from a hovering position. Accordingly, they seem to violate Newton's Second Law of motion wherein any body with appreciable mass, comparable to our aircraft, cannot accelerate to such speeds—let alone almost instantaneously.

Bringing to mind *A-gravity*, reportedly discovered by reverse-engineering a recovered UFO, is U.S. Patent #3,626,605 for an apparatus that involves a force field maximized in an atom's nucleus, the nuclear binding force found in all of nature.

Second, the machines are reported to have a virtual capacity to disappear and reappear, to stop electrical devices and other mechanical artifacts, and to generate sensations of timelessness as well as weightlessness when persons are nearby. These effects and properties are referenced in both military and civilian reports. And the reports find intriguing expression in a now declassified document by US General Nathan Twining of the Air Materiel Command at Wright Field, Ohio, on 23 September 1947. Flying "discs" were stated to be real, to be as large as man-made aircraft, and to generally make no sound

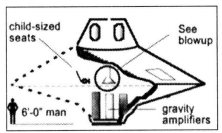

Reverse engineered? The reactor (*below*) is said to use Element 115 as fuel – source of gravity-A waves amplified to distort space-time for interstellar travel. The *Scientific American* (17July 02) now reports interest in stable superheavy elements by Berkeley researchers and scientists abroad.

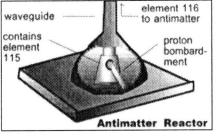

Bringing to mind *A-gravity* is US Patent #3,626,605, METHOD & APPARATUS FOR GENERATING A SECONDARY GRAVITATIONAL FORCE FIELD by Henry Wallace: "Since the dynamic interaction field arising through gravitational coupling is a function of both the mass and proximity of two relatively moving bodies... the resultant force field is predictably maximized within the nucleus of an atom.... Such force fields may... account for a significant portion of the nuclear binding force found in all of nature."

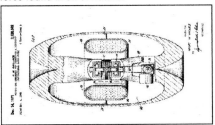

except when under apparent conditions of high-performance. Civilian reports are consistent with these descriptions.

Descriptions in Some Case Studies

Seventy-seven year old Alfred Burtoo stated that he saw a "vivid light" moving soundlessly toward him, then hover and land as he was fishing in the early-morning hours on August 12, 1983.[22] When taken into the craft by two child-sized beings, he noticed that the object was a metallic machine, despite its prior luminescence, and that it had no visible seams, bolts, or nuts. And in an earlier case, a police official saw a metallic object with a lighted bluish-green dome travel silently over her at a 300- foot altitude in the early Fall of 1977.[23] She had no sensation of weightlessness as the craft came to within about 50 feet of her, which is frequently reported when UFOs are nearby. But she did experience a sense of timelessness.

An amazing description, though controversial, comes from Robert Lazar. After being layed off work at a high-security military installation in Nevada, he declared on *Las Vegas* and *Nipon TV News* in the late 1980s, initially incognito since he feared reprisal, that research was being conducted on extraterrestrial spacecraft. One 35- to 40-foot saucer with an aluminum-like finish was alleged to have a 15-foot height at midpoint with 'portholes' on its dome. This craft was said to have the appearance of a large melted piece of wax with no joints, rivets, or bolts.

Contrary to a prevalent notion that superheavy elements were unstable and that gravity was not a wave, Lazar said the disc had a reactor that — emitting antimatter by bombarding a stable superheavy element '115' with protons — was the source of 'gravity waves' *via* an almost 100% efficient thermoelectric generator that amplified them to distort space-time for virtually instantaneous interstellar travel. Controversy over a physics of this travel notwithstanding, creation of a stable isotope with atomic number $Z=118$ at Lawrence Berkeley National Laboratory was announced by scientists in *Physical Review Letters* (1999). Though their claim was later retracted in 2002, the continuing experiments might give us some pause about some of Lazar's 'ridiculous' reports.

Lazar's further reports of the disc's glow and test-flight disappearances, attributed to gravitational fields induced by the generator, echoes physicist Alan Watts' *UFO Quest: In Search of the Mystery Machines* (1995) and NASA engineer Alan Holt's *Field Resonance Propulsion Concept* (1979).[24] Together with top scientists named by Lt. Col. Philip Corso, former Chief of US Army R&D at the Pentagon, Holt and Watts suggest that UFO gravitational fields are caused by either electromagnetic or hydromagnetic propulsion systems. These systems could account for both their partial and total invisibility, power disruptions of cars, and static electricity effects that are often reported.

...the disc's paper-thin metallic skin was impervious to either drilling or 10,000° heat. It had an 18-foot cabin around which rings spiraled with unknown gear ratios.

It needs to be added that many effects of UFOs are related to a wide variety of their shapes and sizes. Sizes range from eight *inches* to a few feet in diameter, seen by pilot Lt. David Brigham in his T-6, to many miles across for shapes which are sometimes described as lighted cities. The most familiar shapes are the crescent, cigar, and disc, although a disc is by far the most common. Some sightings of the cigar or cylinder may themselves be caused by viewing discs from the side. NASA's Alan Holt notes that a disc is most efficient for overwhelming the natural mass energy pattern with an artificial energy (*e.g.* electromagnetic) pattern that permits a higher-dimensional space or hyperspace travel.

Alleged inner features of UFOs are especially enigmatic. Often, witnesses say they felt sedated as if they were suffering akinesia or shock. However, some reports are from those who are amazingly observant and involve either crashed spacecraft or craft whose alien occupants make contact. While contact might be telepathic, it can also be in the native language of a witness. The report of forty-five year old electrical technician Sid Padrick, for instance, stated that he remained calm when encountering a 70 foot diameter saucer on 30 January 1965 near his home about 75 miles south of San Francisco.[25]

Alleged Inner Features of Extraterrestrial Spacecraft
Reportedly, after Padrick entered the craft by invitation from alien beings who seemed human, with a proviso that they were *inhumanly well groomed*, he said one of them spoke pristinely in English. He noticed stationary and movable instruments akin to teletype dots with dashes that moved left to right. Though there were no ordinary screens such as oscilloscopes, there were meters without dials and an oblong lens with a three-dimensional screen. And on it there appeared an enormous cigar-shaped craft in space. In UFO idiom, this sort of craft is often called a mothership or Leviathan.

Another description was gleaned from scientists and key personnel on condition of anonymity by investigator Bill Steinman. He followed up on an Frank Scully's investigation of a UFO crash 25 March 1948 at Aztec, New Mexico. Given plausibility by retired government scientist Wilbert Smith who said Scully's report was "substantially correct,"[26] the disc's paper-thin metallic skin was held to be impervious to 10,000° heat and diamond-tipped drilling. Entry was gained by a fractured metallic 'port hole', said to reveal an 18-foot diameter cabin around which spiraled large rings with gear ratios unknown to our science.

The upshot of this report on a recovered UFO is as follows: There was a tubeless radio, gears with anomalous ratios made out of metal that was not identified after over 150 tests, and an instrument panel with buttons and levers having hieroglyphic-like symbols, and symbols lighted on screens. While there was a control panel with roll-out drawers, there were no nuts or bolts or wires anywhere in the craft.

Sometimes, in other cases, interior surfaces of alleged alien craft are said to be fabricated out of a jointless metallic plastic-like material on which colors are not painted or baked but part of the material. Also, there are claims about an evident lack of propulsion systems, though the latter may stem from either control by 'mother ships' or destruction in crashes. But again, Lazar insisted that, in at least one case, a fuel element was bombarded with protons in a small half-spherical reactor at the craft's bottom level. It yielded another element that, decaying, radiated antimatter up a tubular 'wave-guide' in the ship's center. In colliding with matter, the antimatter produced an extraordinary energy which generated immense gravitational fields.

President Harry Truman on 16 Dec 1950 signing a proclamation to declare a national emergency. The CIA was formed under Truman in the same month and year as a reported UFO crash at Roswell. And after a memo classified *Secret Signal Communications Activities Intelligence* on 24 Oct 1952, this give-em-hell President abolished the Armed Forces Security Agency (AFSA) and created the more expansive National Security Agency (NSA) on 4 Nov 1952. A month later in this year of UFO cover-ups, he issued the following press release: "These unexplained phenomena are neither a secret weapon, neither a rocket, nor a new type of plane under test."

⋀

Microgravity, diminishing the weight of astronauts, was reportedly generated by a plasma-like light in the ceiling that was "acquired as a result of recovered alien ships." These sorts of reports are typically augmented by technologies used outside the ships.

⋎

In one of two modes of travel, the gravitational fields are allegedly directed by antenna-like 'gravity amplifiers' to a space-time target that may be many light years away — to pull it to the ship and retract with it almost instantaneously. Certainly, the claim of a technology that bypasses the speed of light has been clouded by, among other things, the skepticism of nuclear physicist Stanton Friedman. He rejects the notion that the element for the fuel would be produced naturally at another star system since, by an apparent principle of homogeneity, the element would be evenly dispersed throughout the universe and, therefore, would be present in our solar system as well. However, to say that the element is not here is not to say that it is not evenly dispersed in the universe at similar binary star systems where it reportedly occurs.

In addition to the external and internal characteristics, there is reference to duplicated technology from recovered UFOs. Former NASA engineer Robert Oechsler reported that when he was at Ellington Field in 1990 to work on a mechanical arm operating under zero gravity for the space shuttle, he was introduced to a top-secret low gravity room used for experiments. The microgravity was allegedly generated by a plasma-like light in the ceiling "acquired as a result of studying *recovered* alien ships."[27] Such reports are augmented by technologies used outside the ships. The reports range from tubular instruments for inducing akinesia and advanced lasers for cattle mutilations to UFOs near crop circles. Occurring worldwide, the circles led astronomer Allen Hynek to conclude that alien intelligences "know something about the physical world we don't know."[28] That a knowledge of natural phenomena is patently insufficient to explain the circles is shown by a farm woman's report in the summer of 1987 near Dorset, England.

Λ

The Majestic-12 Group Special Operations Manual allegedly describes alien technology for an elite group of officials by order of President Harry Truman. The Manual states that mechanical devices have not been analyzed fruitfully as to method of manufacture.

$\vee$

Reports of Additional Technologies

This British woman said that when a "bright golden star" moved silently towards nearby wheat circles at Westbury, the star turned out to be a round object which projected downward a ball or globe with "a lot of small, brilliant, golden lights on it."[29] Crop-circle researchers Pat Delgado and Colin Andrews, who investigated the circles and interviewed witnesses for many years, suggest that anything associated with UFOs is *so officially taboo* in the scientific community that curious scientists usually inquire about the circles without divulging who they are. But why were the scientists embarrassed? Inquiry into the causes of anomalous phenomena has always been the chief source of scientific progress.

In his former position as Chief of the US Army's Foreign Technology Desk at the Pentagon's R&D, Col. Corso said that anomalous material from the crashed Roswell machine was seeded into private industry. Passed off as curious artifacts gained by foreign intelligence during the Cold War, in countries such as former East Germany and the Soviet Union, select corporations were provided samples of alien artifacts for analysis. Technologies developed from them were allowed to be patented in private industry in order to conceal their top-secret origin. Accordingly, analysis of the material yielded patented discoveries, he stated, that included fiber optics, supertenacity fibers, integrated circuit chips, and lasers.

Though the propulsion system was of paramount scientific interest, it was not reverse engineered successfully — as far as Corso knew. Other anomalies included the craft having no controls, power supply, and fuel. Thus there were inquiries into whether the machine was itself an electrical circuit, if alien flight-suit material had elongated atomic structures to provide directional flows of currents applied to it, and if genetically-engineered aliens controlled the craft by their brains — in terms of head-bands with

Radio telescopes seek communication from ET: "If humanity is not the center of attention of a deity, what does that do to various theologies and religions?" asks astronomer Dr. Steven Dick of the US Naval Observatory. "If it is Christianity, for example," what about "the doctrines of redemption and incarnation?" Does "Christ have to die on other worlds for their sins the way he did here?" Though these questions naturally come to mind for scientists, the latter are often naive about the major theological and philosophical traditions. In addition to the fact that the traditions of Islam, Judaism and Christianity either accept or are consistent with a God of Nature as a First Cause in terms of Natural Theology, as opposed to only Supernatural or Revealed Theology, theologians such as Jesuit priest and astronomer Dr. Christopher Corbally have already entertained a plethora of scenarios about a possible spiritual status of ET, one of which includes that suggested by Dr. Dick. Moreover, the theologians have been open to the existence of ET long before even the secular scientists with the Center for SETI (Search for Extra-Terrestrial Intelligence). Only a few years ago SETI scientists scoffed at UFOs because of distances that cannot be traversed feasibly in terms of our present-day science, in disregard of the Principle of Pessimistic Induction that warrants pessimism about the absolute truth of our theories. Today, SETI is more open-minded: "Many of us dream of... meeting up with a (friendly) alien race. Much could be learned and discovered about each other"

their suits. Was the machine a virtual extension of ET's brain? Was electronic control gained by instructions transmitted from brain to body to suit, *via* the craft's control panels, to the craft itself?[30]

A speed-of-light limit lies behind the relegation of even well-witnessed UFOs, operating like intelligently flying craft, to natural phenomena, when this is patently absurd.

The craft seemed to store and conduct current. What if the craft *were* the engine "with a steady current from another source it stored as if it were a giant capacitor?"[31] Corso likened this to charging the battery of an electric car but controlling the stored current by the pilot. And in regard to propulsion, scientists who saw the Roswell debris at Wright Field in 1947 were reportedly reminded of the German-American antigravity experiments of the 1930s. But "the effort to develop true antigravity aircraft never came to fruition" since aircraft manufactures already had "a perfectly good weapons technology[!]."[32] Thus although there were theories on electromagnetic antigravity propulsion, even if it was not completely understood, our latest power source of a nuclear fission generator was infeasible in contrast to the Roswell UFO which, like a "flying battery," possessed its own electric potential and storage capacity.

A Top Secret/MAJIC Report, April 1954

Finally, there is a 1954 report obtained by nuclear physicist Stanton Friedman, the *Majestic-12 Group Special Operations Manual: Extraterrestrial Entities & Technology, Recovery & Disposal*. Stamped "TOP SECRET/ MAJIC Eyes Only," there is a subtitled area *Description of Extraterrestrial Technology* that allegedly describes UFO technology for an elite group of military and civilian officials by order of President Truman.[33] Descriptions are said to be based on military analyses of crashed alien craft from 1947 to 1953: (i) debris indicates extremely light medium sized craft; (ii) material composition remains undetermined by metallurgic analysis; (iii) the material has far greater strength and heat resistance, proportional to weight/size, than that used on civilian or military craft; (iv) some material has an appearance of aluminum foil but resembles plastic without metallic features; (v) other material includes beams and structures similar to dense grain-free wood, light in weight, but with tensile and compression strengths unknown to industry; (vi) no magnetic properties or residual radiation; (vii) attempts to decipher engraved marks have been unsuccessful; (viii) mechanical devices and gears have not been analyzed fruitfully as to either their function or manufacturing method.

⋀

NEC physicist Lijun Wang stated that "light pulses did indeed travel faster than the accepted speed of light." The report adds, "scientists are beginning to accept... interstellar space travel."

⋎

Reasons the Technology "Cannot Exist"

Disbelief in alien UFOs by most scientists is largely founded on an assumption that ET could not exceed the speed of light. Since Alpha Centauri is the closest star system and *our* safe travel there would take *us* thousands of years, ET's visit from more likely star systems much farther away is rejected out of hand. Thus a presence of alien technology is dismissed *a priori* since it violates a theoretical speed limit. The limit lies behind a tendency to identify even well witnessed UFOs as natural phenomena when this is patently absurd. One absurdity follows another when a premise is foolish. Denying feasible interstellar travel due to a twentieth-century theory is as silly as a denial of crossing the ocean by air in terms of a theory in the fifteenth century. In point of fact the accepted speed of light has already been challenged by NEC physicists at Princeton, New Jersey, with mind-boggling implications:

The experimental study has been performed with great care and repeated

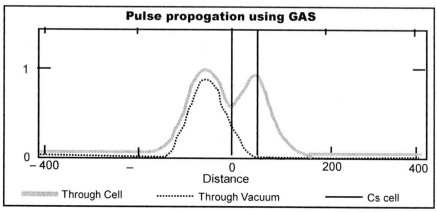

Pulse propogation using GAS

Through Cell ⋯⋯ Through Vacuum —— Cs cell

Time Travel – ET a Savior from the Future?
Diagram by physicists L.J. Wang, A. Kuzmich, and A. Dogariu at the NEC Research Institute in Princeton in which a pulse of light advanced 310 times the "vacuum transit time" (time it takes light to traverse the 6-cm length in vacuum) where the pulse behaves as if a distance is traversed in a negative time. An imaginative inference was drawn in the article "Will Time Travelers Alter Our Future?" Jim Donahue notes that Dr. Wang seems to have proved "the possibility of time travel." Maj. Ed Dames, he adds, believes most "UFOs are time machines visiting us from the future." Thus "we might dream that a *savior* really will pull us out of the soup at the last moment. Only instead of dropping [from] the sky as the Christians believe, this savior will come out of the future."

numerous times... The theoretical model is entirely based on existing physics theories of Electromagnetism and Quantum mechanics... *our experiment does show that the generally held misconception "nothing can move faster than the speed of light" is wrong...* In other words, the net effect can be viewed as that the time it takes a light pulse to traverse through the specially prepared atomic medium is a negative one. This negative delay, or a pulse advance, is 310 times the "vacuum transit time" (time it takes light to traverse the 6-cm length in vacuum).[34]

Physicists are intrigued by light traveling forward in time since information might be sent into the future. While there is controversy over implications of the experiment, it raises a specter of the need to reinterpret a fundamental presupposition of physics — a causal principle in which causes must be prior to effects. And it would clarify, if not undercut, relativistic physics since not all phenomena are necessarily limited to the accepted speed of light. Dr. Lijun Wang noted, "Our light pulses did indeed travel faster than the accepted speed of light."[35] And the report adds that scientists are starting to accept that there can be feasible "interstellar space travel."[36]

It would miss the point to object that interstellar travel and traveling in time will be rendered chimerical if, aside from phenomena which have no rest mass, such as light waves, the accepted speed of light still holds. While Wang says it holds and denies a violation of Einstein's special theory, why is that theory so special? In "Impulse Gravity Generator Based on Charged $YBa_2Cu_3O_{7-y}$ Superconductor with Composite Crystal Structure" (*arXiv:physics, Vol. 2, 2001*), E. Podkletnov and G. Modanese note that their control of gravity cannot be understood in terms of general relativity. Indeed, a patent probability of why these relativistic theories will be violated is noted by Oxford's W. H. Newton-Smith in terms of a principle called Pessimistic Induction. In terms of inductive reasoning, the history of science warrants pessimism about the truth of any present theory since all prior theories were superseded by later ones within two-hundred years of being propounded:

> We may think of... our current theories as being true. But modesty requires us to assume that they are not so. For what is so special about the present?... Indeed the evidence might even... support the conclusion that no theory that will ever be discovered by the human race is strictly speaking true.[37]

A
Dr. Bernard Haisch, Professor at the CA Institute for Physics and Astrophysics, notes that "Scientists are more closed-minded on the subject of UFOs than the general public."
V

Newton-Smith's insight can be refined. The pessimism is not that our theories are false or no better than previous ones. Rather, new theories can do all that previous ones can but more, so that there is a historical sequence of increasingly truer theories. A superseded theory's truth still holds for a given domain as does Newton's theory, despite theoretical developments of Einstein, where Planck's constant is small and bodies do not approach the speed of light. In virtue of an increasing truth-likeness, called "verisimilitude" by Sir Karl Popper, it is rational to believe that a present impossibility of feasible interstellar travel will be possible in terms of our future science. Given that more advanced alien civilizations would have sciences which render *possible* what is impossible for us, some statements by the SETI Institute seem patently irrational. Having its roots in NASA's Ames Research Center in the 1970s, an irrationality is manifest in its information to the public in 1998. SETI stated that if there was "evidence that extraterrestrials were visiting our planet, tens of thousands of university scientists would be busy investigating this idea. They're not."[38]

On the one hand, SETI seems wantonly naive about universities where professors are often chastened for unpopular and politically incorrect ideas. Dr. Bernard Haisch, of the CA Institute for Physics and Astrophysics, noted that "Scientists are more closed-minded on the subject of UFOs than the general public." And University of Illinois philosopher Neal Grossman stated, "One could effectively substitute the three letters 'UFO' for 'NDE' [Near Death Experience].... The taboo against having any interest in the paranormal except for... debunking it has persisted in academia." The "punishment for violating this taboo is to be marginalized by colleagues."[39]

On the other hand, these colleagues are the very scientists to whom SETI circularly appeals for rejecting any evidence of ET's visit because the nearest star is over 4 light years away. "With *our current rocket technology*, it would take around 300,000 years to travel there."[40] Further, in addition to this circularity, the reasoning is both closed to counter evidence in violation of a verification principle, assumed for meaningful truth-claims, and baffling insofar as it appeals to our "rocket technology" for a denial *a priori* of UFOs when UFOs are *not* comparable to that technology.

In short, to indicate why the speed-of-light limit is not reasonable is to indicate the unreasonableness of identifying all UFOs as natural phenomena and rejecting any presence of alien technology. Even the most rudimentary spirit of scientific inquiry beckons open-mindedness. Major contributions to science were made by Faraday, Clark, and Maxwell who were intrigued by even the most *ordinary* phenomena. These included Maxwell's fascination with sparks emitted by rubbing fur on amber. We may well wonder why more interest is not aroused by the *extraordinary* phenomena of UFOs that might be piloted by beings from another world.

Since modern academics are fond of recalling that the medieval Church was anthropocentric, they need to be reminded that they are themselves often more dogmatic with far less excuse.

Behind the Story of Why the Technology "Cannot Exist"

In considering the gravity of questions such as why UFOs do not arouse much curiosity, we may suppose that there are personal factors at work against being open to possible extraterrestrial technologies. On the assumption that most scientists are not as silly as one might believe, the factors may include false pride in reaction to other 'intelligences'. Even equally intelligent beings may enjoy civilizations much older than ours. Or in not being older, the inhabitants may be far more intelligent. Either alternative, not to mention aliens who could be both more intelligent and from older civilizations, would afford more advanced sciences — possibly by thousands or even multiple millions of years — that should interest any scientist worthy of the name.

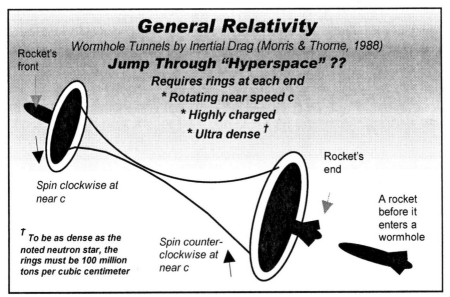

General Relativity

Wormhole Tunnels by Inertial Drag (Morris & Thorne, 1988)

Rocket's front

Jump Through "Hyperspace" ??

Requires rings at each end
* Rotating near speed c
* Highly charged
* Ultra dense †

Spin clockwise at near c

Rocket's end

† To be as dense as the noted neutron star, the rings must be 100 million tons per cubic centimeter

Spin counter-clockwise at near c

A rocket before it enters a wormhole

Wormholes would afford interstellar travel by bypassing the speed of light (c). But the impracticality of creating them is said by NASA to be evident: Collect enough superdense matter, such as that of a neutron star, to construct a ring the size of our Earth's orbit around the Sun. Then make another ring for the worm hole's end at the desired distance. We charge them with incredible voltage and spin them to near the speed of light. But having achieved all that, we would already have had to be where we want to go!

Some of these reasons for overzealous scientific skepticism may not hold, but they are plausible.

Academic skeptics infer surreptitiously that they have a stake in the *status quo* of science in virtue of devoting their lives to professing what they 'know'. Professor of astronomy, strident atheist, and UFO debunker Carl Sagan is a case in point. While his popular TV shows glibly unveiled the vast unbreachable spaces between *billions of stars*, the stellar spaces would have been far less impressive if he had admitted that science cannot exclude their effortless traversal by ET. Since these academics remind us endlessly of the medieval Church's dogmatism, they need to appreciate that they are often more dogmatic with far less excuse—starting with Sagan for whom, as with his forerunner David Hume, religious superstition kept modern physics from being discovered centuries earlier!

One might object that raising personal factors commits an *ad hominem* fallacy in which motives or circumstances are accosted irrationally rather than arguments. But the well worn arguments of UFO skeptics have been painstakingly perused. When dubious arguments are stubbornly repeated by intelligent and well educated persons, a consideration of their motives or circumstances is warranted, indeed begged for, for explaining the anomaly.

An intriguing point on anomalous UFOs, notes Stanford University physicist Peter Sturrock, is if there were nothing to UFOs scientists who investigate them would lose interest. The exact opposite is true.

Notes

[1] Maj. George Filer, USAF (Ret.), *Filer's Files* 20, 23 May 2000.

[2] Timothy Good, *Above Top Secret: The Worldwide UFO Cover-Up* (NY: William Morrow, 1988), pp. 374-375.

[3] See Timothy Good, *Alien Contact: Top-Secret UFO Files Revealed* (NY: William Morrow, 1993), p. 82.

[4] See *Physics Today*, Vol. 52, July 1997. From *Filer's Files* 28, 15 July 1999. Also see D. Deutsch and M. Lockwood, "The Quantum Mechanics of Time Travel," *Scientific American*, Vol. 270, March 1994, pp. 68-74.

[5] See "NASA Breakthrough Propulsion Physics Project," Breakthrough Propulsion Program (BPP), *www.grc.NASA.GOV/WWW/BPP*, 12 Sept. 2003

[6] See J. Brody's interview of neuropsychologist Dr. Nancy Etcoff, Dr. D. Perrett, and Dr. J. Langlois in "Notions of Beauty Transcend Culture," *Themes of the Times* (NY: Prentice Hall, The New York Times, 1994) p. 1.

[7] See Michael Talbot's *The Holographic Universe* (HarperTrade, 1992), David Bohm's *Wholeness & the Implicate Order* (Routledge, 1996), and Rupert Sheldrake's *The Rebirth of Nature* (Inner Traditions, 1994), noted by Michael Hesemann and Philip Mantle, *Beyond Roswell* (Marlowe, 1997), p. 193. My criticism of the determinism is augmented by others. In "Matter is Consciousness," S.C. Malik says that "by confining consciousness to an effect in the brain of an assumed (but ex-hypothesi unknowable) external cause, it excludes from knowledge the very object that the knowledge seeks and claims to embrace." See J.V. Narlikar, *The Nature of Matter*, Vol 4 (DK Ltd., 1995, and ignca.nic.in/ps). And in "David Bohm, Compassion, & Understanding" (*Sunrise magazine, Theosophical University Press*, Aug-Sep 1997), James Belderis notes other implications of the holographic universe: It "has infinite levels while still acting like one indivisible unit, that higher levels determine the nature of what exists at lower levels, and ... everything is intimately interconnected as an organic whole. What happens in one locality affects what is happening on that entire level, and sets up infinitely complex *chains of causation* above and below... [emphasis]."

[8] William Schaaffs, *Theology, Physics, and Miracles*, Tr. by R. L. Renfield (Washington DC: Canon Press, 1974), pp. 35, 41.

[9] Theodore Schick and Lewis Vaugn, *How to Think About Weird Things* (Mountain View, CA: Mayfield Publishing, 1995), p. 6.

[10] Dr. Kevin Randle, Capt. USAFR, and D. Schmitt, *UFO Crash at Roswell* (NY: Avon Books, 1991), pp. 105-6.

[11] Good, *Above Top Secret*, p. 414.

[12] LTC Philip J. Corso (Ret.), *The Day After Roswell* (NY: Simon & Schuster, 1997), p. 32.

Since crashed planes are often destroyed *in toto*, traveling by both land and air would seem quite rational.

[13] *Ibid.*, pp. 89, 90.

[14] David Jacobs, Ph.D., *Secret Life: Firsthand Documented Accounts of UFO Abductions* (NY: Simon & Schuster, 1992), pp. 22, 202.

[15] Good, *Above Top Secret*, p. 344.

[16] See Marti Kheel, "Ecofeminism & Deep Ecology," *Covenant for a New Creation*, Ed. C.S. Robb and C.J. Casebolt (NY: Orbis Books, 1991), p. 142. In regard to gender feminism being called 'politically correct' in higher education, see Wendy McElroy's "A Conscientious Objector to the Gender War" (*Ifeminist.com*, June 17, 2003). And for a 'biodenial' in which gender is regarded as a mere social construct, see Daphnie Patai and Noretta Koertge's *Professing Feminism: Education and Indoctrination in Women's Studies* (NY: Lexington Books, 2003), pp. 135. This study also documents an ideological agenda of academic feminists that, as noted in my work, explains their hostility to irrelevant 'otherworldly' areas of research such as that on religion or UFOs. Dr. Koertge is Professor Emerita in the Department of the History & Philosophy of Science at Indiana University and Dr. Patai is Professor of Literature at the University of Massachusetts. Both Patai and Koertge were professors of Women's Studies for decades.

[17] See A.J.S. Ragl, "Inside the Military UFO Underground," *Omni* 16 (April 94) 50-54. Timothy Good expressed his opinion in a letter to me, on 22 June 1999, that many of Dean's claims are bogus.

[18] Good, *Above Top Secret*, p. 241.

[19] See S. Friedman, *Top Secret/MAJIC* (NY: Marlow & Co., 1996), p. 5 of the document after its introduction on p. 166. The above is a summary.

[20] See Pierre Guerin, "Are the Reasons for the Cover-Up *Soley* Scientific?," *FSR* 28 (1982) 2-6.

[21] Corso, *The Day After Roswell*, pp. 89, 117, 134, 211, 214,

[22] See Good, *Above Top Secret*, pp. 106-11.

[23] *Ibid.*, p. 116.

[24] See Good, *Alien Contact*, p. 185. In regard to element 118, Lazar's claim of stable superheavy elements was not a howler. Despite retracted claims about 118, the *Physics Web* notes: "A team of American and German scientists has discovered two new 'superheavy' elements... 118 and 116 — along with the discovery of element 114 at Dubna in Russia earlier this year — [confirming] the existence of the so-called island of stability... 'We have jumped over a sea of instability onto an island of stability...', said Victor Ninov, the first author on the paper describing the results." The paper was submitted to *Physical Review Letters*, 8 June 99, physicsweb.org. Thus although some of this research may disagree with some of Lazar's claims, are not other seemingly improbable claims generally supported?

[25] Good, *Above Top Secret*, p. 295.

[26] *Ibid.*, p. 379.

[27] Good, *Alien Contact*, p. 231, emphasis added. Tim Good informed me that, with the

exception of a recorded conversation with Adm. Bobby Inman, Oeschler's reports are unreliable.

[28] P. Delgado & C. Andrews, Circular Evidence (Phanes Press, 1989), p. 174.

[29] Ibid., p. 174.

[30] Corso, The Day After Roswell, p. 109

[31] Ibid., p. 109.

[32] Ibid., p. 109.

[33] See Friedman, Top Secret/MAJIC, for p. 6 of the document.

[34] See L. J. Wang, A. Kuzmich, and A. Dogariu, "Gain-assisted superluminal light propagation," Nature 406, 277-9 (2000) and Dr. Lijun Wang, NEC Research Institute, http://www.neci.nj.nec.com/homepages/lwan/faq.htm, 6 May 2003, emphasis added.

[35] See Maj. George Filer (Ret), Filer's Files 22, 5 June 2000, p. 10, and J. Leake, Science Editor, London Times, posted by Dr. Rex C. H. Kim on 4 June 2000 at http://maxmet.com/bbs/messages/46.html

[36] Ibid., p. 10 and http://maxmet.com/bbs/messages/46.html

[37] W. H. Newton-Smith, The Rationality of Science (London: Routledge & Kegan Paul, 1981), p. 14. In terms of his point about current theories not being entirely true, see Evgeny Podkletnov and Giovanni Modanese's "Impulse Gravity Generator Based on Charged $YBa_2Cu_3O_{7-y}$ Superconductor with Composite Crystal Structure," arXiv:physics/0108005, v2, 3 August 2001. Their experiment of generated artificial gravity either violates or is anomalous to Einstein's general relativity. The article lists Podkletnov as a researcher at Moscow Chemical Scientific Research Centre; Modanese at the California Institute for Physics & Astrophysics.

[38] SETI Institute Website www.seti-inst.edu, March 6, 1998. See the 2003 website which seems more open to the prospect of UFOs.

[39] See physics professor Bernard Haisch's "Open Letter to His Colleagues," in Appendix I, and philosophy professor Neal Grossman's article "On Materialism as Science Dogma" at www.ufoskeptic.org, 2003

[40] Cf. Good, Alien Contact, p. 139.

4

Singular People and Strange Places

S INGULAR PEOPLE INCLUDE researchers who have presaged that ET may be ubiquitous in the universe, and link it to an angelic spirituality, as well as those whose allegations about alien propulsion systems begot praise by notable scientists. Strange places range from remote alien bases on earth, about which White House insiders have been concerned, to urban areas in broad daylight where police report not only alien confrontations but unearthly machines. Clearly, standard responses in the social and natural sciences beg for a reexamination of assumptions.

Some Key People

Key people range from Robert Achzehner who investigated UFOs during its golden days in the 1950s to John Altshuler, M.D., Clinical Professor of Medicine in hematology and pathology in Denver, Colorado. Whereas Dr. Altshuler raised public concern over mutilations of cattle, concern for the entire spectrum of UFO phenomena was anticipated by Achzehner.

Engineer Robert Achzehner, c. late 1990s, outside his home in California.

Robert Achzehner's Insight

Though Robert Achzehner was inspired to investigate UFOs and novel technologies after sighting a bizarre aerial phenomenon, he is virtually ignored in the literature. In an interview with the author in 1994, he said that his investigations were sparked by a maneuvering saucer which was observed with several friends at a backyard cookout in 1950 near Louisville, Kentucky. That Achzehner persisted in esoteric inquiries is revealed by the *New Energy News* (March 1997): "Electrical Engineer Robert Achzehner visited T.H. Moray in his Salt Lake City Laboratory during the summer of 1957 [to demonstrate] a new metal that he had invented." Robert was asked to use a 1/4 high speed drill bit to bore a hole. "It was impossible to drill the metal and after several attempts," stated *NEN*, "both the drill bit and... metal remained cool." Bringing to mind reports of recovered UFOs when there were vain attempts to drill their surfaces, there "was no heating due to friction."

𝔸

Having lost patience with the media's hypocrisy and self-exaltation by 1991, the prominent aviator John Lear questioned sarcastically who would report UFOs... "Tom Brokaw?"

𝒴

Having been the National Service Manager for Motorola Automotive Products Division and Quality Control Engineer for Mitsubishi Electronics Company, Achzehner also had experience in electromechanical engineering that made him well qualified for inventing a device to detect the magnetic fields of UFOs and to conduct field investigations for the Aerial Phenomena Research Organization (APRO). Beyond his work for APRO, his lectures and interviews on UFOs anticipated concessions in later years by prominent scientists. In an interview with *The Dallas Morning News* on November 20, 1965, his statement that UFO "Reports are more frequent in... Brazil" suggested a reluctance by the US news media to report the sightings. There were, of course, as many sightings in America as any where else in the world.

Former CIA pilot John Lear, of Lear-Jet family fame, echoed this point in a lecture to the 1st International UFO Conference in 1991 at Tucson, Arizona. And in the same interview Achzehner said "it's a little egocentric... to think of ourselves as the only intelligent beings.... There could be others more advanced, living on some of those millions of other planets." Certainly, a denial of the planets reflected a pre-Copernican egocentric attitude of many scientists, belying their self-avowed humility. In response to this arrogance, astronomer and Air Force consultant Allen Hynek cautioned that "Supercilious attitudes, pontifical *ex cathedra* statements... that authority be worshiped" do not help "to see science as an adventure pursued in humility."[1]

One need only contrast other scientists to Achzehner to appreciate his farsightedness in 1965. For example physicist G. Feinberg and chemist R. Shapiro noted fifteen years later in *Life Beyond Earth* that we may be "alone [in the universe]."[2] Dr. R.C. Pine stated twenty four years later in *Science and the Human Prospect* that there is "no consensus as to whether any star other than our Sun has planets" and it is unlikely "we have been visited [by ET]."[3] Then, after discovering a planet outside our solar system in 1994, the University of California's Dr. John Welch, at the Radio Astronomy Lab, stated that a detection of water at the Markarian-1 Galaxy is "very exciting" because it indicates life-supporting water "everywhere in the universe."[4] Finally, Cardiff University astrobiologist W. M. Napier reported in *Notes of the Royal*

John Lear echoed many of the insights of Robert Achzehner. A captain for a major US Airline, Lear flew missions worldwide for both the CIA and other government agencies. He became interested in UFOs after talking with military personnel who reportedly saw a UFO land at Bentwaters Air Force Base (UK) and three small aliens approach the Wing Commander.

Astronomy Society (Oct. 2003), in agreement with Sir Fred Hoyle and other colleagues, that there is evidence of a chain reaction of life-bearing particles, ejecting at a mean rate of ~10^{20}, that seeded our "Galaxy within a few billion years." He added, "it is unlikely that life originated on Earth."

To object that scientists move cautiously by empirical evidence is to miss the point. As Welch implies, the Cosmological Principle suggests Achzehner's insights apart from the caution. Admittedly, there were others who had foresight and intellectual independence. Independence and the courage to exercise it are especially needed in regard to the UFO question. The question continues to be suppressed by an epistemological totalitarianism in academia, noted by Dr. John Mack, and by a "scientific tar and feathers" lamented by Allen Hynek. But it is noteworthy that Achzehner pursued the issue at a time when intimidation was exacerbated by a greater lack of public awareness of the question.

After decades of UFO research and even some claimed telepathic communication with an alien who informed him where its spacecraft would appear, Achzehner was reluctant to ascribe a physical nature to these creatures or their craft. In an interview on Century-Cable TV's *The Cosmic Egg*, he spoke of favoring a paranormal reality in which aliens are on a spiritual mission. This position situates him more with Strieber or Vallée than with Randle or Hynek who hold a nuts-and-bolts view of UFOs and a biological nature of aliens. Admiration for his contributions is not lessened by questioning the degree it is influenced by an exaltation of science.

The modern ideal of scientific knowledge may lead from astonishment over UFOs to a belief in their miraculousness, similar to 'primitive people' who think the gods must be crazy when they drop a coke bottle from the sky. But make no mistake: Robert, whose death in 2000 left us in shock, was highly intelligent. Unlike many Americans who are increasingly ignorant of science, he was a first rate experimental engineer. At the time of my interview, he was a consultant for DPI Labs and a minister with an "ontological approach to ascending consciousness."

Robin Adair and the Press

Adair was a reporter and photographer with the *Associated Press* from Albuquerque, New Mexico, who was present in 1947 at the newsroom of the *Roswell Daily News* when Bill ('Mac') Brazel retracted his claim about finding debris of a crashed disc on his ranch. He had said the disc was seized by military officials at Roswell Army Air Field who, first affirming the crash, later said it was only a weather balloon.

Robin Adair witnessed how Mac Brazel adopted the new official story, of which other reporters were also being advised, from Gen. Roger Ramey at Fort Worth Army Air Field in Texas. The press in Texas, the reporters at Roswell, and the military press conferences serve to reveal the extent of both public and governmental interest. Yet after the Roswell UFO crash in 1947, American reporters never again heard any official admit of anything like flying saucers, let alone of a saucer that crashed. Indeed, Lt. Col. Philip Corso says that an elite government group, Maj-12, was formed after Roswell to ensure the utmost secrecy. And in alluding to the academic community, he states that after the military's fake balloon story, snitches and skeptics in the university, as well as hacks in the news media, were doing the government's work! This work meant treating "flying disk reports as the delusions of citizens who had seen too many Buck Rogers movies."[5]

Mac Brazel: first witness to the debris of the Roswell UFO crash. This was the first wire-photo sent out after the alleged crash by Robin Adair, during an interview with the *Daily Record* in early July, 1947.

Thus it is significant that Adair boldly reported the Brazel story over the years in contrast to his colleagues in the mainstream news media who flatter themselves as our republic's fourth-estate watchdog with a sacred mission. In a way similar to the university's zealous defense of tenure to pursue truth, the press jealously guards its right to investigate truth about government cover-ups. But like selective inquiries of many academics, investigations of cover-ups such as Watergate

> **Funeral Notice: Robin D. Adair**
> *Albuquerque Journal*
> **(Jan. 27, 1999)**
>
> Robin D. Adair was born on August 17, 1901, in Texas. Robin passed away on January 26, 1999, while in the Albuquerque Manor Nursing Home. Mr. Adair is survived by two sons, Robin D. Adair, Jr. living in Westminster, Colorado, and Richard Weldon Adair living in Boulder, Colorado. Mr. Adair was employed by the Associated Press for approximately 40 years, starting on 8-31-24. His last position was as Traffic Bureau Chief in the Albuquerque office of the AP. He retired from the AP on 9-1-66. A Memorial Service will take place in the Strong-Thorne Mortuary, Palm Chapel, 1100 Coal S.E., Albuquerque, New Mexico, at 2:00 P.M. on Thursday, Jan. 28, 1999. Strong-Thorne Mortuary is in charge of arrangements.

are lauded when a possible 'Whitewater' of cosmic proportion is studiously ignored.

If governments give free reign to notorious mobsters in trade for information, which *is* done, might they not tolerate alien activity on earth in exchange for a hyper-advanced technology?

Other witnesses bear on this point. As opposed to Adair who tried vainly to revive the true story about Brazel with no support from the national press, Col. Corso notes that the press helped the Air Force use Project Blue Book as a façade to "make the general public happy."[6] Tidbits of *genuine* UFO information, with the whole channeled to higher top-secret levels, was reported to distort the truth or to gradually raise public awareness. But the reporters "fell on the floor laughing or sold the story to the tabloids, who'd print a drawing of a large-headed, almond-eyed, six-fingered alien."[7] He adds with irony, "But that's what these things really look like because I saw the one they trucked up to Wright Field."[8]

Question: Where were all the investigative reporters and college professors pursuing truth when "Flying saucers [were] buzz[ing] over Washington, DC, in 1952, [with] plenty of photographs and radar reports to substantiate" the incident?[9] The military denied it, without a whimper from the pursuers of truth, while encouraging films such as *The Man From Planet X* to blow off the pressure "about flying disks. This was called camouflage through limited disclosure, and it *worked*."[10]

Purportedly, the cover-up continues after five decades because a ridicule of witnesses was augmented by a disinterest, and even by a complicity, of the

Members of a NASA Academy Display Group, posing facetiously as Men in Black (MIBs), reveal the extent of the MIB controversy. MIBs are said to be elite government intelligence agents who investigate UFO reports. Certainly, there are reliable cases of men in dark suits, with enigmatic government credentials, who menace witnesses. The cases become bizarre, as depicted in the TV series *X-Files*, when these men are thought to be aliens who materialize mysteriously and do such things as drive outdated black Cadillacs in perfect condition, as if they came at the wrong time in time machines.

US news media. Having lost his patience with its hypocrisy and self-exaltation, the famous aviator John Lear asked sarcastically who would report UFOs? Would it be "Tom Brokaw? Tom wants Sam Donaldson to risk his credibility. No one... is going to risk their reputation on such outlandish ideas regardless of how many people report sightings."[11]

Thomas Adams and Alien Cooperation

Adams is a researcher who, by 1980, had documented hundreds of cattle mutilations. Their link to UFOs was fortified by his other investigations at Dreamland, Nevada, near the Nellis Air Force Range and nuclear test site. He reported that military personnel told him that they had seen other personnel operate alien discs. It was suggested that the discs, seen near the mutilations, indicated a cooperation between aliens and the government. If a government were concealing its possession of alien spacecraft and its collusion with aliens, would it engage in activities that draw attention to itself? Would it permit crimes that provoke state and federal investigators who are unaware of secret projects? The answer may be 'yes': *Once a reality of UFOs is accepted,* which is scientifically viable, *it is a short step to admit of official attempts to keep UFOs secret at whatever cost, for political, military and cultural reasons. And with a likelihood of UFO crashes, it is likely the crashes would yield live as well as dead aliens. Thus it is equally likely that aliens and humans have communicated and that this has resulted in either cooperation or hostility.* But this would not be all.

Since a hostility of aliens would lead to our decimation, given their far superior technology, can not cooperation be supposed? This could include technology traded for 'benign experiments' on our creatures. Detections of the collusion might be an acceptable risk due to discoveries that bear momentously on national security. Mundane analogies may strengthen this prospect. The *Associated Press* (Sept. 10, 1999) reported that US officials conspired with notorious mobsters who, in trade for information, were "given free rein by the FBI to commit crimes short of murder." Might not the virtual murder of citizens be all the more tolerated when national security itself is at stake? In another case, "The US government secretly hired hundreds of private companies during the 1940s and 1950s," reported *Reuters* (Sept. 6, 2000) "to process huge amounts of nuclear weapons material leaving a legacy of poisoned workers and contaminated communities that lingers to this day...."

⋏

Foreign agents in America may have discovered more than the American Congress. Congressional inquiry into UFOs was stonewalled by its own military establishment.

⋎

Logically, there is a slippery-slope reasoning that is *not* fallacious. Crashed aliens who survive could conceivably become liaisons for their species and ours. Their technology could be traded for the government's toleration of alien investigations which range from the earth's composition to a biological nature of its creatures. Notwithstanding a cogency of the reasoning, however, many persons will find it and the reports to Adams much too extreme to accept.

Hugh J. Addonizio and Congressional Inquiry

While not noted in any UFO literature,[12] Addonizio adds credibility to the presence of UFOs since he was a colorful Democratic Representative born in New Jersey in 1914. Raised Roman Catholic, graduating from Fordham University, he rose to Vice President of A&C Clothing Co. and served with honors in World War II. Graduating as a Lieutenant from Army OCS, he served over three years with the 60th Infantry and fought from Morocco to Normandy and Battle of the Bulge. At the time of his discharge as a Captain, his decorations ranged from the American Theater Campaign Ribbon to a Ribbon with the bronze arrowhead and 8 campaign stars for the European-African and Middle Eastern Campaigns. In short, Addonizio was a tough but well educated patriot with enviable business, military, and political experience.

There can be no doubt that he would have risked his life for his country and not have pursued government secrets if they posed a security risk. These facts bear on his input into UFOs after he was elected. Elected to six Congresses from 1949 to 1961, he was one of the rare Congressmen in 1961 who opposed the official withholding of information on UFOs from the public. Congressmen read a plethora of news to be informed from many viewpoints. Thus although Addonizio's denouncement of the secrecy was triggered by his access to

Former army officer, war hero, city mayor, and Congressman who pressed bravely for the release of classified UFO documents, Hugh J. Addonizio lived from 1914 to 1981. This portrait was taken in 1954, during the "golden days of UFOs," in Orange, New Jersey.

studies of the National Investigations Committee on Aerial Phenomena (NICAP), his increasing interest would inevitably have been piqued by specific newspaper reports.

The New York Sunday Times reported, 28 February 1960, on Adm. Roscoe Hillenkoetter's admission that the Air Force had "silenced its personnel" on UFOs. As one of the highest ranking Naval officers and former CIA Director, his remark drew Congressional attention to two succeeding articles in the *New York Times*. On January 9, 1961, one noted that the official communist news acknowledged widespread "flying saucer" reports in the Soviet Union. But on February 8, 1962, the other echoed an incredulous claim by the US Air Force that it found absolutely no evidence of any "flying saucers" in 483 investigations in the United States! Interestingly, KGB files, released after a collapse of Soviet communism, revealed that Joseph Stalin ordered his spies in America to find out about the UFO crash at Roswell in 1947. This was near Los Alamos, Nevada, where the network had previously probed US nuclear secrets with disturbing success. In any case, Stalin may have discovered far more than Congress. Its inquiry into UFOs, in which Addonizio played an important role, was stonewalled by the military. The obstruction was aggravated by a host of other pressing political issues to which attention was demanded by the legislature's constituents.

Captain Robert Adickes' Bizarre Sighting

Robert Adickes and Bob Manning were flying a TWA DC-3 at 8:30 p.m. on 27 April 1950 at 2000 feet towards Gushen, Indiana, when a glowing disc began to pace them at 175 mph. It looked solid, unlike a cumulus cloud, and traveled in a self-directed mechanical way. Suddenly, it accelerated to twice their speed in three seconds or less and then *disappeared*. Although it has been held that this acceleration indicated that the craft must have been unoccupied, Mark Cashman noted that a formula for acceleration (a), where $a = (2d)/(g)(t^2) = 2*293/ (32.2)(3^2) = .09G$, shows that the G-force does not support that finding. Nevertheless, the UFO's *disappearance* suggests an initial acceleration followed by one with a much greater magnitude. Besides a possibility that the craft may have been occupied in virtue of having an

Captain Robert Adickes

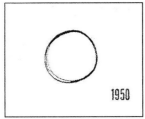

1950

A deceptively simple sketch of the UFO: One witness, the flight hostess, stated that it looked like "a large red wheel rolling beside us." Another witness on the plane noted that it "was the color of a neon sign — just a large red disc. Before, I joked [about] the stories of saucers — but I would not do it any more." After pacing the TWA DC-3 for a while, the UFO doubled its speed and vanished.

advanced technology that negated G-forces, physicist Dr. Robert Sarbacher noted that if aliens had small bone masses like insects, they might endure an acceleration without negating G-forces. Adickes and Manning typify reports after World War II which intrigued a growing minority of scientists. The latter neither spurned all witnesses of UFOs as kooks nor arrogated all UFOs to either illusion or natural phenomena.

Ⱥ

Despite being a paradigm knowledge-yielding enterprise, modern physics cannot be the last word either on the credibility of witnesses or on whether UFOs are extraterrestrial.

Ɏ

Some scientists, not wanting to patronize the crew by denying them the power of their own observation, admitted that the craft might be alien. But they offended the larger scientific community. In dismissing any technology better than their own, they echoed the musical film *Oklahoma!* in which naïve cowboys visit a modern city and sing that it had gone about "as fer as she could go!" The renowned scientist Dr. Stanley Jaki, by analogy, referred dismayingly to his colleagues who disparaged new discoveries. Schrödinger, Bohr, and Planck were similar to those who, being good at physics but bad at metaphysics, naïvely and arrogantly believed that their work was "the last word in physics."[13]

NASA physicist Alan Holt's *Field Resonance Propulsion Concept*, among other research, refers to a possibility of superior electro- or hydromagnetic processes of UFOs that would make them glow and negate internal forces of gravity.[14] And although the object seen by Adickes did not leave an exhaust, its solid physical nature and virtually abrupt acceleration bear a statement by Northwestern University astronomer Allen Hynek. He noted that the association of such craft which exhibit "at times near-zero inertial mass yet able to leave physical traces," is surely a phenomenon that is beyond the reach of "mid-twentieth century physics."[15]

Scientist John Ahrens on a Police Report

Ahrens was a psychologist with the Condon UFO Committee at the University of Colorado. He and physicist Roy Craig investigated a state trooper's report of a saucer hovering 40 feet in front of his squad car during a routine patrol.[16] The object was about six to ten feet off the ground, tilting 15° from the horizontal, emanating blinking red lights from 'windows.' Suddenly, it glowed, rose with a siren-like sound, ejected a fiery exhaust underneath, and shot out of sight. When the trooper returned to his barracks, he felt sick and was unable to account for twenty minutes. Though there was no material evidence, he handily passed hypnotic tests, a polygraph exam, and psychological evaluations. His superiors also testified to his reliability and sober Marine-Corps background.

The Condon Committee's report was published as a book with over 900 pages under an auspices of the US Air Force. There was a sufficient number of credible photos and reliable witnesses to provide compelling circumstantial evidence for countering the dogma that all UFOs can be glibly dismissed as natural phenomena, delusion, hysteria, or popular myth. Contrary to a *myth* propagated in American universities and by Condon's negative conclusions, the cases revealed UFOs to be real and the need for research. Indeed, many scientists after its publication confessed to a revitalized interest in the subject. And interest was clearly strengthened by Dr. Ahrens who went beyond what

Air Force officers discuss Proj Blue Book. *Officially*, it sought 1) to explain witnessed UFOs, 2) to determine if UFOs pose a threat to the US; and 3) to assess whether UFOs have an advanced technology that the US could exploit. *Unofficially*, the Project was to *salve* the public's alarm over an explosion of sighted UFOs. That is, it was advanced to make the public think that their government was "doing something."

physicists could say by lending a credibility of the human sciences.

⅄

A UFO with a fuselage arm at the front and three at the rear, seen by a former British intelligence officer, darted down from the sky with the front arm in the lead.

⋎

This point underscores that the natural sciences, to which Condon's expertise as a physicist was restricted, have limits in assessing the nature of UFOs. Admittedly, natural science bears on both material evidence and observation by providing theoretical interpretations of them. But the philosophy of science poses some nasty problems for the interpretations. In speaking of the interpreted phenomena as data, there is a knotty epistemological problem of the Underdetermination-of-Theory-by-Data Thesis. It holds that any data interpreted by one theory permits an empirically equivalent but logically inconsistent interpretation by another theory.[17]

Leslie Akhurst's Probe of a Most Unusual UFO

Leslie Akhurst was a member of an S4 intelligence unit in the British Ministry of Defense (MoD). With several government scientists, he probed an extraordinary UFO incident on 26 October 1967 at Moigne Downs, England.[18] The incident involved J.B. Brooks, a former RAF intelligence officer and retired flight administrator for British Airways, who reported a UFO when he was walking two dogs. The object was not an ordinary flying disc or glowing light but rather a translucent circular object, with one leading fuselage-like arm at the front and three at the rear, which darted down diagonally from the sky with the front arm in the lead. The strange object reportedly leveled off at an altitude of 200 to 300 feet, stopped abruptly, and hovered silently about a quarter mile away.

Suddenly, two of the rear arms swiveled open to form four arms 90° apart. Brooks estimated they were 75 feet long and with a circular center 25 feet in diameter. The object hovered for over twenty minutes, unaffected by a strong wind, after which the two front arms swiveled to the rear and it ascended out of sight at astonishing speed. Besides its novel features, what assuredly enhanced the report's interest to the MoD was its location between the Winfreth Atomic Power Station and a US Air Force Unit at Ringsted Bay. Akhurst's conclusion that the UFO was attributable to vitreous floaters in the witness' eyes was vigorously denied by Brooks. While having had a corneal transplant operation several years before, he noted that eye floaters generally

77

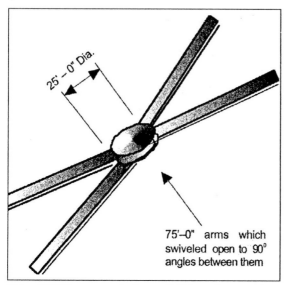

25'-0" Dia.

75'-0" arms which swiveled open to 90° angles between them

This odd UFO was seen by former British intelligence officer J.B. Brooks. The object darted downward from the sky, leveled off at an altitude of 200 to 300 feet, hovered silently a quarter mile away, and then two rear arms snapped open in mechanical precision to form four arms at 90-degree angles. Ominously, the craft stayed poised between an Air Force Base and an atomic power station before its arms folded to the rear and it flew upward at astonishing speed.

move up or down in contrast to the craft.

A relegation of the phenomenon to eye floaters *(muscae volantes)* illustrates a scientific approach in which ordinary explanations have priority *ceteris paribus* over extraordinary ones. But the extraordinary one of a UFO that maneuvered horizontally and vertically outweighed the ordinary one, in this case, which admitted of eye floaters that moved only vertically. Moreover, a preference for the ordinary was offset by an extraordinary witness. As a former intelligence officer, Brooks had an impressive professional background. While his background and these facts may be insufficient to resolve the case in favor of an alien craft, surely the case did not warrant the more mundane conclusion.

In drawing a conclusion from insufficient information, as in the case at hand, one commits the fallacy of an *appeal to ignorance*. And in being an uncharitably negative conclusion at that, one may suppose that this irrational gambit would foster growing suspicions of UFO cover-ups. Presumably, the government wishes to avoid attention to the very things that arouse suspicions of a conspiracy. If a principle of military intelligence proscribes a risk of lying when unnecessary, then, without being facetious, the principle needs to be augmented by not averting the truth by fallacious reasoning.

Daniel Fry and Aerojet-General Corporation

Listed in *Manufacturing USA* as a manufacturer of space propulsion units with 8,000 employees at La Jolla, CA, Aerojet-General entered the UFO controversy in 1954 when Daniel Fry was doing field work at the White Sands Missile Range in New Mexico.[19] A pioneering rocket engineer, Fry made the astonishing claim that he had flown in a pilotless saucer by an invitation beamed down from a mother ship. With this ship as the source of a voice booming out of the landed saucer, an alien reportedly said his race had inhabited earth thousands of years before, that they now live in space, and that their life-spans are over twice those of humans. After his encounter, he was allegedly contacted five years later by this alien who, calling himself as 'Alan' (*Stranger*), said he had acquired a birth certificate and was posing as an international businessman to influence our governments.

Though alien contact with the world's governments seems outlandish, Fry's report of the UFO's propulsion system won some respect. His description of acceleration by a force identical in nature to a gravitational field, in *Steps to the Stars*, led MIT Professor of Engineering Dr. Parry Moon to say that "A more rigorous treatment might be of great value to scientists."[20] The question ensues of why a rigorous treatment was not explored by laboratory infrastructures such as one at Aerojet-General.

Having noted an oddity of Aerojet not exploring Fry's claims — if that is in fact the case, is it not likely that his claims about the propulsion would have been so far beyond our knowledge that scientists would not have been able to grasp it, despite Dr. Moon's interest? And even if there was adequate

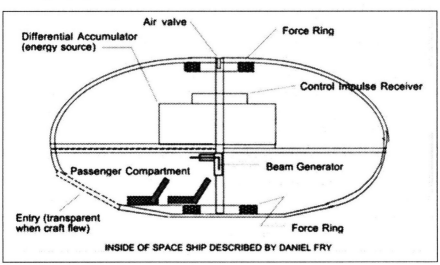

INSIDE OF SPACE SHIP DESCRIBED BY DANIEL FRY

Daniel Fry (1908-1992) drew this cross-section of a UFO, wih no pilot, that allegedly flew him to New York City and back in an impossibly short time

information for experiments, the data may well have been insufficient to garner much financial support due to Fry's incredible story. Finally, even if his story had been accepted and attempts made to exploit the technology, a subsequent lack of any publicized sensational discovery is not as odd as it is reminiscent of many incidents in which stunning new technologies were ignored irrationally.

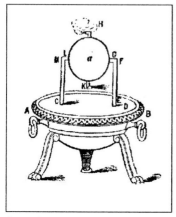

A mechanical device, c. 200 B.C., often described as the first steam engine. Its development was ignored by humans for nearly two thousand years because labor was cheaper. The bent hollow arms cause steam to rotate the globe, just as water-wheel rotations are produced by out-flowing water.

When jet propulsion was discovered, propeller craft were not replaced because of politico-economic reasons. Would these reasons keep UFO technologies from being admitted publicly?

For example, steam engines go back over *twenty centuries* to the scientist Hero in Alexandria who described its ability to open doors. But there was no "need to harness the power of steam because the labor and toil of slaves and animals was sufficient."[21] And the British Luddite workers destroyed new textile machinery in the early 1800s since it threatened their employment. Also, jet propulsion was discovered in the 1930s by Hans von Ohain and Sir Frank Whittle.[22] Leading to a Draper Prize in engineering, it did not result in a moratorium on inferior propeller aircraft, despite a life-costing handicap in World War II, due to economic concerns which influenced a negative reaction by politicians. Though the Nazis put their own discovery of the jet into use in 1944, the allies were fortunate that it came too late to be used effectively.

An even more dramatic case is that of Nikola Telsa who worked with Thomas Edison in 1884. Though he won the support of George Westinghouse, he was studiously ignored by fellow scientists despite his enviable discoveries: a theory of the particle-weapon beam over fifty years before the Strategic Defense Initiative (SDI), development of a cyclotron, and the conversion of "circumscribed small-area antigravity fields out of electromagnetic fields."[23] Forgotten until after the Roswell UFO crash, the antigravity was allegedly noted by scientists who surveyed the debris. "Telsa's name kept bubbling up" because the craft "must have converted an electromagnetic field into an antigravity field."[24] In short, new technologies threaten complacent scientists and undercut vested interests in established industries, evoking political suppression. These points weaken the skeptic's objection that if major recent discoveries had their source in UFOs, then, given their vast superiority, there would be either a moratorium on old industries or sudden technological advances that are made public.

Akkuratov: A Soviet Report

Valentin Akkuratov was a war hero in the Soviet Union, a pilot on arctic expeditions, a member of the Papanin Expedition to the North Pole, a pilot of the first USA-USSR passenger flight, and chief navigator for polar aviation. When flying his Tupolev-4 over Greenland during the Cold War, he saw a pearl-colored disc fly parallel to him. Since he felt it might be an experimental US military craft, he angled into the clouds to avoid any conflict. But on coming out of the clouds, Valentin again saw the strange object and decided to move in. The disc reacted by flying parallel to his aircraft for over fifteen minutes. Suddenly, it accelerated, changed its course, and vanished at an incalulable speed. His descriptions never became more definitive than that the UFO paced his plane, was like a pearl-colored eyeglass lens, had pulsating wavy edges with no exhaust, and accelerated at an impossible rate of speed.

British Parliament member Lord Clancarty is said to have added that five different extraterrestrial space vehicles landed while President Dwight D. Eisenhower looked on.

⩔

Dr. Felix Zigel, of the Moscow Aviation Institute, challenged the assumption that the UFO could be relegated to an atmospheric plasmic phenomenon because of its partial translucency.[25] Only four to five-inch lightening balls, after storms are moving plasmic phenomena that have distinctive shapes. And aside from their spherical shape and vastly smaller size, there was no storm. A more scientifically informed explanation for the large luminescent disc was that it had electromagnetic capabilities — inferable from our own more primitive technology — that caused its fuzzy translucent appearance. And while this appearance is true of balloons, the object's kinematics were intelligently controlled. Thus although it is not *deductively certain* that Akkuratov had observed an anomalous intelligently operated craft from an alien world, that sort of logical certainty is not at all necessary for drawing a reasonable conclusion.

Admittedly, an overzealous conclusion one way or another about the UFO could commit an Appeal to Ignorance. This *ad ignorantiam* fallacy occurs when ignorance is appealed to for drawing a conclusion. For example, although the fallacy would be committed by inferring "The disc *was* an alien device" because "No one proved it *wasn't*," the fallacy would apply equally to inferring "It *wasn't* such a device," since "No one proved it *was*." No one proved it was, or, indeed, that a host of other creditable sightings involved alien craft.

Seal of Soviet Intelligence Agency

Yuri Andropov, former Soviet leader and Chief of the feared KGB, ordered a 13-year program for all military personnel to monitor UFOs over Soviet territory. Igor Sinitsin, his aide in the Politburo, told *The Observer* that in 1977 he discovered that Andropov kept a file on UFOs. At the time there was speculation about a large UFO over Petrozavosk. Sinitsin, now 70, said he monitored the foreign press whereupon he gave Andropov a *Stern* magazine article on UFOs. "I dictated a summary of the piece to my secretary and was sure [he would] express some doubts...." Sinitsin was shocked by the KGB Chief's calm reaction. Andropov gave him an official report from the counter-espionage directorate. "It described a UFO... in Astrakhan that an officer had seen while fishing." Due to Andropov's interest, two committees were formed in 1978 in order to investigate UFOs and four million soldiers were ordered to file detailed reports of incidents. Platov said the program led to hundreds of thousands of recorded sightings before the breakup of the USSR in 1990. *(Report based on Iolik4, 25 Mar 03, Alpha Block: KGB chief orders 4m soldiers to keep watching the skies for UFOs, Akvilon Trident)*

However, evidence can lead to probable conclusions. Inductively, a conclusion is highly probable that the craft was not the product of any terrestrial technology. Given a patent unlikeliness that this technology was terrestrial, can not one infer beyond a reasonable doubt that the UFO was extraterrestrial?

Did President Eisenhower Meet with Aliens?

With Edwin Nourse of the Brookings Institute, Gerald Light of Borderland Sciences Research Associates, Los Angeles Catholic Bishop James McIntyre and newsman Allen Franklin, President Dwight D. Eisenhower is said to have witnessed the an *otherplane aeroform* in April 1954 at Muroc Field (aka Edwards Air Force Base). Reportedly, the aeroform turned out to be a saucer that violated our laws of physics. Even more bizarre is the claim that President Eisenhower, though having arranged to be in nearby Palm Springs to golf, was secretly whisked to Muroc while reporters were fed the cover story that he needed to see a dentist. There, he witnessed the landing of a disc and talked to a *star visitor* who approached him.

Allegedly, this alien *visitor* requested that Eisenhower make public the alien contact, wherein the President protested that humans needed more time to prepare for this stupendous fact. Reportedly, British Parliament member Lord Clancarty said the event was conveyed to him by a former high-level test pilot. "The pilot was one of the six people," Clancarty said, who were "present at the meeting of Eisenhower with the beings. He had been present as a technical adviser because

British Parliament member Lord Clancarty allegedly stated that a high-ranking American pilot told him that five different alien space vehicles landed in California at Muroc Field (aka Edwards AFB). Three were saucers and two cigar shaped. While Eiseinhower looked on, an extraterrestrial *person* emerged to meet him. Purportedly, the aliens resembled humans, but not exactly.

of his reputation and capacities as a test pilot." Clancarty is said to have added that five different extraterrestrial space vehicles had landed. Three were saucers and two cigar shaped. And while Eiseinhower looked on, an extraterrestrial emerged and met with him. The aliens resembled humans. While intriguing, these claims are among the less reliable ones. Still, it is interesting that Lord Clancarty formed a group to investigate UFOs in the House of Lords in 1978. This commission was then directed by Admiral Lord Hill-Norton.

Dr. John Altshuler on UFO-Related Mutilations

In the 1990s John Altshuler, M.D., was Clinical Professor of Medicine, specializing in hematology and pathology, at the University of Colorado Health Sciences Center in Denver. He officially investigated cattle mutilations, amid sightings of glowing discs, which occurred largely in the Southwest and West. His photomicrographic samples and analyses of some of the mutilations resulted in his conclusion that a "surgical procedure... took place quickly, probably in a minute or two," which "utilized high-temperature heat (for example, laser) as a cutting source applied as a fine probe or cutting instrument."[26]

Human technology at that time included lasers, but they would have taken much longer to do the operations and involved heavy cumbersome equipment. In short, given that the equipment was hyper-advanced, that there were no animal intrusions at the mutilation sites, that the sites excluded duplicable surgery by humans, and that there were reliable witnesses who saw glowing discs, Altshuler drew an extraordinary conclusion shared by other scientists: An inability to explain the cattle cauterizations by a technology on this planet, implied that they were administered by the technology of another world.

SOME UNUSUAL PLACES

Places that utilize military surveillance by aerial technologies are obviously the best candidates to detect anomalous phenomena. But there is virtually no

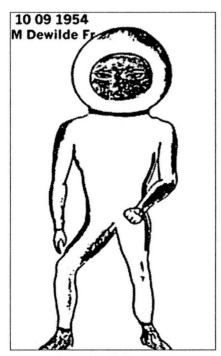

10 09 1954
M Dewilde F

An Odd Mutilation Report

On 10 September 1954 at about 10:30 pm, in the French village of Quarouble, steel worker Marius Dewilde heard his dog howl. After finding the animal crawling near railroad tracks, he saw a dark mass on the railroad and heard strange sounds. Lighting a lamp, he observed two small broad-shouldered beings with helmets and clothing that resembled diving gear. Dewilde sought to confront them, but a green light was flashed that paralyzed him, while leaving him conscious. He saw a machine slowly ascend, suddenly shoot up, and then disappear with a reddish-colored luminosity. There was the aftermath of a hot wind and distinct ozone odor. Later, the Police Chief was convinced of the witness' sincerity. And personnel from the French Ministry of Defense certified that there were five fresh symmetrical prints of an object that was estimated to weigh approximately 30 tons. Afterward, Dewilde suffered various ailments and a loss of breath. His dog continued to behave in an unnaturally nervous way until its premature death in six months. Several days later, three cows died mysteriously and an autopsy revealed that they were devoid of any blood.

place in the world, however inhospitable to humans, which has remained untouched by phenomena that is allegedly impossible. The places discussed here are as intriguing as any and beg for some renewed attention, starting with Afghanistan.

Afghanistan Then and Now

The *Political Handbook of the World* indicates that from 1953 until the Soviet invasion in 1979 there was enough political stability to report UFOs.[27] More feedback came from foreigners, sensitive to worldwide news of the phenomena, than from the introverted nomadic tribes in the vast rural areas where the craft are usually seen. Among foreigners were Soviet and US military advisors who monitored each other and continually watched the sky. Accordingly, it is not surprising that most UFO reports came from those advisors, although some were from US Army intelligence and the Defense Intelligence Agency (DIA).

On 8 November 1959, for example, US army intelligence personnel reported to the DIA that a "huge luminous object" was flying at an extraordinary speed and that it crashed into the Shurad Mountains. They stated that it caused tremors, but no evident loss of personnel.[28]

A second report in November and third in December of a bright circular object that vanished in minutes, led a US Army Colonel to assert sarcastically that UFO reports meant

Made declassified by the Freedom of Information Act, this US Army Intelligence document states that a UFO was seen in Afghanistan on 8 Nov 1959: "a huge luminous object was seen moving at great speed over the sky in Kandohar... it blew up with a loud roar on SHURAD mountains, causing slight earth tremors in the area."

that Afghanistan was catching up to the progressive nations. Intended to debunk the reports, his ridicule illustrates an *ad hominem* attack as well as a fallacy of *false cause* in which UFOs cause societal progress! This species of the fallacy, *post hoc ergo propter hoc* (after this so because of it), is similar to the comical reasoning that since an earthquake struck Kabul after a General's speech, it is imperative for the safety of the people that he keep his mouth shut!

Seen flying over several towns of Balochistan, a UFO landed near a PAF air base in the tribal area of the Rajanpur district on 15 Aug 2000.

Was the UFO monitoring a potential conflict? Was the crash due to its disruption by radar? An advanced technology of extraterrestrial UFOs might have been vulnerable to our peaceful technology but, ironically, invincible in regard to our assault weapons. With a development of laser and particle-

beam weapons, there have been allegations that UFOs are now susceptible to both. These weapons were not employed on 15 August 2000, however, when a fleet of UFOs was seen at 8:15 p.m. in Kandahar. Eyewitnesses told the *Afghan Islamic Press* that missile-looking "flames of fire" flew across the border in neighboring Pakistan. And in Pakistan, a UFO was reported to have landed in the tribal area of Dera Ghazi Khan.

"It was nose diving and after a few moments the sky was lit up," a tribesmen was quoted as saying in *DAWN* (28 August 2000). Sources ruled out casualties since the area was deserted. Uraniumrich Baghalchor is desolate,

Alleged photo of the recovery of a crashed UFO during World War II by troops of the Soviet Union in an unknown area — though perhaps in or near Afghanistan, thought to be sometime between 1940 and 1945.

while Rounghin is thinly populated. A political assistant in the tribal area confirmed the report and stated that a team was sent to ascertain facts. It was the second UFO that landed in Punjab within two weeks. Earlier, on August 15, a UFO flew over towns in Balochistan and *landed* near a PAF Air Base in a tribal area of the Rajan-pur district. The debris was removed by 'sensitive agencies' a few days before, but the divisional administration would not discuss a recovered UFO.

"Instrumentation for UFO Searches"
Ch. 9, p. 1214 , by Frederick Ayer II

V_{nom} = 48.9 km/mm nom

V_{max} = 96.6 km/mm max

V_{min} = 0km/mm

These velocities should be compared with those of the UBO in Haleakala II, that is, ~142 mi/min = ~228 k/min. If the UBO was in orbit, the distance GB is the projection of its path SB making an angle Beta with the line of sight. Assuming that the velocity in Haleakala II is typical, then

$$\text{Sin } ?_{nom} \sim \frac{48.9}{228} \sim 0.214$$

$$?_{nom} \sim 120$$

and the object was in a highly elliptical orbit. Alternatively, the distance OB might have been the projection of the apogee of the ballistic trajectory of a body launched in a retrograde direction.

Investigation showed that no sub-orbital missiles were launched from Vandenberg AFB or Pt. Mugu until one or more hours after this sighting. The Aerial Phenomena Office at Wright-Patterson AFB suggests that it might have been an artificial satellite on which information is not readily available. The object is thus in the unidentified category.

From Fred Ayer's "Instrumentation for UFO Searches," *Scientific Study of Unidentified Flying Objects* by the University of Colorado under contract with the US Air Force – Dr. Ed Condon, Scientific Director *(Final Report originally copyrighted in 1968 by the Regents of the University of Colorado – published in reports of the Air Force & other government agencies. Published commercially by Bantam, OP).*

UBOs V. the APO Office at Wright Patterson

Scientists usually resort to observatories to identify UBOs (Unidentified Bright Objects), but the little- mentioned Aerial Phenomena Office (APO) at Wright Patterson Air Force Base was noted by Fred Ayer in "Instrumentation for UFO Searchers."[29] A physicist at Brookhaven National Lab and research high-energy physicist at the University of Colorado, he described a tracked UBO at the Hawaiian Haleakala Observatory. A magnitude of its brightness was greater than $Mv = +3$, where a five-magnitude difference corresponds to a brightness ratio of 100 such that $Mv-1$ is 100 times brighter than Mv +4. By this convention, the sun is -26.72, the full moon is -12, Venus -3.2 to -4.3, and the faintest magnitude visible to the naked-eye is about +6. The UBO traversed an elliptical orbit with a 228 km/min velocity (8,500 mph). While no suborbital missiles were launched from Vandenberg AFB or Pt. Mugu until an hour afterward, the APO said "it might have been an artificial satellite on which information is not available."[30] Still, Ayer left no doubt about a reasonableness of holding that some UBOs were extraterrestrial devices. The powerful magnetic fields of UFOs, a salient feature of flying discs, may "be investigated by proton magnetometer measurements."[31] For political reasons, no doubt, Ayer deferred paradoxically to the APO by concluding that facilities should not be wasted on tracking UFOs since "these sightings are so infrequent."[32] Even if they were infrequent, sheer scientific curiosity about artificial UFOs of alien origin should far outweigh mundane concerns for the facilities.

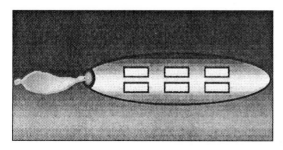

The Alabama UFO Controversy
In July 1948 at 2:45 a.m., pilot John Whitted and copilot Clarence Chiles observed a dull-red object on a collision course with their DC-3 at 5000 feet as they approached Montgomery, Alabama. The UFO veered pass them in ten seconds, stopped abruptly, reversed direction, and paced their plane before suddenly disappearing in overhead clouds. A passenger stated that the UFO, on its initial pass, appeared as if a bright light had flashed by. The pilots, describing it at a slower speed, reported a 100-foot long, cigar-shaped object with two rows of lighted windows, a bluish glow beneath, and a redish-orange exhaust flame about 50 feet long. But the UFO was dismissed as a meteor by astronomy professors Donald Menzel of the Harvard College Observatory and Allen Hynek of Northwestern University.

Though Hynek came to increasingly believe that many UFOs were extraterrestrial,[33] Menzel paraded an unrelenting skepticism before the public. Ironically, he was an alleged member of Maj-12 who doodled at Harvard faculty meetings by drawing pictures of aliens. Was his being one of the few persons who had ever seen aliens related to his being one of the most zealous debunkers? This would explain a legacy of antagonism to UFOs at Harvard where Dr. John Mack was harassed in the 1990s. In the case at hand, Menzel expressed a disingenuous skepticism in *UFOs: The Modern Myth*.[34] He held that the meteor scenario was consistent with the pilots' final report since it denied any rocking effect on the DC-3 as would be expected with an exhaust detected by flames. But this retort was a bad-faith legalism because the pilots' initial description, altered under pressure, referred to a prop "or jet wash rocking our DC-3."[35]

The Alabama incident was used by skeptics to challenge our *seeing* UFOs. Since our perceptions sometimes deceive us, a point made famous in René Descartes' *Meditations*, we cannot be certain that any are true. But this skepticism was countered by St. Augustine who did not appeal, as Descartes, to the proof of a God who would not deceive us. At the dawn of the medieval age, Augustine distinguished our perception of oars being bent in water from judgments that they really are bent. Precisely, their looking bent is how they *should* appear. The appearances corroborate judgments based on scientific theories, say one of optics which explicate and predict the appearance. And since we rely on true theories for our very survival, we can have a general faith in appearances.

⋀

Harvard psychiatrist Dr. John Mack notes that some UFO debunkers are themselves debunked by their own extraterrestrial experiences!

⋁

St. Augustine also presaged another abuse of scientific reasoning. Over fifteen-hundred years before a principle called 'Pessimistic Induction' was formulated in the philosophy of science, he cautioned scientists not to reject anomalous events, such those noted in the Bible, because scientific laws change over time. Scientists should not patronizingly impose current theory on the perceptions *of others* since it excludes the very phenomena that, counting against scientific theories, enhance theoretical developments. These developments now include UFOs whose *impossibility* may be possible in terms of future theories. He suggested that scientists do not know the complete truth and that those who think they do impede science in its name.

Alamogordo, New Mexico
Alamogordo is near the White Sands Proving Grounds where, in the 1940s, nuclear weapons and military craft were tested. These craft included the V-2 and A-9 Rockets. Skeptics say these were confused with flying saucers. But others counter that those devices could not perform as the saucers and that, given their intelligent maneuvers which far exceeded the capability of the rockets, it is reasonable to ponder whether they belonged to aliens who were concerned with military developments. Atomic bomb tests might naturally arouse alien alarm about wholesale terrestrial destruction and a human race posing future threats in space. The old debate continues over experimental craft versus UFOs.

For example, a military report of a crashed disc in 1947 near Alamogordo was almost immediately silenced by the military and said to be a weather balloon. On the one hand, the balloon scenario found some support from local military officers. An officer from Alamogordo, Major W.D. Prichard, held that personnel at his relatively nearby base had sent up the balloons.[36] On the other, two civilians reported to the *Roswell Daily Record* on 8 July 1947 that they saw the object. The latter was said not to be a balloon since it descended faster than a conventional craft, as if it was going to crash, and "looked oval in shape like two inverted saucers."[37]

The saucer-balloon dispute rose again in 1995 when the US Air Force admitted that the balloon story was a cover! But the cover was allegedly for a nuclear test balloon to detect atomic-bomb testing by the Soviet Union.

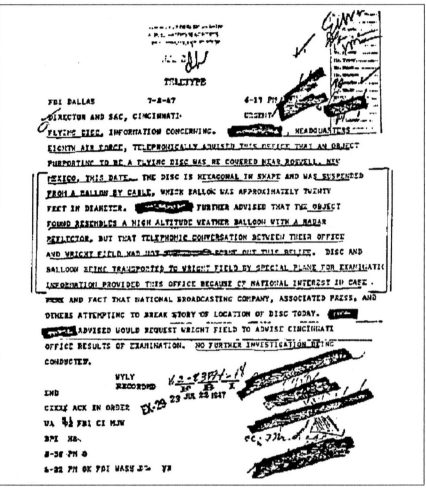

On advice from the military about a crashed disc near Alamogordo, the Dallas FBI teletyped SAC and Cincinnati FBI near Wright Field — where a disc and alien bodies were reportedly sent. It warns in part: "National Broadcasting Company, Associated Press... [are] attempting to break story of location of disc today."

There was also a spectacular claim of former NASA engineer Robert Oechsler. Having interviewed the relative of a deceased government employee, he stated that this person had been part of a UFO retrieval operations team in 1947 that found unknown clear material of a crashed UFO as well as dead and injured aliens. One was rushed to Alamogordo, possibly to a medical station, where *it* later died.[38]

Skeptics say that such allegations play into the hands of the military which exploits false UFO claims as a cover for its secret experiments. This might have been plausible due to a fear of imminent nuclear attack during the Cold

War, from 1945 to 1990, when almost any mode of deception might have been deemed legitimate. Now, however, national security per se may be the cover for UFOs! Dr. Kevin Randle (Capt., USAFR), notes that Alamogordo could headquarter a new UFO project named Aquarius or Bluepaper.[39]

Strange Lights in Alaska

These lights were seen by the late UFO debunker Dr. Donald Menzel, Professor Emeritus of Astrophysics at Harvard and former Director of the Harvard College Observatory. In seeking to show how easily even an astronomer can be deceived by what he *thinks* he sees, Menzel debunked his own UFO sighting in an Arctic zone near the Bering Strait on 3 March 1955.[40] When he was flying in a commercial aircraft at an altitude of 20,000 feet, he saw a bright UFO with flashing green and red lights shoot toward the craft. Having a diameter of about one-third the moon, the object stopped at an estimated distance of 300 feet away, executed evasive movements, disappeared and then reappeared over the horizon.

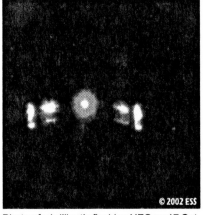

© 2002 ESS

Photo of a brilliantly flashing UFO on 17 Oct. 2002. Bringing to mind the Alaskan sighting, this UFO was recorded by a judge near Spokane, Washington. This and similar objects appeared to both land and leave the ground, as well as to interfere with commercial air traffic.

Menzel did not doubt his perception but its interpretation. The 'UFO' was a fuzzy image of the bright star Sirius that appeared to move because of a 'random walk' of light beams through irregular atmospheric layers. Its disappearance and reappearance were due to a mountain that momentarily cut off its light. Dr. James McDonald, Professor of Atmospheric Sciences at the University of Arizona, noted that this account was absurd because it required detailed refractive measurements through hundreds of miles of atmosphere tangent to earth. That this earthly phenomenon was more unlikely fact that no other interviewed astronomer ever saw anything like it. In short, the famous debunker of UFOs may really have witnessed a UFO!

Intriguingly, McDonald's assertion that a UFO debunker had actually seen a UFO bears on Harvard psychiatrist John Mack, M.D. He noted good naturedly that a debunker, who supposedly lied about an abduction, had really been abducted by aliens! At a debate sponsored by the Committee for the Scientific Investigation of Claims of the Paranormal (SICOP) in June 1994,

Mack was criticized by a former patient who claimed she had fabricated her story to expose the chicanery of the "UFO community."[41] But in sticking to his diagnosis that she was indeed deeply troubled and had been abducted, Mack and McDonald suggest that some UFO debunkers might themselves be debunked by the misinterpretations of their own authentic UFO experiences!

Bizarre Places: Alien Bases

A bizarre possibility of extraterrestrial bases on earth is related to the suspicion that many nations are engaged in UFO cover-ups. The cover-ups might have their origin in crashes of alien craft, such as the alleged crashes at Roswell in 1947 or Del Rio in 1950, because living as well as dead aliens and their technology could have been recovered. The recoveries, spawning contact by either the surviving extraterrestrials or attempts by their species to rescue them, might lead to their communication with us and our cooperation with them. This evokes the idea that governments might permit aliens to establish bases in exchange for the aliens not making their visits public, say by landing openly in Paris, until an agreed time.

Alien bases allegedly concerned the White House. An Army officer at the Pentagon stated that an elite unit tried to get the President to set up posts "to report on any UFO activity."

That is, despite a ridicule of UFOs by both the mainstream news media and academic community, *the only way official secrecy could be kept in confidence is if a government knew in advance that aliens would not suddenly reveal themselves in a public way.* In other words, a cover-up suggests that at least some officials have advance knowledge. And this knowledge, in turn, invites a specter of cooperation as well as communication. The fantastic idea of aliens collaborating with a government, and thus of alien bases, was directly supported by Wilbert Smith, senior radio engineer for the Canadian Department of Transport.

Based on contacts with top scientists and his government sponsored work on duplicating extraterrestrial technology, Smith stated unequivocally that "every nation... has been officially informed of the existence of the space craft and their occupants from elsewhere."[42] A more immediate relation of the occupants to alien bases comes from the report of a university professor and two colleagues. On a hunting trip in December 1989 at 4:30 a.m., they reported that a metallic disc had ascended silently from a camouflaged opening at a remote American-Australian Defense Space Research Facility at Pine Gap, Australia.[43]

Another link is a glowing egg-like craft that radiated Betty Cash and Vickie Landrum in December 1980 at Huffman, Texas, when they stopped their car to watch the UFO descend and hover beforethem.[44] The increasingly astonishing reports are that over twenty unmarked helicopters escorted the craft, that it was an alien machine operated by government pilots, and that it was part of a military project called 'Snowbird' which involved a government agreement with aliens. In trade for the machine, the aliens were allegedly allowed to operate it from designated sites if they did not publicly disclose their existence and otherwise disrupt society.

Moreover, authorities in the Laguna-Cartagena Cabo Rojo area of southwest Puerto Rico have been besieged historically by UFO sightings. Along with reports of various alien species, there were also allegations of extraterrestrial bases. A 1956 report of glowing discs leaving and entering a lagoon was followed by others. Tall blond humanoid aliens were reported in 1964, for instance, followed by other claims in August 1990 of three to four-foot tall aliens with slanted eyes, long arms, three fingers on each hand, and three toes on each foot.[45] Usually, the UFOs are elusive. In some cases, there are reports of alien craft flying with US military planes as well as claims of aliens on the ground near of high-ranking officers.

A base at Pine Gap was established by the US and Australia to track satellites, with apparent involvement of the CIA, NSA and NRO. Underground facilities are rumored to extend below ground 12 stories or more. And in addition to tunnels laid out like spokes of a wheel that extend several miles from the center, there is reportedly a bore hole 20,000 feet or more below the base. It allegedly contains an ultra low frequency antenna used for secret experiments such as ones related to Nikola Tesla's resonance theories, and for low worldwide frequency communication. Ostensibly, this is used by US intelligence agencies and by the ultra secret Australian Defence Science & Technology Organization (ADSTO) that deals with crashed UFOs, test flies them and allegedly even harbors aliens themselves in the facility. Three hunters, one of whom was a US college professor, reported that in 1989 at 4:30 am they saw a large camouflaged door open on a hill side in the security compound. The metallic disc ascended from a black hole, tipped on its edge, and disappeared at a tremendous speed. The door slowly shut into camouflaged surroundings, after which they were unable to see anything unusual about the area. If the report is true, was the UFO flown by alien or human pilots? Other UFO reports occur repeatedly.

Finally, it is said that there are cases in which alien bases seem hostile to humans. Bringing to mind the film *The Thing* (1952), the former Chief of the US Army's Foreign Technology Division stated that the existence of alien

Photo of the Laguna Cartagena Refuge, said by some researchers to harbor an alien base — based on repeated reports of people native to the area. "The present lagoon," states the US Government, "is a remnant of what was once a large open expanse of water." Moreover, there are uplands, abandoned sugar cane fields, and 263 acres in the foothills of the Sierra Bermeja. Geologically, the hills are some of the oldest in the whole Caribbean area.

bases concerned the White House. Lt. Col. Philip Corso said that when he was there, the working group (Maj-12) tried to get President Eisenhower to set up "formal listening posts — electronic pickets staffed by army and air force observers... to report on any UFO activity."[46] This group, under Gen. Nathan Twining, held that if extraterrestrial biological entities had plans for semipermanent bases, they would be locations such as the polar regions, "in the middle of the most desolate surrounding they could find, or even underneath the ocean."[47] That is, the UFO becomes an unidentified underwater object (UUO).

UUOs bear on a report by *The Associated Press* (August 27, 2000). Titled "Putin Honors Crew as Probe Launched," it states, "As military prosecutors began a criminal investigation into allegations that the submarine Kursk collided with an unidentified vessel, Russian President Vladimir Putin conferred posthumous state honors Saturday on the 118 men who died on the sub." The article notes that "Military prosecutors believe an unidentified vessel violated safety rules and was directly responsible for the sinking of the Kursk...." Curiously, "Military officials claim the most likely scenario was that the Kursk collided with... a foreign submarine" but "Britain and the United States denied their submarines collided with the Kursk." Since the mere five-year old sub was constructed of "immensely strong steel," the question arises of how another sub could have caused the calamity without also destroying itself. And although former intelligence officer Maj. George Filer says that Russia may have been hiding a mechanical failure, he admits that the collision with an alien UUO could have destroyed the submarine."[48]

Maj. Filer adds that "Naval personnel who operate sonar... often report high speed comparatively large objects. Ivan T. Sanderson wrote 'Many submarines have just plain vanished, and not only in wartime. Two of them disappeared at the same time in the Mediterranean Sea only a few weeks before the Scorpion — one Israeli, the other French. Also, we have lost other

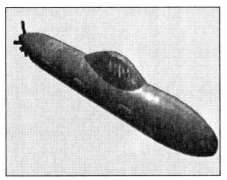

Cigar-shaped UFOs, described by many witnesses, are similar to submarines. Whereas the subs do not yet have airborne capabilities, this image depicts a sensor-coated submarine, of the new "virginia class," with an improved stealth that brings to mind either stealth aircraft that may someday have underwater capability or a future capability of subs to go airborne.

nuclear-powered subs such as the Thresher'.... We can assume these [alien] craft have the capability of operating under the ocean or in the atmosphere with equal ease. *Underwater bases* provide ideal locations.... Throughout the Cold War years, there were numerous chases of mystery subs.... Naval forces would corner the mystery sub in a fjord, but inexplicably the underwater craft would vanish."[49]

Some Concluding Thoughts on Strange Places

Next to ET introducing himself by landing in Washington DC, a specter of alien bases is one of the most mind-boggling possibilities in history. Most investigators would not even entertain the idea except for the inextricable steps of reasoning from credible reports of UFOs *to* UFO crashes *to* recovered aliens *to* communication *to* cooperation. Given a cooperation in which alien bases are permitted in exchange for advanced technologies, there are novel political concerns. What would justify such an unparalleled secret collaboration in free societies? These societies are rooted in citizens being ultimately responsible for the laws of the land. In being entrusted to the highest elected officials, whose *raison d'être* is to serve the people, the laws apply to government officials who are answerable to the citizens.

It is unlikely that contact with ET would uproot all of our traditions. Our own anthropologists seek to preserve the rituals of 'primitive people'. In this case, however, we are those people!

On the one hand, citizen inquiries might be legitimately resisted because of threats to law and order. First, alien bases could reflect a coerced government cooperation wherein to admit of them is to admit that elected officials cannot protect their citizens against superior technologies. Second, in addition to the technologies posing a danger to the world, most citizens might suffer a cultural shock of unprecedented historical proportions. These possibilities could wrongly lead to a loss of faith in our political institutions and spiritual

Monarca Merrifield's stunning depiction of the Red Sea miracle suggests some startling theological implications. At the same time while miracles are an important part of Scripture, many great thinkers such as Ludwig Wittgenstein did not find the Bible profound merely because of miracles but because it explained a spiritual sense of human guilt and need for redemption by a messiah. Still, unprecedented questions arise such as whether aliens suffered a 'Fall' and if so, whether there is a messiah peculiar to *their* species.

traditions, the very infrastructure for our laws that roots present to past and renders cultural stability.

On the other hand, if aliens are known to be friendly, the reports may *not* be entirely suppressed in order to prepare us for the inevitable announcement that ET is here. Films such as *E.T.* and *Close Encounters of the Third Kind* may have been encouraged. Some of them were in fact produced with the aid of former consultants to the US Air Force. Even scientists, theologians, and philosophers may be tacitly countenanced to speculate on the far-reaching implications of extraterrestrial contact. Presumably, counterintelligence operations could prevent things from getting out of hand by, say, feeding some false information to the public. A possibility of misleading the public, addressed later, raises serious ethical questions.

However, dealing with the import of an alien presence at theoretical levels would have a practical payoff. We would be spared unprecedented shocks if the idea of contact were perused before the event and if the event's possibility filtered into popular culture. Our culture would not suddenly suffer a shattering displacement of anthropocentric ideas by more cosmic ones. The latter could needlessly subvert viable traditions that should be retained. Our own anthropologists, afterall, seek to preserve the rituals of a 'primitive people.' In this case, we *are* those people!

A

A fondness for noise and sexy streamlined machines would *alienate* us culturally from UFOs. These silent, odd-shaped craft collide head-on with not only our primitive technology but ideals of beauty and power.

Y

This point does not discredit human religion. It has a profoundness that was upheld by Ludwig Wittegenstein who is widely considered to be the greatest twentieth-century philosopher. Held in a virtual religious awe himself by many secular philosophers, he appreciated how the Bible gave singular expression to his own sense of guilt and need for redemption.[50] The Scripture was not admired merely for its reference to miracles. The latter alone are typically noted by some UFO researchers when they try to relegate the major Biblical figures to either superstition mongers or those endowed by aliens with super-advanced scientific abilities.

The prospect of an alien contact would animate an appreciation of our religion, a re-evaluation of the scientific enterprise, and how our limited technologies influence our popular culture. We need only consider how this culture is challenged by UFOs. Their typical disc and spherical shapes are unsuited to our propulsion systems. The systems of UFOs are silent and fumeless, yet profoundly powerful. These properties are foreign to our aerodynamic understanding of an atmospheric interaction with moving objects. Historically, our mechanical objects have been streamlined, noisy, and smoky. A socially conditioned value of these traits would *alienate* us culturally from UFOs since they collide head-on with not only our technology from shapes of autos to missiles to spacecraft but even our ideals of power and beauty.

One thinks of beauty in terms of 'macho men' who, paradoxically, have an almost effeminate love of their sleek autos. Traditionally, fuss over style is a distinctively feminine trait. And power is related to phallic symbols of a male-dominated culture by many feminists who say that the symbols influence our aerodynamic technologies.[51] But both genders often relate power to noxious noises of internal combustion engines. This fact led a university colleague, in her irritation with traffic, to suggest an inverse relation between intelligence and loudness: The louder a car, the lower a person's IQ! While her remark may seem hyperbolical, it and the other insights reveal how technological norms are infused into popular culture. These cultural nuances would, at least subconsciously, increase our intransigence to accepting a superiority of the objects that maneuver so elegantly in our skies. A specter of open skies and closed minds beg for attention to official UFO reports.

Notes

[1] Dr. Allen Hynek, *The UFO Experience: A Scientific Inquiry* (Chicago: Henry Regnery Co., 1972), p. 207.

[2] Dr. G. Feinberg and Dr. R. Shapiro, *Life Beyond Earth* (NY: Morrow, Publishers 1980), p. 214.

[3] Dr. Ronald C. Pine, *Science and the Human Prospect* (Belmont, CA: Wadsworth Publishing Co., 1989), p. 261.

[4] Dr. Jack Welch, RAL Director—University of California at Berkeley, "Water Discovered In Another Galaxy," *Associated Press*, in *TCE*, June 3, 1994, A6.

[5] Lt. Col. Philip J. Corso, *The Day After Roswell* (NY: Simon & Schuster, 1997), pp. 74, 75.

[6] *Ibid.*, p. 78.

[7] *Ibid.*, p. 78

[8] *Ibid.*, p. 78

[9] *Ibid.*, pp. 78, 79.

[10] *Ibid.* 79, emphasis added.

[11] John Lear, "The UFO Cover-Up," *International UFO Congress*, Tucson, Arizona, 1991. I am grateful to the late Robert Achzehner for a copy of this paper.

[12] See "Addonizio, Hugh J.," *Biographical Directory Of The American Congress 1794-1961* (Washington, D.C.: U.S. Government Printing Office, 1961), p. 461.

[13] See Dr. Stanley L. Jaki, "The Last Word in Physics," *Philosophy in Science* V (1993) pp. 9-32.

[14] A. Holt, *Field Resonance Propulsion Concept* (Houston, TX: NASA, Lyndon B. Johnson Space Center, 1979), p. 3.

[15] Hynek, *The UFO Experience: A Scientific Inquiry*, p. 232.

[16] "Case 42, North Central, Fall 1967, Investigators: Craig, Ahrens, Staff." From D. S. Gillmor,

ed., *Final Report of the Scientific Study of Unidentified Flying Objects:* Conducted by the University of Colorado Under Contract to the US Air Force (NY: Bantam Books, 1969), pp. 389-391.

[17] The UTD thesis refers to Underdetermination-of-Theory-by-Data.

[18] *Cf.* Timothy Good, *Above Top Secret: The Worldwide UFO Cover-Up*, With a Foreword by the Former Chief of Defense Staff, Lord Hill-Norton, G.C.B. (NY: William Morrow and Co., 1988), pp. 64, 68, 69.

[19] See *Manufacturing USA: Industry, Analyses, Statistics, & Leading Companies*, Ed. by A. J. Darnay (London: Gale Research, 1994), p. 2034.

[20] See Daniel Fry's *Steps to the Stars* (Lakemont, GA: CSA, 1965). From Timothy Good, *Alien Base: The Evidence for Extraterrestrial Colonization of Earth* (London: Arrow Books Ltd., 1999), p. 100.

[21] Kendra Bolon, "The Steam Engine," University of Dayton, Spring 2001, at http://www.udayton.edu/~hume/Steam/steam.htm.

[22] See H. Petroski's "The Draper Prize," *American Scientist*, Vol. 82, March-April 1994, pp. 114-117.

[23] *Cf.* Col. Corso, *The Day After Roswell*, p. 110, 238.

[24] *Ibid.*, p. 110.

[25] Good, *Above Top Secret*, p. 569, fn. 11.

[26] See Linda Howe, *An Alien Harvest* (Huntingdon Valley, PA: Linda Moulton Howe Productions, 1989), pp. 94-103.

[27] A. E. Burk, ed., *Political Handbook of the World* (New York: CSA Publication, SUNY, 1993), pp. 3-8.

[28] From a declassified DIA report on 3 December 1959. See Good, *Above Top Secret*, p. 308.

[29] Fred Ayer II, "Instrumentation for UFO Searchers," *Final Report of the Scientific Study of Unidentified Flying Objects*, ed. Gillmor, pp. 761-804.

[30] *Ibid.*, p. 781.

[31] *Ibid.*, p. 793.

[32] *Ibid.*, p. 804.

[33] Hynek allowed for alien UFOs even before *The UFO Experience* in 1972: A Large segment of our citizens hold "that superior extraterrestrial civilizations might visit the earth... as if to make periodic checks on a tribe of aborigines. The theory finds support throughout history in accounts of strange apparitions in the sky.... The scientific community... has labored under the misconception that UFO reports are necessarily made by untutored, untrained, and gullible people." *Encyclopedia Britannica* (Chicago: William Benton, 1969), p. 500.

[34] Donald Menzel, "UFOs: The Modern Myth," *UFOs — A Scientific Debate*, Ed. By C. Saga and T. Page (NY: Cornell University Press, 1975), p. 135.

[35] Good, *Above Top Secret*, p. 264.

[36] Kevin Randle, Capt., USAFR, and D. Schmitt, Dir. Special Investigations, Ctr. UFO Studies,

UFO Crash at Roswell (NY: Avon Books, 1991), p. 215.

[37] *Ibid.*, p. 215.

[38] Good, *Alien Contact*, pp. 101-2.

[39] Kevin Randle, *The UFO Casebook* (Warner Books, Inc., 1989) p.175.

[40] Menzel, "UFOs: The Modern Myth," in Sagan and Page, *UFOs – A Scientific Debate*, pp. 133-4.

[41] P. Monaghan, "Encounters With Aliens," *The Chronicle of Higher Education*, 6 July 1994, p. A10.

[42] *Ibid*, p. 205.

[43] *Ibid*, p. 110.

[44] Good, *Alien Contact*, p. 137.

[45] *Ibid*, pp. 239-40.

[46] Corso, *The Day After Roswell*, p. 229.

[47] *Ibid.*, p. 229.

[48] Maj. George Filer, US Air Force Intelligence (Ret.), *Filer's Files* 33, 21 August 2000, pp. 1, 2. It is now thought, of course, that a missile exploded in the Kursk.   [49] *Ibid.*, emphasis added.

[50] See N. Malcolm, *Ludwig Wittgenstein: A Memoir*. With a Biographical Sketch by G. H. von Wright. NY: Oxford University Press, 1984, p. 60.

[51] See Professor I. Gotz' *The Culture of Sexism* (CT: Praeger Publishers, 1999), p. 67: "airplanes, autos, skyscrapers, rockets, pens... stand up like so many phallic tributes to male creativity." Also, the author heard a feminist philosopher present a paper at the APA Conference in Chicago, in the Fall 1985, which held that the male's sexual organ has determined all technological designs in our male-dominated culture. That our culture panders to egregious feminist distortions of science which promise to politicize any study of anomalous phenomena — much more phenomena that is conventional — is further revealed by the sarcasm of Professor Emerita Noretta Koertge at Indiana University. Having an M.S. in Chemistry, a Ph.D. in the Philosophy of Science from the University of London, and being Chief Editor of the journal *Philosophy of Science*, she is well qualified to criticize a "female-friendly science" whose supporters want to start out girls in physics on the study of wave phenomena or fluid mechanics rather than "those darned old rigid bodies." See Koertge and Daphne Patai's *Professing Feminism: Education and Indoctrination in Women's Studies* (NY: Lexington Books, 2003).

5

Official Reports and Odd Encounters

OFFICIAL UFO REPORTS have allegedly included close encounters with aliens. The reports are intriguing when serious analysis supports their possible veracity. Admittedly, the possibility is virtually nil that many of the fantastic reports of New Age cultists are plausible. But based on the caliber and shared experience of others, the plausibility ranges from serious consideration for well regarded citizens to a difficulty of dismissing scientific summaries by the military. Whereas the citizens often insist that they were anesthetized by tubes and blinding spheres of light near flying discs, the military admits that these discs are singular machines which are "not visionary or fictitious."

Official UFO Reports

Some of the more dependable reports of UFOs, ranging from spheres and cylinders to discs — although discs dominated reports in the late 1940s, are rooted historically in declassified military records which have been compiled by the armed forces since that time. Accordingly, we begin with an official report of the Air Materiel Command.

Reports of the Air Materiel Command
The Air Materiel Command (AMC), commanded by Lt. Gen. Nathan F. Twining, was a division of the US Army Air Forces, now the US Air Force, with headquarters at Wright Field. This field is now known as Wright-Patterson Air Force Base near Dayton, Ohio, where the AMC figured most prominently in the UFO controversy on 23 September 1947. Twining, at that time, filed an astonishing report, with a later declassified copy for The National Archives, in response to an AC/AS-2 request for findings on 'flying discs.'

"The unidentified lenticular aerodyne that has been designated ULAT-1, has been evaluated as a non-air breathing aircraft of unknown origin... designed to operate outside the earth's atmosphere."

Twining's report from the Air Materiel Command states that the discs are real; as large as our aircraft; make extraordinary maneuvers; are evasive, thus revealing manual, automatic, or remote control by some intelligence; have metallic reflecting surfaces; show no exhaust; are circular/elliptical with flat bottoms and domed tops; often fly in formation; make no sound unless exercising high performance; and have horizontal speeds usually above 345 statute miles per hour.

Gen. Twining *(left)* and SoAF Harold Talbott

The AMC report especially stirred controversy in the 1980s when it spoke of a "lack of physical evidence in the shape of crash recovered exhibits which would undeniably prove the existence of these objects."[1] A reality of alien craft was *affirmed unequivocally* but, paradoxically, downplayed because there was no material evidence of crashed discs! British investigator Timothy Good notes that Twining's words were exploited by skeptics to reject the Roswell UFO crash, announced and then retracted by the US Army Air Forces, less than three months earlier in July 1947. That Twining made no reference to the much publicized crash, if only to lament it did not occur, seems extremely puzzling.

Good and others suggest that Twining may have filed his report *as if* there was no UFO crash because the personnel to whom he was reporting may not have had the requisite top secret clearance.[2] Lack of clearance is likely, since Twining refers to a possible high security project for which most AMC and AC/AS-2 personnel would not be cleared. Thus he admits of a top secret clearance without a denial of his own access to it. His access would mean he knew the crash was either alien or only military debris. Given it was *not* this debris, as implied by its not being mentioned, the craft *would* have been alien. What is the import for science?

Doubt among scientists about alien UFOs, since they violate our laws of

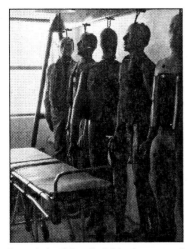

In reply to claims that it had recovered alien bodies in 1947, the Air Force said they were test dummies. Since this was ludicrous because dummies were not used until the 1950s, former Air Force Intelligence Officer Maj. George Filer noted that the Air Force apparently sought to create doubt about whether aliens *really did crash*. Is doubt one way the government seeks to raise our awareness to the fact that ET *is* visiting us?

physics, reveals the significance of new technologies and a military concern for their threat by ET. If there *was* material evidence of a crash, various scenarios bear on its denial: (1) The crash had occurred only two months earlier and was reportedly sent to Wright Field. Thus the debris may not have been sent until *after* Twining's report. This might have led to a secret follow-up memo. (2) The debris may have been delivered on time for his report but insufficiently analyzed for concluding that it was extraterrestrial. Either the destruction or unusual nature of the propulsion system may have excluded a conclusion. (3) Alien debris could have been analyzed on time but Twining's report served as a counterintelligence disclaimer for fear of future disclosure. The third possibility agrees with Timothy Good's view and a truth of the memo as far as it went.

Finally, Col. Corso, former Chief of US Army's Foreign Technology, states that Gen. Twining saw the material and, before returning to Wright Field, conferred with the scientists who were "his brain trust at Alamogordo."[3] His disavowal revealed a need for lower security channels. While a cover-up was required even *within* the military, Twining's memo layed a basis for data exchange with other commands that did not have proper security clearance. Adding further credibility to Corso beyond his reliable background, to take seriously many of his claims, he explains Twining's anomalous report while agreeing with the third scenario for needed secrecy.

Right: Stamped MAJIC EYES ONLY, dated 9

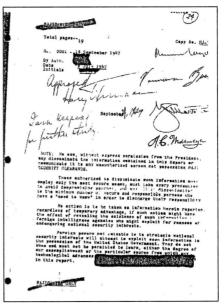

July 1947, and signed by Harry Truman andother top officials who include his scientific advisor Dr. Vannevar Bush, is this reported top secret document by Gen. Twining on a recovered "unidentified lenticular aerodyne" at Socorro, New Mexico, designated ULAT-1. This first page (0020-120) states:

"Note: No one, without express permissionfrom the President may disseminate the information contained in this Report or communicate it to any unauthorized person not possessing MAJ SECURITY CLEARANCE. Those authorized to disseminate such information must employ only the most secure means, must take every precaution to avoid compromising sources, and must [keep ?] dissemination to the minimum number of secure and responsible persons who have a 'need to know' in order to discharge their responsibility. No action is to be taken on information herein [reported on?] regardless of temporary advantage, if such action might have the effect of revealing the existence of such information to foreign intelligence agencies who might exploit [it?] for reasons of endangering national security interests. Foreign powers not amicable to the strategic national security interests will attempt to exploit such information in the possession of the United States Government. They do not know and must not be permitted to learn, either the degree of our accomplishment or the particular source from which any technological advances [blocked out] in this report."

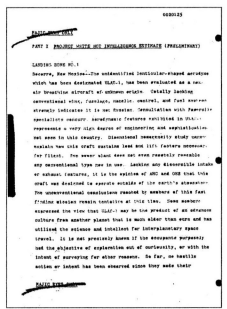

Left: Titled "PART I PROJECT WHITE HOT INTELLIGENCE ESTIMATE (PRELIMINARY)" and "LANDING ZONE NO. 1," page 0020125 addresses the recovered UFO:

"Socorro, New Mexico —The unidentified lenticular-shaped aerodyne which has been designated ULAT -1, has been evaluated as a non-air breathing aircraft of unknown origin. Totally lacking conventional wing, fuselage, nacelle [enclosure for crew or cargo], central, and fuel system strongly indicates it is not Russian. Consultation with [...?] specialists concur. Aerodynamic features exhibited in ULAT-1 represents a very high degree of engineering and sophistication not seen in this country. Dimensional homogeneity study cannot explain how this craft sustains load and lift factors necessary for flight. The power plant does not even remotely resemble any conventional type now in use. Lacking any discernible intake or exhaust features, it is the opinion of AMC and ONR that this craft was designed to operate outside the earth's atmosphere. The unconventional conclusions reached by members of this fact finding mission remain tentative at this time. Some members expressed the view that ULAT-1 may be the product of an advanced culture from another planet that is much older than ours and has utilized the science and intellect for interplanetary space travel. It is not precisely known if the occupants purposely had the objective of exploration out of curiosity, or with the intent of surveying for other reasons. So far, no hostile action or intent has been observed since they made their [from page 0020125 to 0020126] presence known."

Opposite Page, Top: Titled PART III SCIENTIFIC PROBABILITIES, page 0020133 states in part:

"1. Based on all available evidence collected from recovered exhibits currently under study by AMC, AFSWP, NEPA, AEC, ONR [?], NACA, JRDB [?], RAND, USAAF, SAG, and MIT, are deemed extraterrestrial in nature. This conclusion was reached as a result of comparisons of

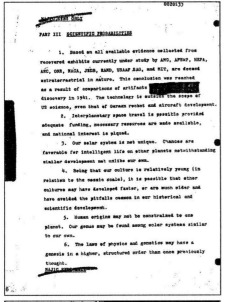

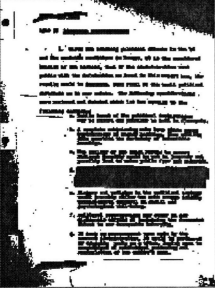

artifacts [blocked out] discovery in 1941. The technology is outside the scope of US science ... 3. Our solar system is not unique... 4. Being that our culture is relatively young (in relation to the cosmic scale), it is possible that other cultures may have developed faster, or are much older... 5. Human origins may not be constrained to one planet. Our genus may be found among solar systems similar to our own. 6. The laws of physics and genetics may have a genesis in a higher, structured order than once previously thought."

Below Left: Titled PART IV <u>POLITICAL CONSIDERATIONS</u>, page 0020-134 states in part:

"1. Given the existing political climate in the US and the unstable conditions in Europe, it is the considered opinion of the members, that if the Administration went public with the information as found in this report now, the results would be damaging, even fatal to the world political structure as it now exists... [Part] e. History and religion in the political context would probably suffer the most damage causing unprecedented upheaval in social and psychological well being.

Reports of the Aeronautical Information Service (AIS)

The AIS Military is a unit at the London Air Traffic Control Center, UK, where reports of UFOs are forwarded. This typifies official government interest in UFOs because they reflect vastly superior technologies that bear on air control and R&D. In being more than a unit for military/police reports, the AIS is listed in the civilian manual of Air Traffic Services for air traffic controllers to contact when UFOs are seen or reported. The relevance of UFOs to national security is indicated by the requirement that descriptions be sent to the British Ministry of Defense.

The MoD's Royal Armament R&D division conducted an inquiry that revealed the British government's interest in even the most bizarre reports related to security. In December 1980 near a hospital redevelopment site,

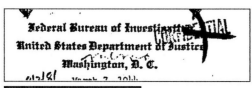

Newspaper image of Sir Winston Churchill, from FBI Files *via* the Freedom of Information Act. Did his alarm about a UFO crash in the US lead to both an investigation of information given him and *his* investigation of the AIS? Churchill asked Lord Cherwell to check into UFOs in 1952: "What does all this stuff about flying saucers amount to? What can it mean?"

for example, forty persons witnessed a triangular object at an estimated 50,000-foot altitude.[4] The object hovered, moved slowly, traversed vast distances in seconds, split into parts several times, and then abruptly shot upward out of sight. These sorts of reports explain politico-military motives in debunking UFOs. Any alien science that affords such 'miraculous' maneuvers, not to mention interstellar space travel, would poignantly reveal a country's absolute vulnerability. If such objects are extraterrestrial, are their occupants demonstrating the defenselessness of terrestrial nations? Since the alien pilots are not exploiting our weakness, are they revealing their benevolent intentions? Or are these beings simply exhibiting their vastly superior technology for some other mysterious purpose?

Given that science is esteemed in modern western culture as an ideal knowledge-yielding enterprise—as a quintessential *rational* endeavor, one of the most intriguing questions is whether an extraterrestrial *intelligence* that has exercised that endeavor, to an almost unimaginable degree, would be moral. Since the time of Plato (c. 400 BC), an increase in moral behavior has been typically held to proceed logically with an increase in rationality and knowledge. Presumably, answers are still being sought by the AIS, its numerous international counterparts, and by various international alliances which would include NATO.

⋏

Since science is a paradigm *rational* endeavor, there is an intriguing question: Would an alien intelligence which exploits that endeavor to an almost unimaginable degree be moral?

⋎

Reports of the Air Force
Office of Special Investigations (AFOSI)
Having its headquarters at Andrews Air Force Base, Maryland, the AFOSI is a counterintelligence service for the US Air Force. Its relationship to UFOs,

A poignantly patriotic plaque at Wright-Patterson Air Force Base, outside the Air Force Museum, for those who served their nation in the US Air Force Office of Special Investigations (AFOSI). The AFOSI has been a field operating agency, with headquarters at Andrews AFB, MD, since 1 Aug 1948. Does its mission include threats from other "worlds" as well as from this world? *One* of its missions is to "Detect and provide early warning of worldwide threats to the Air Force."

including any unidentifiable natural phenomenon or artificial flying object, concerns investigations of potential and actual security threats to the Air Force by such things as anomalous power loses, radar interference, and communication difficulties.

There was an alleged document on 17 November 1980 which stated that results of an 'Aquarius Project' were still restricted to the top-secret level of 'MJ Twelve.' Also, green fireballs were seen traveling in a controlled way at high velocity near both the former Atomic Energy Commission's Sandia Base and Los Alamos, New Mexico, from 1948 to 1950. Further, the associated activities of Jim and Coral Lorenzen, who founded the Aerial Phenomena Research Organization (APRO) in 1952, were discovered to have been monitored. Still, skeptics and some UFO researchers respond that if they were covertly watched, it might have been due to their seeking security clearances.

Further, there were mysterious radar jammings and blackouts at Kirtland Air Force Base in 1980 caused by an extremely bright object that had incredible speeds and sudden stops at odds with any known terrestrial craft. Finally, Project Sigma in 1954 had the alleged purpose of communicating with aliens. For example, there was the report by former military intelligence officer Richard Doty that one alien taken alive, although a mere "common mechanic," was examined by experts who certified that *its* IQs was over "200."

However, in addition to a questionableness of leaked information from such organizations, there is a problem that is especially relevant to counterintelligence units. Any information that is obtained legally or illegally from these units could be intentionally skewed. Precisely, their *raison d'être* is to counter misbegotten intelligence. At the same time, UFO investigators know that if there is distorted information, it invariably contains some degree of truth in order to be believable. And when all of the information in a given document is bizarre, investigators may cogently believe that even the partial truth, among undifferentiated information, becomes significant as well as titillating. Undoubtedly, there is a well founded hope that the truth will be

The US Air Force Security Service was established on 20 October 1948 at Arlington Hall Station, Virginia. The HQ USAFSS moved to a new Headquarters building in August 1953. The AFSS Logo symbolizes the Security Service's multifaceted mission.

ferreted out over time by, among other things, comparing documents in either the same or different government organizations.

Reports of the US Air Force Security Service (AFSS)

Later named the "Air Force Electronic Security Command," the AFSS was responsible for air security within the US National Security Agency (NSA). In directing the security and intelligence branches of the government, the NSA would naturally have turned to the AFSS for threats of UFOs to national security. What make its reports so intriguing, even if they are few, is that the highly trained military personnel who generated them were further restrained by both bureaucratic procedures for reporting and their reports having to ascend an authoritarian chain of command.

In the midst of a public perceived to be 'shrill' at the prospect of extraterrestrial visitors, this command might expect due tactfulness as a matter of military professionalism. That is, the AFSS reports may have deleted buzz words such as either 'UFO' or 'saucer'. However, the reports cannot omit reference to potential threats as could other Air Force organizations with different special services. These services afford discussion of some novel reasons for a possible government secrecy, if not a conspiracy, though they are by no means peculiar to the Air Force Security Service.

⋀

Officials can tangibly assess alien intentions only in light of a *human history* with which they are familiar. But an alien species might have both a vastly higher and more malignant intelligence than the human race.

⋁

With these considerations in mind, the AFSS entered the controversy. It issued reports in early June 1955 that referred to both Unidentified Airborne Phenomena (UAP) and electronic disruptions of military craft. The craft included Boeing RB-47 reconnaissance planes in the initial report. But on June 4 there was a more descriptive report of a plane's gun-warning light

Sgt. Joe Patterson of SBCCOM models a "future warrior" ensemble for the new "Army of One" recruiting campaign.

This "future warrior" (c. 2025) shown by the US Army Soldier & Biological Chemical Command (SBCCOM) is similar to "cyber-warrior" displayed by the Air Intelligence Agency (AIA). The display suggests that future threats will come from both outer space and 'other worlds'. The threat revitalizes the words of Gen. Douglas MacArthur, in a speech at West Point on 12 May 1962, for which he received the Thayer Award. He stated: *"We deal now not with things of this world alone, but with the illimitable distances and as yet unfathomed mysteries of the universe* [emphasis]." On 24 May 1948, less than a year after the reported Roswell UFO crash in New Mexico, the Air Force informally activated the Air Force Security Group (AFSG). It was formally established on 23 June 1948 with officers and enlisted personnel on loan from the Army Security Agency. On 20 Oct 1948, the US Air Force Security Service (AFSS), a forerunner of the Air Intelligence Agency, was established at Arlington Hall Station, Virginia, as a major air command for cryptologic and communications security missions for the Air Force.

flashing intermittently with radar contact of a UAP 7,000 yards distant near Melville Sound, in northwestern Canada. Visual contact was eventually made of a flying object described as being 'glistening' and 'silver metallic'.

After nine minutes, the UFO abruptly accelerated to an inestimable speed and veered north out of visual and electronic contact.[6] While the gun-camera film was useless, it is significant that it was analyzed anyway and that the anomalous object was reported to top security officials. These facts weigh against the object's being a natural phenomenon, rocket, or any sort of wing-shaped plane. Also, the pilots were trained to recognize conventional and experimental craft.

If human beings are seen as being greatly inferior, morally and intellectually, might intelligent aliens view our elimination as being morally justified — given our increasing development of weapons?

Also, it can be reasonably supposed that the AFSS linked a third report on June 7 to a similar object in virtue of another RB-47 encounter involving electronic problems and the object's proximity in time to the previous ones. In this case, the plane was headed toward Elison Air Force Base in Alaska when radar contact was made at 3,500 yards. In addition to a small rectangular appearance on the radar's scope which was interpreted as radar jamming, the plane's target-warning light flashed on and off three times in as

Examined closely by the Air Force Security Service, expert opinion supported an authenticity of this now wrinkled photo by Rex Heflin. A Highway Accident Investigator for the LA County Highway Commission, he took the photo in 1965 near Santa Ana, California.

many minutes. The intermittent flashes caused by the target bring to mind typical descriptions of UFO disappearances and reappearances, their acute high-speed turns, and their almost instantaneous stops and starts. Since RB-47s were often used for manipulating enemy radars, NSA would naturally be interested in *UFO manipulations* of Air Force craft.

Importantly, UFOs are often said to be hostile since they probe military installations and disrupt power sources of our craft. Their attacks are frequently depicted in science fiction as super-advanced 'laser beams.' Admittedly, there are cases of such attacks being reported. And there are reports of physical assaults by UFOs themselves. But can these cynical interpretations be better understood as defensive disruptions? In most instances thus far, there has been little evidence of malign alien intentions according to Nick Pope, a former official in charge of UFO investigations at the Air Secretariat 2-A of the British Ministry of Defense (MoD):

> If aliens have harmed us, perhaps we have brought this on ourselves. In the paranoia of the Cold War, American pilots certainly, and probably Russian pilots, had orders to intercept and, if necessary, shoot down unidentified flying objects... What if some of them were actually alien craft?[7]

While an absence of immediate threats weakens the case for keeping UFOs secret, there can be no doubt that the AFSS held that the long-term motives of aliens were unclear. Admitting of unclear intentions would cause public panic because the government would be unable to protect its citizens should more sinister intentions materialize. And there was official concern about panic over the unknown *per se*. The sheer unknown in the unchecked imagination is as likely to induce a horrid specter of insect-like aliens as it is to evoke an image of the amusing Mr. Spock.

Certainly, policy on secrecy was not set by top authorities of the AFSS but rather by the highest echelon of the NSA or another esoteric intelligence unit and, ultimately, certain high elected officials who may not have known subsequent details. The officials can only assess alien intentions that may

Men in Black (MIBs): Bearing on the Air Force Security Service, the following is a NORAD document on UFO sightings – released by the Freedom of Information Act (FOIA) – that refers to the elusive MIBs at the time Rex Heflin photographed a saucer:

REPLY TO ATTN OF: AFCCS

TO: ADC AFSC HQCOMD USAF SAC AFCS ATC CAC TAC AFLC AU MAC USAFSS [emphasis added]

Information, not verifiable, has reached Hq USAF that persons claiming to represent the Air Force or other Defense establishments have contacted citizens who have sighted unidentified flying objects. In one reported case an individual in civilian clothes, who represented himself as a member of NORAD, demanded and received photos belonging to a private citizen. In another, a person in an Air Force Uniform approached local police and other citizens who had sighted a UFO, assembled them in a school room and told them that they should not talk to anyone about the sighting. All military and civilian personnel and particularly Information Officers and UFO Investigating Officers who hear of such reports should immediately notify their local OSI offices.

initially seem innocent in light of a *human* history with which they are familiar. Given a firsthand familiarity with military attacks, holocausts, and ethnic cleansings throughout the twentieth century, despite 'everyone professing to love everyone', the officials would be not only be naive but derelict in their primary duty of protecting the public if they did not assume that an alien species might be vastly more malignant in its cunning than the human race. Are intelligent races in the universe, who evolved from biophysical systems, protected naturally by Nature's God? Is there a divine intelligence in Nature that both governs evolution and ensures their benevolence? Good will would be necessary to survive their own self-destruction at a critical mass of technology. Akin to a dilemma faced now by humans, I call this understanding the *Self-Destructing-Evil-Species (SES) Thesis.* This Thesis is assumed by UFO enthusiasts who build landing pads as well as by many scientists who hope for contact with a superior alien intelligence by such things as the radiotelescopic *Search for Extra-Terrestrial Intelligence* and *Pioneer 10* Project.

Scientists who hold the SES Thesis assume that *Pioneer 10,* laden with instructions to earth, will be intercepted by an alien species whose intelligence and technology pale those of humans. The hope is rooted in utopian ideas of doing away with poverty, curing disease, and obtaining a quantum leap in scientific progress. These ideas were aired in a variety of films such as *Starman* (1984), *Contact* (1997), and *Mission to Mars* (2000). But the 1950's cult classic *The Thing*, starring James Arness of the *Gunsmoke* TV series no less, depicted the idealistic scientist who is contemptuously devoured by an alien after *it* was naively approached in good will to obtain new knowledge of the universe.

Å

Former MoD Minister Harold Watkinson stated that his signed promise to never reveal certain facts about investigations included "the subject of flying saucers."

Ɣ

Science fiction notwithstanding, the *Self-Destructing-Evil-Species Thesis* may fail to distinguish a species which destroys *itself* from one that terminates a different species. By analogy, some human political ideologies eschew violence for some races and classes but encourage it for others. Also, units which gather UFO data, such as the AFSS, would acknowledge a possibility that humans are viewed as being greatly inferior, intellectually and morally. And although we pose only an immediate threat to ourselves, there might be a future danger to the cosmic environment by such things as nuclear weapons. Thus even if the SES Thesis applies, an elimination or restraint of humans may well seem morally warranted.

Reports of the British Air Ministry (BAM)

A division of the Ministry of Defense (MoD) until 1964, when it was unified into a new MoD, BAM began its investigation of UFOs in 1947 at about the time of the Roswell UFO crash in the America.[8] Its US connection is significant since British military intelligence had information on the crash that was unavailable to US citizens. BAM's inquiries were said to have been made by the Deputy Directorate of Intelligence in room 801 at Northumberland Ave, London. This room for BAM's secret work had *no* name. Other officials heard only rumors about it and former MoD Minister Harold Watkinson said he signed a promise to never reveal facts about flying saucers.[9]

However, many saucer reports became public before they were classified. And others were uncovered by civilian investigators despite their classification. On 19 September 1952, for example, Royal Air Force officers at RAF Topcliffe saw a silver disc tailing a Meteor Jet as the latter approached to land at Yorkshire.[10] The possibility that the disc was a fallen engine cowling,

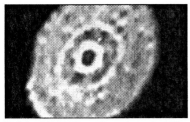

Notched UFO filmed by woman in UK

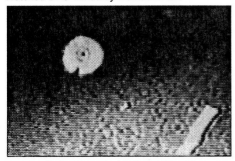

Notched UFO filmed by NASA's STS-75

Top: Multicolored disc flipped over and later vanished in a red flash. Published in virtually all major British newspapers, it was filmed by Sharon Rowlands of Bonsall after hearing an eerie noise on 5 Oct 2000. *Bottom*: NASA reportedly asked to examine the tape because it was similar to a UFO filmed by the STS-75.

jet exhaust, or balloon were quickly discounted. The UFO continued in flight, then slowly descended with a pendulum motion and hovered in a rotating state before suddenly accelerating out of sight at an astonishing speed. Lt. John Kilburn said it was about the size of a Vampire Jet.

This incident bears on Air Minister George Ward who confided in 1953 that his official admittance of UFOs without material evidence would lead to ridicule. His remark cast light on a dilemma of foreign officials. While Gen. Twining's AMC report left no doubt that US officials acknowledged the discs, it sent the covert message that 'due consideration' be given to a lack of physical evidence. He affirmed their *unquestionable* reality by military sightings but, paradoxically, a *questionable* reality because of a lack of material evidence. Ward suggests that most nations do not see a need for crashes to affirm that alien UFOs are real. But any official declaration of the reality would invite an unwelcome attention of the academic and scientific communities which labored under the psychological denial of a cover-up.

The American report affirmed an unquestionable reality of some UFOs as mechanical devices. But any official affirmation would invite the unwelcome attention of the academic and scientific communities.

The apparent cover-up extends to today's MoD. The *BBC News World Edition* (3 December 2002) notes that the British Parliament's Ombudsman Ann Abraham accused the MoD of suppressing documents on the "Rendlesham File" with details about the sighting of a "glowing triangular object" by US Air Force police in Rendlesham Forest near RAF Woodbridge in Suffolk. It turns out that in the early hours of 27 December 1980, US Air Force personnel witnessed the object hover, transmit blue pulsating lights, and send nearby farm animals into a "frenzy." In the report "Unexplained Lights," Lt. Col. Charles Halt, Deputy Base Commander at RAF Bentwaters, adjacent to Woodbridge, told how he saw an object emitting a "red sun-like light" moving through the trees. Former Defense Minister Peter Kilfoyle stated that "The Ministry of Defense is bedevilled by a culture of secrecy."

Reports of Lt. Col. Philip J. Corso

Lt. Col. Philip J. Corso (Ret.) was as an intelligence officer on Gen. MacArthur's staff in Korea, served on President Eisenhower's National Security Council, and was Chief of the Army's Foreign Technology Division (FTD). His position at the FTD provided knowledge of highly classified sightings by the military. Intriguingly, he enumerated military responses to threats of extraterrestrial craft as well as an array of esoteric UFO reports. Some of the reports need to be highlighted.

First, the military admitted of alien spacecraft which passively, though tangibly, disrupted both US defenses and worldwide communications by either electrical or magnetic-field interference. The authorities sought to neutralize the interference by developing circuitry which was hardened against it.[11] Second, his superior in Foreign Technology, Lt. Gen. Arthur Trudeau, as well as others in the Pentagon charged with strategic planning, were concerned about the aggressive behavior of the alien craft. Corso notes that the craft did not just "surveil our spacecraft in orbit" but also "tried to create such havoc with our communications systems that NASA... had to rethink astronaut safety in the Mercury and Gemini programs."[12] In this vein, he also states that there was later speculation among some in Army Intelligence that the "Apollo moon-landing program was ultimately abandoned because there was no way to protect the astronauts from possible alien threats."[13]

⋀

Given that ET's presence might pose an unparalleled threat to humanity and that a raison d'être of government is the security of its citizens, as poignantly illustrated by '9/11', there may be justification for a cover-up. A cover-up morally warranted?

⋁

Third, this seasoned army officer, privy to military intelligence, states that extraterrestrial craft were aggressively buzzing US defenses on the Eastern European frontline in terms of either "looking for blind spots or weaknesses,

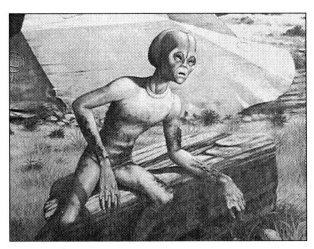

Picture of an alien pilot, based on witness reports, who is alleged to have survived the UFO Roswell crash in New Mexico. Though two aliens were said to have survived, one of whom was killed at the site and the other interrogated, a former Chief of the Army's Foreign Technology Division affirms only that the crash did in fact occur and that he was in charge of having the debris assessed technologically.

or... probing our radar to see how quickly we responded."[14] Blips were seen to shoot across radar screens which could not be identified and that suddenly disappeared. Soon, the blips caused by UFOs would reappear even closer to sensitive installations such as missile-launchers and airfields. Typically, when intercepting aircraft were dispatched and had the UFOs almost in sight, the latter would shoot up into the atmosphere at speeds greater than 7,500 miles per hour.

The only defenses against UFOs in the 1950s and 60s was to get a firm-tracking radar lock, which seemed to interfere with their navigation. Though radar locks signaled to targeted objects that aircraft missiles were locked onto them, some aircraft were apparently able to boost the signal before the alien craft escaped and hits were scored. Intriguingly, at Ramstein Air Base in Germany, an army air defense battalion allegedly succeeded in such a strike with an antiaircraft missile in May 1974. Without noting if aliens were taken dead or alive, Corso states that the alien "craft was retrieved and flown back to Nellis Air Force Base in Nevada."[15]

Then there were sightings of abductions and animal mutilations. Next to a direct assault on our defense installations, these were the most worrisome interventions by ET. Col. Corso cites UFO debunkers who sought to downplay them as hoaxes, natural predators, false childhood memories, and fabrications of the news media.

However, the military took seriously such things as an advanced laser surgery of aliens used in animal mutilations and a troubling similarity among the descriptions made by abductees and other eye witnesses of UFOs and EBEs (Extraterrestrial Biological Entities). The general consensus of military intelligence, relevant to an ethics of secrecy, has three likely scenarios.[16] First,

Thermodynamics Experiment at the Air Force
Institute of Technology
The NAIC and Col. Corso

Graduates of the Air Force Institute of Technology (*above*) would make good candidates for the National Air & Space Intelligence Center (NAIC). It admits of UFO investigations by the Air Force at the time of the Roswell UFO crash. The NAIC had roots in the Foreign Technology Division and in T-2 Intelligence, established in July 1947, and states about its history: "Initially called Project Sign [redesignated Project Grudge in 1949], the UFO program brought T-2's successors their greatest public visibility." Was this visibility to deflect public attention from more secret military inquiries, *i.e.* those of the *Army's* Foreign Technology Division headed by LTC Corso?

the most benign one is that the EBEs, with no intention of causing much disruption, are merely conducting scientific experiments and collecting specimens of earth's life forms. Second, a more vexing scenario is that our life forms and other data are being analyzed to judge if the earth is hospital to inhabit — possibly with humans in a peaceful way? But they are unconcerned about their disruptions (that earthlings could do little about). Third, coupled to our military defenses being tested by UFOs, the most sinister scenario is that the experiments and specimen collections presage an outright invasion of earth.

Given these scenarios and that security is a *raison d'être* of government, there may be justification for UFO cover-ups. Cover-ups morally warranted? A breached secrecy could induce a loss of faith in government and diminish national defense — a critical consideration during the dawn of the Cold War in the late 1940s. A cover-up could only be lifted by adequate responses to the disturbing second and third scenarios. Corso thinks these scenarios were averted by weapon developments of the Strategic Defense Initiative (SDI). Precisely, SDI was the ostensive reason he revealed the cover-up in his book *The Day After Roswell*, in 1997.

Encounters with Extraterrestrials

The expression 'encounters of the third kind,' made famous by the 1970's film *Close Encounters of the Third Kind*, is generally used to denote first-person contacts with intelligent extraterrestrials. The following claimed contacts range from those which seem fanciful to others that have garnered serious scientific attention. This attention has most recently been brought to bear on abductees.

∧

Women often report being impregnated by beings who are half or pure alien. Tulane University philosopher Dr. Michael Zimmerman, in response, issues a cogent challenge to skeptical scientists.

∨

Claims of the Abductees

Some persons allege that they freely chose to board UFOs by 'invitation'. But the term 'abductee' generally refers to those who claim they were forced aboard the craft. Some abductees report they were physically hauled aboard by robotic devices or aliens themselves. Incredibly, some aliens are said to be the size of dwarfs. Other abductees state they were put into a semiconscious dreamlike state — a state of akinesia discussed later — in which they were unable to resist. Reports of abduction are worldwide and range from those of children and senior citizens to unskilled laborers, teachers, and scientists.

Despite some notable exceptions, there are uncanny similarities in the descriptions of alien abductors. They tend to have robotic behavior, are usually said to have either a pale or gray skin tone, thin torsos, disproportionately large heads, and a height of three to four feet — although these features are said to hold more in some parts of the world than others. Having said this, their clothing is typically uniform, if any is worn, and is often described as being similar to skintight jump suits and scull caps. The creatures rarely communicate except by motioning with their arms, by nods, and by conveying messages through mental telepathy.

Typical descriptions by abductees bring to mind a report of child-sized aliens in "jump suits" encountered in Cennina, Italy, who tried to forcibly haul a housewife into their craft on 1 Nov 1954.

Furthermore, these beings tend to ignore questions about either their extraterrestrial cultures or terrestrial missions and appear patronizing in the sense that they *seem* to smile and say 'Don't worry about that.' They also evidence no interest in human societies in regard to religion, philosophy, science, ethics, politics, and art. Finally, the aliens are said to perform physical examinations and genetic experiments on the abductees with enigmatic advanced technologies. And the 'medical' technologies are frequently applied

117

LA DOMENICA DEL CORRIERE

Attack on the woman in Cennina, Italy (1 Nov 1954).

under blinding lights when the abductees are lying down and while others, waiting their turn, are seen to be in a trancelike state.

Without seeking to evoke prurient interests, there needs to be some mention of sexual encounters. Frequently, men report that their semen is extracted mechanically by an apparatus which resembles devices used for breeding animals.

The abducted men frequently say that sexual sensations, induced involuntarily, are isolated in the genital area in an unnatural manner. At other times men report that they are seduced to have intercourse with either women or female aliens who resemble attractive women. Women often report mechanical impregnations with a semen of aliens who are half or pure alien. Reports of half-aliens are often based on women who saw *hybrid babies*. Also, women say they are forced to mate with abducted men and aliens themselves. In the latter case it usually occurs with taller and more humanlike, but thin, creatures. And they usually seem to be in charge of the smaller less humanlike beings. Repeated kidnappings make sense for an alien recovery of fetuses in impregnated women. Some abductees say that their multiple kidnappings are aided by implanted microscopic transmitting devices for follow-up checks, similar to honing devices used by wildlife researchers to relocate wild animals.

Much information on nightmarish sexual encounters was obtained by regressive hypnosis such as that used in the case of Betty and Barney Hill. Yet the memories of abductees are *as* questionable as the recollections of those who underwent other traumas such as Near-Death-Experience (NDE). This sort of experience was criticized by University of Washington psychologist Elizabeth Loftus at a meeting of the American Association for the Advancement of Science because leading questions are easily imposed, facts can be subject to conflicting interpretations, and there may be a willingness to recall what *should* have happened, even if it *didn't*.[17]

In regard to leading questions, the famous psychologist Ronald E. Shor states in *Hypnosis* that impersonal objective research cannot obtain when the very inferences of hypnosis require, circularly, a personal involvement of the therapist.[18] And with respect to interpreting behavioral data, philosophers are poignantly aware of an Underdetermination-of-Theory-by-Data (UTD) Thesis.[19] This Thesis holds that, in principle, all data — understood as so-called facts — permit equally viable but inconsistent theoretical interpretations in a way similar to forensic explanations of a crime inferred by police.

All things equal, as only criminal confessions may resolve alternatives, only alien *confessions* may settle if persons are UFO abductees or deluded. But just as the law convicts criminals on enough circumstantial evidence, the latter is sufficient *prima facie* for a belief in some abductee reports. Psychiatrists such as Harvard's Dr. John Mack say that delusion is eliminated for all practical purposes in many cases.

Having said this, some cases beg for a *reality checklist*. Tulane Professor of Philosophy Michael Zimmerman challenges scientists to use a liberal principle of verification, integral to a rationality of science, in terms of which evidence can count for or against abuductee claims. On the one hand, he says of claims about alien impregnations that "If it could be clearly demonstrated that a pregnancy was never initiated," doubt would be cast "on the 'hybridization program' hypothesis."[20] On the other hand, Professor Zimmerman notes that alien abductions can take a step closer towards acceptance "if one could verify some of the information and predictions that abductees are often given by their abductors."[21]

Cultural tracking means that ET *keeps track of* developing technologies in human history. This historical conformity would both ease integrations into our culture and facilitate ET's investigations.

Professor Zimmerman's conclusion that there is sufficient evidence for "a strange phenomenon that merits funded research by the scientific community"[22] is a laudable exception to the academic community. Indeed, his conclusion highlights a neglect of the entire issue of UFOs by most philosophers whose discipline demands a greater epistemological open-mindedness than other areas of study. At the same time, all studies are challenged by questions of interpretation. Are human interpretations skewed

Tulane University Professor of Philosophy Michael Zimmerman is one of the few scholars open to the possibility that ET is here. He states that we need to provide an adequate, non-reductive account for the traditional "soul" realm that lies between the realms of spirit and matter. Many apparently sane people have encountered angels, devils, elves, fairies, and possibly alien abductors, that are said to be denizens of the soul realm. Moderns, who oppose religious dogmatism and otherworldly beliefs, depict the soul-realm beings solely as products of either superstition or hallucination. But according to Dr. Zimmerman a more inclusive and adequate exploration of the nature of consciousness will eventually make sense of the soul realm, without regressing to premodern cultural beliefs and without omitting the findings of natural science.

culturally by concepts imposed on extra-cultural experience? If experience of UFOs rivals that of Near-Death-Experience, wherein some experiencers say angelic beings reveal a universe rife with intelligent life and that transitions to heaven are eased by culturally familiar settings,[23] then aliens may sometimes utilize biological facsimiles and employ a *cultural tracking* with replications of our craft.

The reports of 'blimps' with unearthly speeds and maneuvers, in the nineteenth century, are augmented by classified twentieth-century messages. One message was from a US Air Force Commander to the Directorate of Intelligence at Carswell Air Force Base in Texas. Tower personnel confirmed that an unidentified 'plane' passed over at 3,000 to 4,000 feet with a conventional fuselage, wings, and stabilizer. But the UFO had no sound or visible means of propulsion.[24] And a civilian reported on 25 August 2000, that a saucer with a top strobe light and a rotating orange band around its circumference was hovering about 300 yards from his car, 1000 feet over Loveland, Colorado. What amazed him was that the strobe and rotating orange lights were suddenly turned off and transformed into conventional lights of a small plane.[25]

Cultural tracking suggests that aliens *keep track of* and conform to our cultural technologies at different historical times. This would make aliens more culturally acceptable, ease integration into our culture, and facilitate their investigations. The investigations would be reminiscent of anthropologists who adopt similar strategies for studying 'primitive people'. If this is the case with bizarre experiences that abductees report, and if their reports are true, one may well wonder: How truly exotic are alien civilizations and UFOs?

Encounters of the Aetherius Society

This society had its inception in George King, a thirty-five year old taxi driver in London, when he was told by a disembodied voice in 1954 that he would speak for an Interplanetary Parliament. Initially, King's response to the voice involved engaging in meditative trances. Several days later, in one of his trances, an Indian Yoga master allegedly appeared and told him that "cosmic intelligences" in spacecraft would use him as a medium for conveying messages to earth. The messages, he said, eventuated in his contact with Aetherius who was a cosmic master on Venus. Thus, King established the Aetherius Society in August 1956. And besides tapping into less scientific aspects of the 1950's flying-saucer craze, the Aetherius Society appealed to almost every rage of the coming decades.

Multicultural diversity found expression in King's counterculture meditative trances, in his notion of karmic evil, and in flower-laden yoga masters who presaged the San Fransisco flower children. Quasi-Christian cultists were equally attracted by King's super Cosmic Masters who included a 'Jesus' on Mars. Pagan ritualists and certain persons with racially sensitized identities found their niche as well in a secret spiritual order of reincarnated perfect beings who were called the 'Great White Brotherhood.' And since this Brotherhood had women priests, despite the 'sexist' name, the society foreshadowed the new feminist church.

The New Age movement was anticipated in the magazine *Cosmic Voice* which touted a healing metaphysics, psychic energy, catastrophe from tests of the atomic bomb, and Jesus' Aquarian-Age Bible for Mother Earth. Quixotically, there was even a yuppy high-tech element for dissolving virtually all unhappiness. Rooted perversely in the eighteenth-century Enlightenment and d'Holbach's utopian *System de la Nature, pace* Professor Zimmerman, the technology was applied to radionics equipment for beaming healing energies to trouble spots of the world. These attractions did not spare the leader's ill health. Retiring in the 1980s to a restful luxurious estate at Santa Barbara, California, King kept himself busy with the questionable future of Aetherius and the earth's fate.

UFO researchers, many with Ph.D. and M.D. degrees, have now become numb to their inevitable fate of being associated with a New Age metaphysics. They have come to humorously accept that their reasoned cases for UFOs will be dismissed by an *ad hominem* ridicule of other puerile beliefs. Those beliefs also bear on fallacious *straw man* arguments in which their reasoning, bearing on good science, is perniciously recast as either mythology or Aetherian mysticism.

UFOs with no visible means of propulsion, but looking like conventional planes, bring to mind this illustration of a rocket-powered supersonic research aircraft, a forerunner of the X-1, drawn for the US Army Air Forces by MSgt. Alex S. Tremulis.

Encounters at Alternative Landscapes

Reports of alien encounters may involve landscapes that are out of place and which variously seem real, fantastic, or fabricated. One case involves a thirty-eight year old French medical doctor who said he was struck by an intense white beam from a large silver-topped disc.[26] A year later whistling noises began to occur in head for which self-therapy was of no avail. The noises were related to odd 'impulses' that directed him to where he would meet a well dressed and strikingly blue-eyed tall man. The man enabled the doctor to experience teleportation and time travel.

Though the doctor experienced the travel as being real, it included a distressing incident in which he saw alternative landscapes on a nonexistent road. The interest of investigators in this so-called 'Dr. X Case' was undoubtedly increased by confirmations of other medical examiners. There was no question about his sincerity or soundness of mind, and they certified a cure of the doctor's previous ailments as well as the existence of an mysterious triangular mark on his body where he claimed to have been struck by a beam.

After a short flight, the craft's landing eventuated in his entering a feudal stone room where there were dwarfs. They had long red hair with lengthy beards and troll-like faces reminiscent of European mythology.

Another case involves Whitley Strieber's *Communion* in which it is claimed, about one UFO experience, that yellow flowers were witnessed where none should have existed. By the same token, a British citizen wrote to the former Defense Department's Dr. Jacques Vallée that pilots at Wethersfield US Air Force Base told him that when they tried to intercept some UFOs,

MSgt. Alex Tremulis, who drew the above ET scene, was given Top Secret clearance to enhance his depictions for the US military. In 1944 the Air Technical Intelligence took him under tight security to a restricted room at Wright Field (later Wright-Patterson Air Force Base). Despite his report that he saw debris of a Nazi V-2 Rocket, later reports of flying saucers inspired this famous drawing of humans interrogating aliens. An officer (Gen. Hoyt Vandenberg?), holds a microphone linked to a machine from which emerges a tape that is examined by an alien with a helmet on which are antenna. Does the translation device transmit voice communication from the men to the helmet on ET? — ET's brain waves transcribed into human language by the machine? Apparently, a cigarette-smoking intelligence officer reads the tape as ET passes it along. "By March 1950, when public interest in saucers was peaking, both of Tremulis's illustrations were in wide circulation in newspapers."

they "found themselves flying in illusory landscapes."[27] These seemed intended, by the UFO occupants, to disorientate them. One pilot thought he was navigating by known terrain, but in fact "was headed straight out to sea."[28]

A medieval theme provided another environment for alien contact, which was deemed to be located either on earth or another planet. In May 1969, at Belo Horizonte, Brazil, Jose Antonio da Silva reported his abduction by four-foot tall masked creatures in an upright cylinder.[29] After some G-force accelerations and what seemed to be a short flight, the craft's landing eventuated in his entering a feudal stone room where there were dwarfs with long red hair, lengthy beards, and the gnome faces of European mythology. The room was equally macabre since shelves held four human bodies and there were wall paintings of animals and other earth scenes. Da Silva's nightmarish experience ended with a fight among the dwarfs and a Christ-like entity who stopped it. Taken back to the cylindrical UFO and released after it landed, he found himself about 200 miles from where he had departed. In agreement with time-travel distortions related to many alternative landscapes, da Silva claimed he was shocked that four days had elapsed.

A fantastic experience involved Elsie Oakensen in November

Alien Drawings by Sutton Family

Sketch of dwarflike aliens encountered in 1955 near Hopkinsville, Kentucky, are similar to those seen by Jose da Silva. After a glowing object was descending nearby (reported later by neighbors) two Sutton men went to their cabin door and saw the approach of a 3 1/2 foot tall entity. It had a disproportionately large head, huge luminous eyes, and talon tipped arms that almost hung to the ground. They opened fire with a shotgun and .22-caliber rifle, whereupon it somersaulted and disappeared. Another alien was seen in a tree and shot, but instead of killing the creature, it floated to the ground and ran away with a lopsided gait. Suddenly, another entity rounded the house and was shot at point blank range. The pellets sounded as if they hit the inside of a metal pot, but the entity was uninjured and retreated. The men went back into the house to care for the frightened children, as the aliens were peering into the windows. Finally, the family ran to their cars and raced to the local police department.

1978, in Northamptonshire, England.[30] But she experienced a created environment that was eventually linked to a dumbbell-shaped craft with bright red and green lights which hovered in front of her car as she was driving home in broad daylight. After passing under the 50-foot wide UFO about 100 feet over her car, Oakensen found herself in total darkness and was forced to stop. Brilliant glowing circles of white light came from the object, which was about a yard in diameter, and approached her automobile as if they were scouting. In the most astounding part of her report, one of the glowing spheres floated to a cottage where it glided up its front and over the top.

The reductionist scientist would reject psychotherapy as a means of accepting alternative realities. But the realities are accepted by a world renowned psychiatrist who dismisses the dismissals of paranormal phenomena.

Oakenson suddenly found herself again driving down the road in daylight. She could not recall the car starting, noticed its being controlled since she was not accelerating or steering, and realized subliminally that a long period of time had elapsed during which she seemed to have been physically examined. At least that part of her story about the dumbbell-shaped object was supported by four other women in a single car who claimed they were paced by such an object at about the same time near that place. But the balance of her report was not corroborated by further testimony and did not admit of greater detail by more questions.

Bearing on the misgivings of Tulane philosopher Michael Zimmerman, who criticizes present-day reductionist assumptions of the Enlightenment, is an overzealous caution of scientists about hypnosis and therapy. Admittedly, the treatments can reinforce false memories about alternative realities. But many of the realities have been entertained seriously by psychiatrist Dr. John E. Mack, noted for his openness to the paranormal, in virtue of his renowned clinical experience in using the therapy on thousands of abductees. To accept their sincerity and a truth of the reports in some cases is, *via* a scientific realism of mainstream science itself, to accept the realities. Thus although the realities may include landscapes created by aliens for disorientation, the question ensues of why the phenomena would be any more unbelievable than the alien UFOs reported by legions of reliable witnesses.

Akinesia in the Encounters

In psychiatry and general medicine the phenomenon of akinesia includes either an absence of motion or a loss of voluntary activity. These tend to be caused by a muscle paralysis that is induced physically or psychologically. For example, one thinks of the deathlike trances of animals that are produced by fright. Frequently, deer are seen frozen in their tracks and staring into the headlights of an automobile. In the reports of alien encounters, akinesia involves a loss of a conscious person's ability to resist ET's intentions.

It seems obvious that if aliens are precocious enough to have hyper-advanced spacecraft and genetic technologies that are virtual mysteries in terms of our science, then they are capable of imposing their intentions on us as well as neutralizing our interference in whatever investigation they undertake. In the famous Valensole event on 1 July 1965, for example, farmer Maurice Masse was in his field of lavender plants near the Basse Alps of France. Shortly after hearing a whistling sound, he came across a strange object that looked similar to a compact car resting vertically on six legs. His first thought that it was a helicopter or experimental craft was dispelled when he saw two child-sized beings in tight-fitting suits who were putting some of his plants aboard the craft.

When these small creatures noticed Masse, for whom they would be no physical match, they pointed *tubes* at him. They seemed good humored and to sort of smile at one another at the time. While he felt a strange sense of peace and remained conscious, whatever struck him caused his immobilization. The long-term effects caused by the tubes were not monitored by either government officials or UFO investigators who took an interest in the case.

However, there was little doubt among most of them that the effects included an extraordinary, if not extraterrestrially related, mode of akinesia which was soon followed by Masse's complete collapse and inordinate need for sleep.

ᴧ

She was aware of being walked to an 'examination room' — not freely choosing to walk, feeling later as if a piece of metal had been implanted inside her.

ᴠ

In another incident at Copely Woods, Indiana, a source of involuntary behavior was even more unusual. On June 30, 1983, Kathie Davis reported an egg-shaped object on legs from which emerged a two-foot diameter ball of blinding light. It "moved up and down her body, painfully irradiating her as she stood paralyzed and unable to respond."[31] Subsequently, in undergoing hypnosis, she said her head hurt, she felt tingly and hot, and everything looked white during the light's inspection. Though Davis could not walk or touch her ear, which felt as if it were injured in the examination, she appeared to be fully conscious and to have retained her cognitive faculties. She recalled being aware of her bodily position and predicament as well as wanting to know what the thing was.

Moreover, in a book foreworded by Harvard psychiatrist John Mack, a more subtle case of akinesia involved sixteen-year-old Barbara Archer in 1982.[32] Just before going to bed, Archer was scared by a light in the window which illuminated the entire room and eventuated in *somebody* standing by her. With the exception of recalling that he was small, she does not remember what he looked like. When he touched her arm between elbow and wrist, she felt no fear as they moved together, passing through a shade-

In 1954 at Nouatre, Indre-et-Loire, France, George Gatay was a foreman with an eight-man crew who encountered a domed craft. It was hovering three feet off the ground with a "man" standing before it who held a weapon-like rod. Gatay tried to run after first seeing this strange creature and object, but he was "paralyzed" during the sighting. Peculiarly, his seven coworkers were also in a sort of a daze during the entire incident.

drawn window! She had no control as they levitated to the bottom of a dark hovering gray oval-shaped craft. Archer remembered that she was not alarmed even though she was frightened by heights, that she saw tree tops and tops of houses, and did not feel the cold to which she was ordinarily sensitive.

Given that passing through solid objects by some aliens *is* physically possible by a superior technology, are some aliens *less* advanced technologically than others — but more than us? When seventy year-old Alfred Burtoo was fishing in the early morning hours of August 1983 in Alder-shot, England, he saw a strange 'bright star'. Shortly thereafter he found himself trailing two four-foot tall aliens dressed in green suits with face visors. But one of them passed *under* a fence, not through it, as they walked to a disc-like object.[33] Typically, he felt no fear. And as most others, it seems doubtful whether he moved entirely by his own volition.

However, no discussion of akinesia would be complete without noting that the most dramatic cases involve abductees who report physical exams and genetic experiments. "I feel as if I'm walking, but... I don't feel as if I can resist...," said nine year-old Jill Pinzarro. She added, "There's something that's lit up, or glowing" and I climb up a "metal, ladder-type thing.... Now it's dark."[34] Her remark is typical of those who report that, without being physically manhandled, they are forced aboard UFOs. Temple University Professor David Jacobs sums up repeated and predictable scenarios among persons of different ages, genders, and vocations when he notes that this young girl said she was confused, dazed, and visually impaired when entering a UFO. Ranging in size from thirty to hundreds of feet in diameter, larger UFOs often have entry chambers which eventuate into larger rooms where creatures seem to be in a mist or fog.

A general preordained doom is induced which led Dr. Jacobs to assert that these "Beings have an agenda to carry out and, once the event begins, nothing can stop it."

Other typification was her awareness of walking without her will into a blindingly illuminated clinical room, feeling tingly while being helped to undress, prodded by long 'fingers', having instruments inserted into body cavities, being scanned by heteromorphous machines, and feeling as if a piece of metal was being implanted. Implantations are common and comparable to tracking devices used by humans on animals. Again, adult female abductees speak of being impregnated by either tubular instruments or tall thin humanlike

Though one could not even dream that 2 + 3 is not equal to 5, said philosopher René Descartes, an all-powerful evil demon might deceive us about even that. Were medieval demons replaced by nightmarish aliens who cause akinesian dreams and make us doubt physical impossibilities known by modern science?

males. And if the abductees are male, they say either that they are seduced to mate with attractive well endowed humanoid females or that their sperm is extracted mechanically. Sexual sensations in the latter event are limited unnaturally to their groins.

Ultimately, the akinesia induced by ET seems to evoke a troubling subconscious sense of fate for humans, individually and collectively. A feeling of preordained doom is brought on, integrated into the lives of the abductees, that brings to mind cases such as that of thirty-year old public relations executive Karen Morgan. Stating that she had been guided aboard an extraterrestrial craft without being able to resist, though she was not intimidated, Morgan described abductions since she was a child by skinny beings with oversized heads and eyes who were swathed in brilliant light. Their gynecological operations when she was a woman eventuated in her bearing a half-alien half-human (hybrid) baby. A study of her case and fifty four other persons led Dr. Jacobs to state that these "accounts display a predictable routine common to most abductions. The Beings have an agenda to carry out and, once the event begins, nothing can stop it."[35]

These cases bring us to enigmatic government, military and civilian organizations that study this phenomena as well as alleged alien intentions that range from altruistic plans for helping humanity to benign indifference and outright hostility associated with a specter of invasion — on a time scale that may be incalculable to humans. The extraterrestrial civilizations in question may not only have reached our stage of primitive technological fire millions or billions of years before recorded human history, or even before the beginning of life on Earth, but the unimaginable sciences of those civilizations would surely produce medical technologies that impart an equally incomprehensible sense of time in virtue of alien life spans that may be longer than ours by multiple factors. Thus although twenty years is 'only yesterday' to elderly persons, by analogy, it is 'ancient history' to youth. The comparison of ET to a virtual 'divinity' by Dr. John E. Mack begets another analogy: A moment to God is like a thousand years, and a thousand years is but a moment.

Notes

[1] See D. S. Gillmor, ed., *Scientific Study of Unidentified Flying Objects*, Conducted by the University of Colorado Under Contract to the United States Air Force (NY: Bantam Books, 1969), p. 895.

[2] T. Good, *Above Top Secret: The Worldwide UFO Cover-Up*, With a Foreword by the former Chief of Defense Staff, Lord Hill-Norton, GCB (NY: William Morrow,1988), p. 477.

[3] Col. P. J. Corso (Ret.), *The Day After Roswell* (NY: Simon & Schuster, Pocket Books, 1997) p. 59.

[4] Good, *Above Top Secret*, p. 78.

[5] See Good, *Above Top Secret*, pp. 260, 266, 352, 406, 425, 522.

[6] Reports 18 July 1955 of the Air Force Special Security Service, HQ, Northeast Air Command, APO 862, New York. *Ibid.*, pp. 285-6.

[7] Nick Pope, *Open Skies, Closed Minds: For the First Time a Government UFO Expert Speaks Out*, With a Foreword by Timoth Good (NY: Dell Publishing, 1998), p. 246.

[8] Though Timothy Good noted that the Air Ministry did not begin its investigations of UFOs in 1947, in correspondence to me 22 June 1999, his *Above Top Secret* (p. 49) specifies that the Air Ministry analyzed the reports from 1947 to 1957. The discrepancy may be due to new information he has obtained, but I have not come across it.

[9] Good, *Above Top Secret*, pp. 49-58.

[10] *Ibid.*, pp. 31-2.

[11] Corso, *The Day After Roswell*, p. 126.

[12] *Ibid.*, p. 126.

[13] *Ibid.*, p. 126.

[14] *Ibid.*, pp. 126, 127.

[15] *Ibid.*, p. 127.

[16] *Ibid.*, p. 128.

[17] See interview of Dr. E. Loftus in D. Haney's "Mind Found Fertile Soil for False Memories," *Associated Press* 16 Feb. 1997.

[18] R. E. Shor, "The Fundamental Problem Viewed From Historic Perspective," *Hypnosis: Research and Perspectives*, Ed. by E. Fromm and R. E. Shor (Chicago: Aldine-Atherton, 1972).

[19] See R. Trundle,"Extraterrestrial Intelligence and UFOs," *Method and Science* 27 (1994) 73-97, for reference to a UTD Thesis. It vitiates appeal to conventional theories for dogmatically asserting what is true about reality and what is real.

[20] Michael Zimmerman, "The 'Alien Abduction' Phenomenon: Forbidden Knowledge of

Hidden Events," *Philosophy Today* 41 (1997) 235-254.

[21] *Ibid.*, p. 250.

[22] *Ibid.*, p. 250.

[23] In my class Philosophy 201: Death and Dying in the Fall 1987, former atheist Professor Howard Storm spoke of his near death experience (NDE). He stated, with relevance to the UFO issue, that it was revealed to him by angelic beings that the universe was well populated by other biological beings who were intelligent. Lest debunkers glibly express pedantic skepticism about NDE, it is not merely the case that he was taken seriously from a clinical standpoint by psychiatrist Raymond Moody, M.D., Ph.D. In addition, the world-renown atheist and positivist philosopher A. J. Ayer, who was notorious for 'demonstrating' that claims of life after death were meaningless pseudo-statements, did himself experience a Near-Death-Experience which he took to be veritable—with apologies to his humanist students and colleagues!

[24] Classified Message from Commander, Carswell AFB, Texas, to the Directorate of Intelligence; Commander, Air Defense Command, Ent AFB, Colorado; ATIC; 8th Air Force, Carswell AFB. From Good, *Above Top Secret*, pp. 281, 282.

[25] Maj. George Filer, *Filer's Files* 36, 11 September 2000, p. 4.

[26] J. Vallee, *Confrontations: A Scientist's Search for Alien Contact* (NY: Ballantine Books, 1990), pp. 102-5.

[27] *Ibid.*, p. 108.

[28] *Ibid.*, p. 108.

[29] J. Spencer, *The UFO Encyclopedia* (NY: Avon Books, 1991), p. 40.

[30] *Ibid.*, pp. 229-31.

[31] B. Hopkins, *Intruders: The Incredible Visitations at Copely Woods* (NY: Ballantine Books, 1987), p.55.

[32] David M. Jacobs, Ph.D., *Secret Life: Firsthand Documented Accounts of UFO Abductions*, with a Foreword by John E. Mack, MD, Professor of Psychiatry at the Harvard Medical School (NY: Simon & Schuster, 1992), p. 52.

[33] See Good, *Above Top Secret*, p. 107. Though Good quotes Burtoo as stating that "The 'form' in front of me went through the railings...," he informed me on 22 June 1999 that this was a misquote — that the alien did not 'pass' through them. Which ever scenario is correct, the point remains that some alien species may not be as advanced as others; for that matter, as advanced as humans. But in this case, of course, these alien species would not be here at this time!

[34] See Jacobs, *Secret Life*, pp. 81-86, for most of the following.

[35] *Ibid.*, p. 86.

6

Outlandish Organizations and Intriguing Issues

THOUGH THERE ARE many discussions of UFO organizations in the popular literature, our analysis raises novel questions about some of the chief groups. Are some military groups public fronts for relegating UFOs to a scientific absurdity? If so, is it ethical to misinform our citizens? Is there an elite government body, privy to extraterrestrial data, that has confused science with religion and made religious decisions for citizens without their awareness? Some tentative responses are considered.

Some Organizations

With various degrees of openness to the public, the following organizations reflect elite groups, small committees, or larger associations which are composed of civilians, military personnel, or government officials. Admittedly, these groups are a mere sample. But they typify the enormous talent and energy that is invested in the young history of UFO research.

Aerial Phenomena Research Organization (APRO)
Founded by Jim and Coral Lorenzen in 1952 at Tucson, Arizona, five years after the Roswell crash, the APRO is the oldest UFO organization in the world. Since 1953 when the CIA formed the Robertson Panel to assess the APRO, the evidence suggests that its legal research has been *both* monitored and hindered by that intelligence agency. Headed by the CIA's Dr. H.P. Robertson who directed the Weapons Systems Evaluation Group, this Panel revealed its hostility to the APRO's research by concluding that the government should debunk UFOs. Given a collusion of intelligence agencies, a suspicion about them is based on APRO member experiences. There is the experience of Ray Ingraham who, with member M. Kellenbarger, investigated UFOs in 1967

Coral and Jim Lorenzen in the early 1950s at home in Tucson, Arizona, founders of the Aerial Phenomena Research Organization.

"They (the US Air Force) spotted six 'saucer shaped flying objects at an altitude of about 7000 feet at the base of a cloud bank about 3:50 p.m.," read the *Los Angeles Times* (Nov. 6, 1957). Having been tracked from the Gulf of Mexico. Mr. and Mrs. A.A. Burnand Jr. reported: "When [the UFO] flashed the rays shot out in all directions and it made you think of an explosion without sound." Frank Altomonte noted that "Whatever happened on the beach in Playa del Rey that morning so long ago was definitely a part of a larger flap. The tantalizing details can be found in the research notes of Jim and Coral Lorenzen now kept at some Midwest University, according to one report, and not readily accessible.... They were the best researchers... of UFO cases that ever lived."

at San Luis Valley, New Mexico. If Kellenbarger had been spied upon, that would explain how US Army Intelligence officials knew about the trip when they questioned him in 1970, after he tried to join the army's intelligence section. Was its interest not in alien technology but a military secret that the Valley was being used for B-52 mock-bomber runs during the Vietnam War? But that War did not evoke concern about spies as did other conflicts. Here, the spying on legal activities of citizens by their own government raises troubling ethical questions. The questions also relate to anomalous UFO phenomena having more credibility than admitted by many elite scientists who might have been privy to the government's involvement.

Given that this involvement would bear on one of the most dramatic events in human history and that mainstream scientists avow a pursuit of objective truth, one would think that the much publicized government interest would arouse their curiosity. Philosophers, if not scientists, are surely prominent for questioning a propriety of government secrecy and coercion. Herein, some of the questions are addressed in 'Above Top Secret and Ethics'. Furthermore, ethics is studied by philosophy. And philosophy is also the discipline whose members should be most able to appreciate the astonishing import of an advanced extraterrestrial intelligence for science, theology, and culture in general.

Accordingly, there is much to excite philosophers and scientists about their government's interest in organizations such as the APRO. This holds all the more for a certain Blue Book Committee.

Ad Hoc Committee To Project Blue Book

Formed in March 1966 by the USAF Director of Information Maj. General E. B. LeBailly to review the Air Force study of UFOs by Project Blue Book, this committee was headed by physicist Dr. Brien O'Brien. Given an increasing dubiousness of the Project's credibility, he recommended that "Contacts be negotiated with a few selected universities to provide scientific teams to investigate promptly and in depth certain select sightings of UFOs."[1] Dr. Edward Condon, former Head of the US National Bureau of Standards and Physics Department at the University of Colorado, stated in 1968 that the investigations of Blue Book by the Ad Hoc committee led to his selection for leading the first official scientific study of UFOs on behalf of the US Air Force.

What became known as the Air Force's "Condon UFO Report" led to notorious debates about its own conclusions. There was documentation, noted herein, that neither Condon nor the Air Force Blue Book investigators were trying to amass information about creditable UFO evidence. But any evidence that was significant was passed on, out of public scrutiny to ultrasecret elements of military intelligence. Precisely, the Ad Hoc Committee was formed to publicly phase out government UFO reports by commissioning the report of an "objective" third party. This party was to be a university whose traditional mission, in the eyes of the public, is to pursue the truth. But the truth was skewed by Condon's predetermined conclusion, in accord with the Committee, that UFOs did not warrant any further scientific study!

This study was biased in advance by the Ad Hoc Committee which reported that "In 19 years and more than 10,000 sightings recorded and classified, there appears to be no verified and fully satisfactory evidence of any case... outside the framework of presently known science and technology."[2] Yet there was *no* critical reference to the official announcement in July 1947 about a Roswell UFO crash and *no* mention of the report by Gen. Twining that acknowledged the reality of flying discs. After a study by the US Air Materiel Command, Twining's report stated that the discs had unknown properties. Also, there was no reference to Project Sign in July 1948 that the UFOs were most likely "interplanetary vehicles" of unknown origin.

Though it was known that the vehicles were beyond any known science of the time, Gen. Hoyt Vandenberg is the one who officially killed the extraterrestrial explanation. And while the astronomer Dr. Allen Hynek *is* referenced as a consultant to the Ad Hoc Committee, Hynek scathingly criticized Condon's anti-UFO remarks. The remarks, contrary to scientific objectivity, were made even before Condon began his inquiry. Hynek dedicated his chapter "Science Is Not Always What Scientists Do" in *The UFO Experience* to enumerating the inconsistencies and confused methodology of the investigation headed by Condon.[3] The investigation was also troubled by Condon's harassment of serious scientists. In the end, the Committee got what it wanted.

Benefits of contact with aliens? Our being informed politically by ageless trials and errors, radically new medical cures born of a vast clinical experience over millions of years, and novel discoveries gained by time travel and crisscrossing multiverses.

Gen. Spaatz (*left*) turning over Air Force reigns to Gen. Hoyt Vandenberg (*right*) with AF Sec. Symington (*middle*) on 2 April 1948. Since Vandenberg is the one who 'vetoed' the Extraterrestrial Hypothesis for UFOs, he is the one who also ultimately steered the Ad Hoc Committee and military away from the view that ET is here.

Admittedly, several of the report's case studies attested to some UFOs being artificial craft that were far beyond any technology on Earth, in virtue of the studies being conducted by independent scientists despite Condon's influence. But his written conclusion vindicated the skepticism of Project Blue Book. The report's conclusion, influenced by the Air Force, found that "there is no evidence to justify a belief that extraterrestrial visitors have penetrated our skies and not enough evidence to warrant any further investigation."[4]

A point about such investigations is beholden to Sir Karl Popper. He argued positively by an Inverse Probability Thesis for a relation between *improbable* predictions and a *probable* truth-likeness of theories:[5] The more *unlikely* a new theory's predictions in terms of well established theories, the more *likely* its truth if the predictions obtain. It is more rational to believe the bizarre predictions obtain because the theory is true than that they occur by sheer chance. But no one has noticed that if Popper's reasoning is accepted, we can argue negatively that a *likelihood* of well-established theories being false or partly false is suggested by an *unlikelihood* of phenomena that are either unpredictable or impossible in terms of the theories!

If alien UFOs were a physical impossibility in terms of current theories, even minimum evidence should arouse maximum scientific curiosity. Scientists should not be most curious about the ordinary but rather, about the extraordinary, since it demands the novel theoretical responses required for a 'Popperian' progress. The spirit of this progress is suggested, though understated, by physicist Fred Whipple, former Director of the Smithsonian Astrophysical Observatory. Scientists, he says, would certainly "devote the rest of their lives to the... [UFO] subject if they had reason to believe that there was even a 30 percent chance for extraterrestrial intelligence."[6]

Alien sun IRC+10216, of interest to the Smithsonian Astrophysical Observatory, noted below, blazes through its death throes — vaporizing surrounding comets that release huge clouds of water vapor. The discovery is from the Submillimeter Wave Astronomy Satellite (SWAS), launched by NASA in 1998. The new sightings provide the first evidence that extra-solar planetary systems contain water, an essential ingredient for *known* forms of life. Is this part of a series of steps in locating ET? Discovering 1) stars with planets, 2) planets with water, and 3) water and alien life? These steps do not even address possible intelligent life forms, however, *which* or *who* may not need water.

Given the benefits pursuant to alien contact, a 30 percent chance of its existence for interest is absurdly conservative. This is partially offset by Whipple's concession that, with a proper percent, the energy now diffused into other research would be dedicated to UFOs. Contact could lead to learning the 'impossible': From a new cultural calculus for near utopian societies, informed by virtually ageless trials and errors, to radically novel medical technologies born of a vast clinical experience over millions of years to unimaginable discoveries gained by time travel and crisscrossing multiverses.

In light of the astonishing implications of ET visiting Earth, it is disheartening that the author tried vainly to convey his excitement to some university colleagues when Lt. Col. Philip Corso published *The Day After Roswell*.[7] The only curiosity one colleague could muster was whether or not aliens would be 'sexist'! Another colleague had previously expressed skepticism about the reality of UFOs, but since Corso confirmed that a UFO had, in fact, crashed in New Mexico, and as he had been a career intelligence officer as well as the former Chief of the Army's Foreign Technology Division

135

This 1994 letter to government official Robert Barrow from Congressman Steven Schiff reveals that Schiff was trying to get to the bottom of the Roswell UFO mystery. Thanking Barrow for his support in getting the General Accounting Office (GAO) to investigate the Roswell incident, he expresses his intention to inform the public. Rep. Schiff died shortly thereafter. The GAO discovered that all of the records, contrary to the law, had been taken out of their files during the months before, during, and after the incident. Below is the GAO's report: *Government Records: Results of a Search for Records Concerning the 1947 Crash Near Roswell, New Mexico.*

at the Pentagon, there now seemed to be better corroboration of the reality of the phenomenon. However, this time my colleague switched from skepticism to saying, as he walked away, that he had no interest in "this area." His new reply was either disingenuous in revealing that nothing would count against his disbelief in ET's visit, or at odds with a life of the mind of his profession. In terms of his profession, mindboggling implications of extraterrestrial civilizations visiting Earth would evidently amount to one more boring area of scholarship!

Amid a sea of top-secret organizations, Rep. Steven Schiff announced to the press that he was being stonewalled by one government agency after another. He demanded accountability to taxpayers: taxpayer records that bore on UFOs.

Aerospace Defense Command

The Aerospace Defense Command (ADC) is often used to make the case that the US military did not, as it often says, abandon its investigations of UFOs in the late 1960s. Indeed, *after* the Air Force ceased its Blue Book Project in 1969, since UFOs are virtually always 'deluded witnesses' and 'natural phenomena', though sometimes not explained, the Pentagon admitted that the "ADC is responsible for *unknown aerial phenomena* reported.... [The] provisions of joint Army-Navy-Air Force publications (JAN AP-146) provide for the processing of reports received from nonmilitary sources [emphasis]."[8]

Many civilian researchers viably suppose that the ADC used the services of the Air Force Cambridge Research Laboratories and Office of Scientific Research. Their facilities were never utilized by Blue Book personnel who, says Lt. Col. Corso, were merely public-relations spokesmen. "At least if we weren't going to do anything about UFOs publicly, we had to have a way to salve the public's fear about UFO sightings."[9] Blue Book "was that salve."[10] Therefore, civilian researchers think that the most credible reports on UFOs bypassed Blue Book and were submitted to the ADC.

In suggesting that many scientists were unqualified to investigate UFOs since a mere mention of the topic evoked anger, official Dr. Tom Ratchford implied that the Air Force was dead serious about the possibility that ET *is* here!

One indication that Blue Book would lack any significant data was the low military rank of its directors. They were largely junior-grade officers, and the staff was reduced to a handful of enlistees. In short, Blue Book was just a front for behind-the-scenes analyses of the ADC. In turn, the ADC's own scientific staff — in consultation with the Cambridge and Scientific Research Labs, subjected the flawless data for UFOs to intense scientific scrutiny before, during, and after the US Air Force claimed that all of its inquiries had ceased.

Air Force Office of Scientific Research (AFOSR)

The AFOSR was located at an office in the Pentagon when it operated in concert with the Ad Hoc Committee. It advised that continued Air Force research on UFOs hinge on evaluations of scientists at a research university. The University of Colorado was selected in 1966 with physics Professor Edward Condon as scientific director of the project. He later noted that AFOSR's Dr. J. Thomas Ratchford had approached him to head the project. Condon admitted that he had recommended other more qualified scientists who *were* interested in the UFO phenomenon. In replying that those scientists "were essentially disqualified for having already taken sides on the UFO question" — a euphemism for saying that they had reacted with anger or sarcasm to the subject,[11] Dr. Ratchford clearly implied both that the Air Force

Logo for the AFOSR that manages all basic research conducted by the U.S. Air Force through a solicitation of proposals from a general Broad Agency Announcement (BAA) in physics, propulsion, the biological sciences and so on. The AFO SR states that the general BAA also has a number of Researcher Assistance Programs that include the *University Resident Research Program.* Presumably, the latter would bear on the AFOSR's choice of the *University of Colorado* at Boulder to scientifically investigate UFOs.

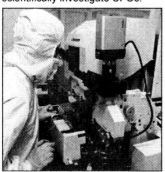

A Star Team Award for excellence, at the Air Force OSR, is performed by several units: the Electronic Device Materials team, the Collisional Plasma & Discharge Physics team, and the Electronic Conjugation in Polymers team.

was dead serious about a possibility that ET is here and that it sought scientific objectivity.

The word 'objectivity' begs for clarification. There should be caution about, *but not necessarily exclusion of,* investigators if they are on record as taking a side but express an openness to new evidence. The same caveat holds for those who are aware of their skepticism even if it is not expressed. Degrees of taking a side range from skepticism that admits of reasonable counter evidence to an irrational bias. For instance, this bias would be manifest in political ideology or in favored outcomes, because of career demands or peer pressure, that permit nothing to count against a position. Short of rejecting a position out of hand, pro-and-con views should be encouraged. Far from being a handicap, these views can enliven points and counterpoints that would otherwise be disregarded in inquiries that can then also cease to arouse curiosity—if this is possible on the issue of ET visiting earth.

Since these points would be obvious to AFOSR officials, the quest for objectivity was disingenuous. The Air Force, and by implication the AFOSR, had already taken sides by a superficial treatment of UFOs in the earlier Blue Book Project. All salient data or material evidence, unacknowledged by the Project and AFOSR when it approached Condon, was given to higher authorities in the CIA, National Security Agency, Department of Defense or other units. Precisely, in this vein Rep. Steven Schiff announced to the press in 1994 that he was being stonewalled. He demanded accountability to taxpayers, taxpayer-related

records bearing on UFOs.

Air Technical Intelligence Center (ATIC)
In an Air Force letter on 29 Apr 1952 classified "RESTRICTED SECURITY

INFORMATION," the ATIC, at Wright-Patterson Air Force Base is said to be "the Air force activity responsible for conducting analysis of all information and material received." When "items of material pertinent to any unidentified flying object incident comes into the possession of any Air Force echelon, two actions will be

Third from the left, Col. Harold Watson was twice the CO of the Air Technical Intelligence Center (ATIC).

taken without delay: a. Safeguard the material carefully to prevent any defacing or alteration which would reduce its value for technical analysis. b. Notify ATIC immediately and request shipping instructions or other special instructions as may be appropriate." And of interest to those who seek the military's secrets on UFOs *via* the Freedom of Information Act, the letter adds, "Reports will be forwarded immediately by electrical means and confirmed and elaborated upon by a written report within three days. a. The symbol FLYOBRPT will appear at the beginning of the text of electrical messages and will be used as subject of written reports to facilitate identification. b. Security. *Reports should not be classified higher than 'Restricted' unless inclusion of data required by 'c' and 'd' below mandates a higher classification* [emphasis]." And importantly, "publicity concerning this reporting and analysis activity is to be avoided."

A higher classification in 'c' would include the "Existence of any physical evidence such as fragments, photographs and the like," as well as "Interception and identification action taken." And 'd' specifies, "These reports will be submitted on AF Form 112, 'Air Intelligence Information report'," and will expand on "photographs, sketches, and signed narrative statements of observers." Intriguingly, the letter was written by USAF Chief of Staff Gen. Hoyt Vandenberg, who had ordered the destruction of an ATIC document in August 1948, *Estimate of the Situation*. It stated that the flying discs were interplanetary extraterrestrial spacecraft! Also, Gen. Vandenberg had been CIA Director — the CIA established in July 1947 in the same month and years as the Roswell UFO crash, and an alleged member of Majestic-12. This group was said to have been formed by President Truman to assess crashed UFOs.

In 1993 the ATC became the National Air Intelligence Center (NAIC). Its coordinator for the Freedom of Information Act, Msg Gery Heulsemen, reportedly stated that the NAIC had about 3,000 boxes of Air Force archives, some on UFOs. In Report TR-AC-47 on 6 March 1998, sent to one citizen by NAIC, Project Silver Bug is said to research "aircraft having a circular platform" that excludes "conventional aerodynamic control surfaces." Does science become science fiction in another letter received by John Greenewald, Jr., in the 1990s? Apparently, the NIAC would be privy to Project Tobacco, so called since it would *burn up* if uncovered, for the public's gradual adoption of futuristic engines—developed by contact with extraterrestrials—to replace the engines on Earth without a collapse of the economy.

⋏

A scientist who claimed to work for the National Reconnaissance Office spoke of a mountain in New Mexico where aliens from the Cygnus Ra Star System have allegedly helped humans upgrade their technology. The latter is said to require a"debriefing" of the public.

⋎

He claims to have received a letter from a man who worked for the National Reconnaissance Office. Referring to a mountain called Baldy in New Mexico where aliens from the Cygnus Ra Star System allegedly helped humans upgrade their technology, Greenewald paraphrased the man's claims. There is a Gravity Catapult which focuses on similarities of Einstein's General Theory and the Maxwell theory of electromagnetism. The bottom line is that microphysical particles are purportedly moved to form a mass current, which generates a new field that is the gravitational equivalent of a magnetic field, for an antigravity engine. The government is said to have had this engine for several years, which is why NASA has been gradually down-sizing to smaller and cheaper engines in its space program for several decades. Transportation on Earth will radically change but the change must be piecemeal.

Bearing on the West's dependence on oil, and thus on the very nations who fuel terrorism, the new transportation will use no fuel, have no sound and can never break. Using the freedom of Information Act (FOIA), John Greenewald wrote to Ms. Rhonda Jenkins, Manager of the FOIA at the Pentagon, to request information on Project Tobacco apart from any reference to UFOs or extraterrestrial issues. Why then, he asks, was there mention of them unless there *was* a Project Tobacco that was related to them? Did the Air Force furtively admit of Tobacco's connection to UFOs by its standard disingenuous reply that UFOs were investigated only from 1947 to 1969 by the AITC, now called the National Technical Intelligence Center?

The Elite Enigmatic Group: Majestic-12

Reportedly, Majestic-12, also denoted Maj-12, MJ12 and MAJIC, is composed of elite US officials who oversee crashed UFOs. Reportedly formed after July 1947 when a UFO allegedly crashed in Roswell, New Mexico, it is often called the 'Group' since its name may change. The Group was first notably referenced in Tim Good's *Above Top Secret* (1988, with a foreword by Britain's former chief of the defense staff), first written about most thoroughly in nuclear physicist Stanton Friedman's *Top Secret/MAJIC* (1996), and first discussed with an apparent authority to know by former US Army Lt. Col. Philip Corso in *The Day After Roswell* (1997).

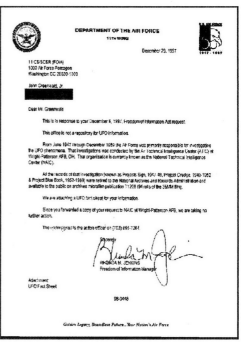

Does this official government reply to John Greenewald's FOIA request admit of Project Tobacco? A development of a hyper-advanced antigravity transportation that requires a public debriefing process?

Since this Group unveils itself in Maj-12 documents, some are highlighted. They were ostensibly prepared in 1952 for retired five-star Army General and President-Elect Dwight D. Eisenhower by Adm. Hillenkoetter. TOP SECRET/ MAJIC EYES ONLY is stamped over a typed TOP SECRET at the top and bottom of each page along with EYES ONLY and COPY <u>ONE</u> OF <u>ONE</u>, OF <u>TWO</u> and so on.

> "An analysis of the four dead occupants was arranged by Dr. Bronk. It was the tentative conclusion of this group (30 November 1947) that although these creatures are humanlike in appearance...."

With an exception of the first page at the top, the following quoted material excludes these monitory words in featuring the most significant part of the contents:

[1]

TOP SECRET / MAJIC
EYES ONLY
* * * * * * * * *

* **TOP SECRET** *

* * * * * * * * *

EYES ONLY   COPY <u>ONE</u> OF <u>ONE</u>

BRIEFING DOCUMENT: OPERATION MAJESTIC 12
PREPARED FOR PRESIDENT-ELECT DWIGHT D. EISENHOWER :
(EYES ONLY)
18 NOVEMBER, 1952

<u>WARNING</u>: This is a TOP SECRET - EYES ONLY document
containing compartmentalized information essential to
the national security of the United States. EYES ONLY
ACCESS to the material herein is strictly limited to
those possessing Majestic-12 clearance level...

[2]

OPERATION MAJESTIC-12 is a TOP-SECRET Research and
Development/Intelligence operation responsible
directly and only to the President of the United
States. Operations of the project are carried out
under control of the Majestic-12 (Majic-12) Group
which was established by special classified executive
order of President Truman on 24 September, 1947, upon
recommendation by Dr. Vannevar Bush and Secretary
James Forrestal. (See Attachment "A".) Members of the
Majestic-12 Group were designated as follows:

> Adm. Roscoe H. Hillenkoetter
> Dr. Vannevar Bush
> Secy. James Forrestal*
> Gen. Nathan F. Twining
> Gen. Hoyt S. Vandenberg

```
        Dr. Detlev Bronk
        Dr. Jerome Hunsaker
        Mr. Sidney W. Souers
        Mr. Gordon Gray
        Dr. Donald Menzel
        Gen. Robert M. Montague
        Dr. Lloyd V. Berkner
```

* The death of Secretary Forrestral on 22 May, 1949, created a vacancy... until 01 August, 1950, upon which date Gen. Walter B. Smith was designated as permanent replacement.

---

[3]

On 24 June, 1947, a civilian pilot flying over the Cascade Mountains... observed nine flying disc-shaped aircraft traveling in formation at a high rate of speed.... Hundreds of reports of sightings of similar objects followed. Many... came from highly credible military and civilian sources. These reports resulted in... efforts by several different elements of the military to ascertain the nature and purpose of these objects in the interest of national defense.... Public reaction bordered on near hysteria.

In spite of these efforts, little of substance was learned about the objects until a local rancher reported that one had crashed... approximately seventy-five miles northwest of Roswell Army Air Base (now Walker Field).

---

[4]

A covert analytical effort organized by Gen. Twining and Dr. Bush acting on the direct orders of the President [Truman] resulted in a preliminary consensus (19 September, 1947) that the disc was most likely a short range reconnaissance craft. This conclusion was based... on the craft's size and the apparent lack of any identifiable provisioning. (See Attachment "D".) A similar analysis of the four dead occupants was arranged by Dr. Bronk. It was the tentative conclusion of this group (30 November, 1947) that although these creatures are human-like in appearance, the biological and evolutionary processes responsible for their development has apparently been quite different from those observed or postulated in homo-sapiens. Dr. Bronk's team has suggested the term "Extraterrestrial Biological Entities," or "EBEs," be adopted as the standard term....

Since it is virtually certain that these craft do not originate in any country on earth, considerable speculation has centered around what their point of origin might be and how they got here. Mars was... a possibility, although... Dr. Menzel considers it more likely that we are dealing with beings from another solar system entirely.

Numerous examples of... a form of writing were found in the wreckage. Efforts to decipher these have remained largely unsuccessful. (See Attachment "E".) Equally unsuccessful have been efforts to determine the method of propulsion or... transmission of the power source involved. Research... has been complicated by the complete absence of identifiable wings, propellers, jets, or other conventional methods....

[5]

A need for as much additional information as possible about these craft... led to... U.S. Air Force Project SIGN in December, 1947. In order to preserve security, liaison between SIGN and Majestic-12 was limited to two individuals within the Intelligence Division of Air Materiel Command whose role was to pass along

certain types of information through channels. SIGN
evolved into Project Grudge in December, 1948. The
operation is currently being conducted under the code
name BLUE BOOK, with liaison maintained through the
Air Force officer who is head of the project....

Implications for the National Security are of
continuing importance in that the motives and ultimate
intentions of these visitors remain completely
unknown.... It is for these reasons, as well as the
obvious international and technological considerations
and the ultimate need to avoid public panic at all
costs, that the Majestic-12 Group remains of the
unanimous opinion that imposition of the strictest
security precautions should continue.... At the same
time, contingency plan MJ-1949-04P/78 (Top Secret -
Eyes Only) should be held in continuous readiness
should the need to make a public announcement present
itself. (See Attachment "G".)...

---

[6]
Page 6 lists the Attachments

Whereas Good initially accepted an authenticity of these MJ-12
documents, his later *Alien Contact* (1993) expressed doubt with a caveat that
they are "essentially factual."[13] And he acknowledges that the group exists,
and its existence was confirmed by the Air Force Office of Special
Investigations.[14] Also, after meticulous analysis, physicist Stanton Friedman
defended their authenticity. The carefulness of his evaluation over several
years, resulting in his certification of their likely genuineness, is captured by
the following remark:

A detailed proposal to validate this document by investigating its internal
references, the appropriateness of the procedures enumerated, the
relationship to the other MJ-12 documents, and the occurrences of any
possible anachronisms has been made by myself, Dr. Wood, Dr. Bruce
Maccabee, and another researcher....[15]

Other considerations support a general factualness, if not genuineness, of the MJ-12 documents: (1) Having received them from Jaime Shandera in 1984, William Moore underwent an intense FBI investigation as confirmed by six pages of FBI records not exempt from disclosure.[16] If they were laughably inauthentic, why the FBI inquiry? (2) The members' whereabouts fit the time of the documents and the latter specify extraordinary phenomena in accord with the unique expertise, association, and security status of these men. (3) Lt. Col. Corso, who was in a singular position to know, admits of Maj-12, calls it the 'working group', and identifies the members in agreement with the critical research of Friedman.

Since it is beyond present purposes to inspect other MJ-12 documents

---

TOP SECRET
EYES ONLY
THE WHITE HOUSE
WASHINGTON

September 24, 1947.

MEMORANDUM FOR THE SECRETARY OF DEFENSE

Dear Secretary Forrestal:

As per our recent conversation on this matter, you are hereby authorized to proceed with all due speed and caution upon your undertaking. Hereafter this matter shall be referred to only as Operation Majestic Twelve.

It continues to be my feeling that any future considerations relative to the ultimate disposition of this matter should rest solely with the Office of the President following appropriate discussions with yourself, Dr. Bush and the Director of Central Intelligence.

Harry S Truman

noted by Friedman, there is attention to Corso. While he avoids discussing data to which he was not privy, he confirms this top secret group in accord with prior research. This and various FOIA documents led to its apparent members whose biographies are listed in *Marquis Who's Who in America* (1948-51). Though these biographies have never before been presented in any literature on this subject, its information reveals some of the most enviably educated and experienced individuals ever assembled in politico-scientific history: those who first sought to comprehend the nature and meaning of a crashed 'flying saucer' and its occupants from another world:

BERKNER, Lloyd V(iel), research scientist; born Milwaukee, Wis., Feb. 1, 1905; s. Henry Frank and Alma Julia (Viel) B.; B.S. in elec. engring., U. Minn., 1927; studied physics, George Washington U., 1933-35; m. Lillian Frances Fulks, May 19, 1928; children — Patricia Ann, Phillis Jean. Engr. airways div., U.S. Dept. of Commerce, 1927-28; engr. 1st Byrd Antarctic Expdn., 1928-30, research engineer, Nat. Bur. Stds, 1930-33; research physicist dept. of terrestrial magnetism. Carnegie Inst. Washington, 1933-41; head electronics material branch, Bur. of Aeronautics, U.S. Navy Dept., 1941-46; exec. sec. research and development bd., Nat. Mil. Establishment, 1946-47; chmn., sect. on exploratory geophysics of the atmosphere. Dept. of Terrestrial Magnetism, Carnegie Instn. of Washington, 1947-48, acting dir. since 1948;

Dr. Berkner: Theoretical Physics – Terrestrial Magnetism

became spl. asst. to sec. of state to direct work relating to Mil. Assistance Program, 1949. Capt., U.S.N.R. Fellow American Phys. Society. Institute of Radio Engrs., Am. Inst. of Elec. Engrs.; mem. Nat. Acad. of Sciences, Am. Internat. Sci. Radio Union (exec. council), U.S. Nat. Com., Internat. Joint Commn. on Ionosphere, Am. Inst. of Elec. Engrs. (exec. council; Washington sect.), A.A.A.S. (Centenniel planning com.), Democrat. Conglist. Contbr. to prof. jours. Home: 7213 Bradley Blvd., Bethesda 14, Md. Office: Dept. of Terrestrial Magnetism, 5241 Broad Branch Rd., N.W., Washington 15, D.C.

BRONK, Detlev W., physiologist, physicist; b. N.Y. City, Aug. 13, 1897; s. Mitchell and Marie (Wulf) B.; A.B. Swarthmore, 1920; studied U. of Pa., 1921; M.S., U. of Mich., 1922, Ph.D., 1926, hon. Sc.D., Swarthmore. 1937, Princeton, 1947; m. Helen A. Ramsey, Sept. 10, 1921; children — John Everton Ramsey, Adrian, Mitchell Herbert. Exec. sec. Phila. Food Administrn., 1918; asst. power engr. Phila. Electric Co., 1920; instr. physics, U. of Pennsylvania, 1921; same, U. of Michigan, 1921-24. instr. in physiology, 1924-26; asst. prof. physiology and biophysics, Swarthmore, 1926-27, asso. prof., 1927-28, prof. 1928-29, dean of men, 1927-29; fellow Nat.

Dr. Bronk (*third from left*): Medicine and Biophysics

Research Council, at Cambridge and London, 1928-29; Johnson Foundation for Med. Physics, U. of Pa., since 1929, dir. Inst. of Neurology 1936-40 and since 1942; prof. physiology Cornell Med. Coll., 1940-41; coordinator of research, Air Surgeons Office, Hdqtrs. Army Air Forces, 1942-46. Weir Mitchell lecturer, Phila. Coll. Physicians, 1938; Hughlings Jackson lecturer, McGill U., 1938; Vanuxem lecturer, Princeton, 1939; Priestley lecturer Pa. State Coll., 1941; Herter lecturer, N.Y. Univ., 1943; Colver lecturer, Brown U., 1946; bd. scientific dirs. Rockefeller Inst. Med. Research, since 1946. Mem. Naval Research adv. com; mem. sci. adv. bd. A.A.F.; Nat. Commn. for UNESCO. Mng. editor *Jour. Cellular & Comparative Physiology*; *Aviation Physiol. Bull.*, asso. editor *Proc. Soc. Exptl. Biol. and Medicine* 1935-41, *Am. Jour. Physiol.*, 1938-43, Biol. Abstracts, Rev. Sci. Instits. Mem. Nat. Research Council (div. physics, 1935-38, chmn. since 1946, div. internat. relations; com. on aviation medicine; chmn. subcom. on oxygen; chmn. subcom. on visual problems). Chief. Div. of Aviation Medicine, com. on Med. Research, Office Sci. Research and Development, 1944-47. Award for Exceptional Civilian Service, 1946. Ensign U.S. Naval Aviation Corps, 1918-19. Fellow A.A.A.S., Am. Physical Soc.; mem. Nat. Acad. Scs. (fgn. sec. since 1945), Am. Phil. Soc. (councillor 1940-43), Am. Physiology Soc., British Physiol. Soc., Optical Soc. Am., Soc. Exptl. Biology & Medicince (councillor 1939-42), Phila. Neurol. Soc., Am. Neurol. Assn., Soc. Naturalists, Sigma Xi, Phi Kappa Psi, Delta Sigma Rho, Sigma Tau, Phi Betta Kappa, Alpha Omega Alpha, Alpha Mu Pi Omega; hon. mem. Harvey Soc., Soc. Anesthetists, Aero-medical Assn.; corr. mem. Soc. Philomath de Paris. Baptist. Clubs: Century (New York), Rittenhouse (Phila.); Cosmos (Washington, D.C.). Contbr. to Am. and British scientific jours.... Office: University Hospital, Philadelphia, Pa.; amd 2101 Constitution Av., Washington, D.C.

BUSH, Vannevar, administrator, electrical engr.; b. Everett, Mass., Mar. 11, 1890; s. Rev. Richard Perry (D.D.) and Emma Linwood (Paine) B.; B.S., M.S., Tuft's Coll., 1913, hon. Sc.D., 1932; Eng. D., Mass. Inst. of Tech., and Harvard, 1916; Lt.D., Brown U., Middlebury Coll., 1939, Johns Hopkins U., 1940, U. of Pa., Yale U., 1942; Washington Univ., Univ. of Buffalo, 1946; Rutgers College, 1942; Sc.D., Harvard Univ., Williams Coll., 1941; Sc.D., Stevens Inst. of Tech., 1943; Trinity Coll., 1946; West Virginia Univ., 1947; married Phoebe Davis, September 5, 1916; children — Richard Davis, John Hathaway. With test dept. Gen. Electric Co., 1913; with inspection dept. U.S. Navy, 1914; instr. mathematics, Tufts Coll., 1914-15. asst. prof. elec. engring., 1916-17; research on submarine detection, with spl. bd. on submarine devices, U.S. Navy, 1917-18; assoc. prof. elec. power transmission, Mass. Inst. Tech., 1919-23, prof., 1923-32; v.p. and dean of engring., 1932-38; pres. Carnegie Instn. of Washington since 1939. Trustee Tufts Coll., Johns Hopkins Univ.; life member Massachusetts Inst. Tech. Corp.; regent, Smithsonian Instn. Levy medalist, Franklin Institute, 1928 (hon. mem. 1947); Lamme medalist, American Inst. Electrical Engring., 1935; Research Corp. Award, 1939; Ballon medal, Tufts Coll., 1941; Holley medal of Am. Soc. Mech. Engrs.; John Scott medal, Phila. City Trusts, 1943; Edison medal, Am. Inst. Elec. Engrs., 1944; gold medal, Nat. Inst. Soc. Scis., 1945; Marcellus Hartley award of National Academy of Sciences; Wash. award, Western

Dr. Bush: *Best* of the best and brightest?

Soc. of Engrs., 1946; Hoover Medal of Asso. Engineering Scientists, 1947. Mem. Nat. Adv. Com. for Aeronautics (chmn. 1939-41). Bus. Adv. Coun. Dept. Commerce, 1939-41; chmn. Nat. Defense Research Com., 1940-41; dir. Office Sci. Research and Development since 1941; chmn. Joint Com. New Weapons and Equip. of Joint U.S. Chiefs of Staff, 1942-46; chmn. joint Research and Development Bd. since 1946. Fel. Am. Inst. E.E., Am. Acad. Arts and Sci., Nat. Acad. Sci., Am. Physical Soc.; member A.A.A.S., Am. Society for Engring. Edu., Am. Philos. Soc., Am. Math. Society; Alpha Tau Omega, Sigma Xi, Tau Beta Pi, Phi Beta Kappa. Clubs: St. Bololph (Boston); Century (New York); Cosmos (Washington). Author: (with W.H. Timbie) *Principles of Electrical Engineering*, 1923; Opera. Circuit Analysis, 1929. Builder of differential analyzer (machine for solving differential equations). Contbr. numerous articles to Am. Inst. E.E. and other scientific pubs. Address: Carnegie Instn. of Washington, Washington 5, DC.

FORRESTAL, James (for'res-tal), sec. of defense; born at Beacon, N.Y., Feb. 15, 1892; son James and Mary A. (Toohey) F.; student Dartmouth College, 1911-12, Princeton U., 1912-15; LL.D. (hon.) Princeton, 1944, Williams Coll., 1946; married Josephine Ogden, October 12, 1926; children — Michael, Peter. With New Jersey Zine Company, Tobacco Products Corp., N.Y. City (1915-16); with Dillon, Read & Co., 1916-40, pres., 1938-40; under sec. of navy, 1940-44; sec. of

SoD Forrestal (*left*) receiving a medal for distinguished service from Pres. Truman.

navy, 1944-47; sec. of defense since July 1947. Served as Lt., U.S. Naval Air Service,

1917-19. Democrat. Clubs: Racquet and Tennis, Links, River. Century Assoc. (New York); Meadowbrook (L.I., N.Y.); Chevy Chase (Washington, D.C.). **[AUTHOR'S NOTE:** The Nuclear Age Peace Foundation (NAPF) states: "James Forrestal committed suicide in May 1949. His death remains shrouded in controversy due to his leadership on military issues and his involvement in covert operations, which he reflected upon in his secret diary." Reportedly, he was replaced by Gen Walter Smith.**]**

GRAY, Gordon, govt. official; b. Baltimore, Md., May 30, 1909; son of Bowman and Nathalie Fontaine (lyons) G.; A.B., U. of N.C., 1930; LL.B., Yale Law Sch., 1933; m. Jane Boyden Craig, June 11, 1938; children — Gordon, Burton, Craige, Boyden, Bernard. Admitted to N.Y. bar, 1934, N.C. bar, 1936; asso. Carter, Ledyard & Milburn, 1933-35, Manly, Hendren & Womble, 1935-37; pres. and publisher Piedmont Pub. Co., Winston-Salem Jour., Twin City Sentinel, operator radio station WSJS, 1935-47; license of radio station WMIT since 1942; asst. sec. of Army, Dept. Defense, 1947-49; sec. of Army since June 1949; mem. N.C. State Senate, 1938-42, 1946-47. Served as pvt., U.S. Army, commd. 2d Lt., Inf., discharged capt., Inf., 1942-45. Mem. Phi Beta Kappa, Delta Kappa Epsilon, Phi Delta Phi.

Gordon Gray: Psychological Operations (Pres. University system North Carolina)

Democrat. Methodist. Clubs: University (New York); Old Town and Rotary (Winston-Salem, N.C.)... Office: Pentagon Bldg., Washington. **[AUTHOR'S NOTE:** President Truman established the Psychological Strategy Board on 20 June 1951, with a directive to the Sec. of State, Sec. of Defense, and Dir. of Central Intelligence for "more effective planning, coordination and conduct, within the framework of approved national policies, of psychological operations." In regard to these operations, Gray was the architect. He was also heir to the R.J. Reynolds tobacco fortune, *via* his father and brother Bowman who were its successive CEOs, and President of the University system of North Carolina, 1950 to 1955 — revealing one of many ways that academia could be influenced by MJ-12.**]**

HILLENKOETTER, Roscoe Henry, naval officer; b. St. Louis, Mo., May 8, 1897; son of Alexander and Olinda (Deuker) H.; B.S., U.S. Naval Acad., 1920; m. Jane E. Clark, Nov. 21, 1933; 1 dau., Jane G. Commd. ensign U.S. Navy, June 1919, and advanced through grades to rear admiral, Nov. 29, 1946. Dir. Central Intelligence Dept. since May 1947. Decorated Comdr. Legion of Honor (France). Mem. U.S. Naval Inst. Office: Navy Dept., Washington 25, D.C. **[AUTHOR'S NOTE:** The Central Intelligence Dept., *aka* Central Intelligence Group

Adm. HillenKoetter: Intelligence (First CIA Director (1947-1950)

(CIG), was replaced by the CIA *per* The National Security Act of 26 July 1947 — the same month and year of the alleged Roswell UFO crash. Adm. Hillenkoetter was well suited to be the first CIA Director in 1947 in virtue of his previous appointment as Officer in Charge of Intelligence on the staff of Commander in Chief, Pacific Ocean Area (Adm. C.W. Nimitz), from Sept 1942 to Mar 1943. Adm. Hillenkoetter avoided suspicion of any personal involvement in a UFO cover-up, at a superficial level, by publicly proclaiming his belief in UFOs and by his later membership on the Board of Directors of the National Investigations Committee on Aerial Phenomena (NICAP). *Cf.* entry below on MENZEL, Donald.]

HUNSAKER, Jerome Clarke (hun'sa-ker), aeronautical engr.; b. Creston, Ia., Aug. 26, 1886; s. Walter J. and Alma (Clarke) H.; grad. U.S. Naval Acad., 1908; M.S., Mass. Inst. Tech., 1912, D.Sc., 1916; D.Sc. (hon.), Williams Coll., 1943; Eng.D., Northeastern University, 1945; married Alice Porter Very, June 26, 1911; children — Mrs. Sarah Porter Swope, Jerome Clarke, James Peter Mrs. T. A. Bird. Officer, later advancing to commander, Constr. Corps, U.S. Navy, 1909-26; Instr. of aeronautic engineering, Mass. Inst. Tech., 1912-16; in charge aircraft design, Navy Dept., Washington, D.C. designed airship Shenandoah, and flying boat NC4, (1st to fly Atlantic), 1916-23; asst. naval attache, London, Paris, Berlin, Rome, 1923-26; asst. v. pres.

Dr. Hunsaker: Aeronautical Engineer (Chief of Aircraft Design, US Navy).

Bell Telephone Laboratories (wire and radio services for airways), 1926-28; vice pres. Goodyear-Zeppelin Corp., 1928-33; head of dept. aeronautical engineering, Mass. Institute of Technology. Chairman Nat. Advisory Committee for Aeronautics. Dir., McGraw Hill Pub. Co., Sperry Corp., Shell Union Oil Co., Goodyear Tire and Rubber Co. Fellow Am. Phys. Society, Am. Acad. Arts & Sciences; hon. fellow Inst. of the Aeronautical Sciences, Royal Aeronautic Soc. of Britain; hon. mem. Am. Soc. Mech. Engrs., Inst. of Mech. Engrs. (London); mem. Am. Soc. Naval Architects, Society Automotive Engrs., National Academy Sciences, Delta Kappa Epsilon, Sigma Xi. Awarded Navy Cross (U.S.); Daniel Guggenheim medal, Franklin medal. Clubs: Century (NY); Army and Navy (Washington); St. Botolph (Boston). Contributor to journals of professional societies. Home: 10 Louisburg Sq. Boston, Mass. [AUTHOR'S NOTE The *Biographical Memoirs* of the National Academy of Sciences (NAS) remarks that Dr. Jerome "Hunsaker was never an enthusiast for rockets." On one memorable occasion at the peak of the Apollo Space Program and faith in rocketry, he was reminded of this by Dr. Charles Draper (after whom the prestigious Draper Engineering Prize is named). Dr. Hunsaker commented, "and I may still be right."]

MENZEL, Donald H(oward) (men-zel), astrophysicist; b. Florence, Colo., April 11, 1901; s. Charles Theodore and Ina Grace (Zint) M., A.B., U. of Denver, 1920,

Dr. Menzel (*left*) next to Dr. Ed Condon – Head of a UFO study for Air Force at the Univ. of Colorado.

A.M., 1921; A.M., Princeton, 1923, Ph.D., 1924; hon. A.M. Harvard, 1942; m. Florence Elizabeth Kreager, June 17, 1926; children — Suzanne Kay, Elizabeth Ina. Field Scout Exec., Denver, 1920; summer and part time teacher of mathematics, Denver Pub. Schs., 1921; instr. in mathematics U. of Denver, 1919-21; asst., Princeton, 1921-22, Thaw fellow, 1922-23, Procter f ellow, 1923-24; summer asst., Harvard Observatory, 1922-24; instr. astronomy, U. of Ia., 1924-25; asst. prof., Ohio State U., 1925-32; exchange prof. U. of Calif., 1928, and 1931; asst. prof. astronomy, Harvard, 1932-35, asso. prof. astrophysics, 1935-38, prof. astrophysics since 1938, leave of absence, 1942-46. Chmn. dept. astron. since 1946; asso. dir. Harvard Observatory for solar research since 1946. Commd. Lt. comdr. U.S.N.R., 1942; comdr. 1945; active duty, 1943-45. Mem. Lick Observatory-Crocker eclipse expdns., Calif., 1930, Me., 1932; dir. Harvard-Mass. Inst. Tech. eclipse expdn., U.S.S.R., 1936; dir. U.S.-British eclipse expdn. Canada, 1945. Corpl. R.O.T.C., 1917-18. Mem. Am. Philos. Soc., A.A.A.S., Am. Astron. Soc. (councillor 1938-41, v.p. since 1946), Am. Assn. Variable Star Observers, Astron. Soc. Pacific, Am. Acad. Arts and Sciences, Sigma Xi (pres. Harvard 1941-42), Received A. Cressy Morrison award N.Y. Acad. Sci., 1926, 1928. Author: *A Study of the Solar Chromosphere* (Lick Observatory Publs.), 1931; *Star & Planets*, 1st edit., 1931, ed edit. 1938; *Story of the Starry Universe*, 1941; *Manual of Radio Propagation*, 1947. Editor of *The Telescope*, 1936-40; member board editors of the *Sky and Telescope* since 1940. Research work in fields of astrophysics, solar atmosphere, eclipses, planetary temperature, absorption and emission lines, theory of gaseous nebulae, wave mechanics, spectroscopy, ionosphere, radio propagation. [**AUTHOR'S NOTE:** Dr. Menzel, with W.W. Salisbury, first made calculations for radio contact with the moon in 1946. Oddly, Professor Menzel was the most vocal *debunker* of UFOs. In his book *UFO: Fact or Fiction* he stated, "it is very possible that intelligent Life" exists in outer space. But "the time required to reach the earth — *even at speeds comparable with that of light* — range in hundreds if not thousands of years for our near neighbors (emphasis)." See author's criticism of his reasoning in "Reasons the Technology 'Cannot Exist' " in Chapter 3. At the same time, Professor Menzel's Harvard colleagues recalled his doodling at faculty meetings by drawing aliens. Indeed and ironically, if he belonged to Maj-12, he may be one of the few persons on earth who ever saw aliens from a crashed UFO. Were there *good cop—bad cop* members? Whereas Menzel might have played *bad cop*, Adm. Hillencoetter may have acted as *good cop*. *Cf.* entry above on HILLENCOETTER, Roscoe.]

MONTAGUE, Robert Miller (mon'ta-gu), army officer; b. Portland, Ore., Aug. 7, 1899; s. Charles David and Effie (Miller) M.; student U. of Ore., 1916-17; B.S., U.S. Mil. Acad., 1919; grad. F.A. Basic Sch., 1920, F.A. Sch., Adv. Course, 1933, Command and Gen. Staff Sch., 1938; m. Mary Louise Moran, June 21, 1921; 1 son, Robert Miller. Commd. 2d Lt., F.A., U.S. Army, Nov. 1, 1918, and advanced through grades to brig. gen. (temp.), June 26, 1942; assigned to Operations & Training Div., G-3, July 1941; Art. Comdr. 83d Inf. Div., June 1942-Mar. 1946; dir. of Antiaircraft and Guided Missiles Br. of the Art. Sch. since 1946. Mem. Delta Tau Delta. Club: Army & Navy Country (Arlington, Va.)... Address: A.A. and G.M. Branch, care AS, Ft. Bliss, TX.

Gen. Montague: Operations / Defense (former G-3 in US Army and Dir. Anti-Aircraft & Guided Missiles).

Adm. Souers: Secretary and Business Input (Former CEO).

SOUERS, Sidney W(illiam) (sou'erz), life insurance; b. Dayton, O., Mar. 30, 1892; s. Edgar D. and Catherine (Ricker) S.; student Purdue U., 1911-12; A.B., Miami Univ., 1914; married Sylvia Nettell, May 28, 1943. President Mortgage & Securities Company, New Orleans, 1920-25, Piggly Wiggly Stores, Memphis, Mar. 1925-Oct.1926; Exec. V.P. Canal Bank & Trust Co., New Orleans, La., 1924-30; v.p. Fourth & First Banks, Nashville, Tenn., July-Nov. 1930; financial vive pres. Mo. State Life Ins. Co., St. Louis, Aug. 1930-Aug. 1933; v.p. General Am. Life Ins. Co., Sept. 1933-Jan. 1937, exec. v.p. since Jan. 1937, now also dir.; pres. Southern Ginning Co., Kenneth, Mo.; vice pres. Missouri Merc. Co., Kenneth, Mo., Delta Realty Co., Catron, Mo., member first board of directors Aviation Corp.; chmn. bd. and chmn. finance com., National Linen Service Corp., Atlanta, Ga.; chmn. bd. and chmn. finance com., Linen Service Corp. of Texas, Dallas; dir. Denver (Colo.) Joint Stock Land Bank. Mem. Life Underwriters Assn., U.S. Naval Researve Officers Assn., Army and Navy Council of St. Louis, Navy League of United States. Delta Kappa Epsilon. Rear admiral U.S.N.R.; active duty July 1940-46; director Central Intelligence Agency Nat. Intelligence Authority. Democrat.

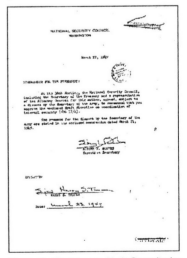

Memo to Truman on Natl. Security by Adm. Souers, 22 arch 1949, with his title noted as Executive Secretary.

Methodist. Clubs: Boston (New Orleans); Mn. Athletic Assn., Farmers (St. Louis)... Office: 1501 Locust St., St. Louis, Mo.

Gen. Twining: Initial Overseer of Roswell UFO Crash (Rose from Nat. Guard and 2d Lt. Inf., to US Air Force Chief of Staff and Chairman of Joint Chiefs)

TWINING, Nathan Farragut (twin'ing), army officer; b. Monroe, Wis., Oct. 11, 1897; s. Clarence Walter and Maize (Barber) T.; grad. U.S. Mil. Acad., 1919; student Inf. Sch., 1919-20, Air Corps Tactical Sch., 1935-36, Command and Gen. Staff Sch., 1936-37; rated command pilot. m. Maude McKeever, March 9, 1932; children– Richard McKeever, Nathan Alexander, Oliv Barber. Served in Ore. Nat. Guard; commd. 2d Lt., Inf., Nov. 1918; transf. to Air Corps, 1924; promoted... to lt. gen., 1945; chief of staff to commdr. gen., South Pacific, 1942-43; commdg. gen., 13th Air Force, Solomon Islands, 1943, 15th Air Force, Italy, 1944-45, 20th Air Force, Pacific, since 1945. Awarded Distinguished Service Medal (Army and Navy), Legion of Merit with oak leaf cluster, Bronze Star (U.S.); Order of Partisian Star first class (Yugoslavia); Companion Brit. Empire.

VANDENBERG, Hoyt Sanford, army officer; b. Milwaukee, Wis., Jan 24, 1899; s. William Collins and Pearl (Kane) V.; student Columbian School, 1918-19; B.S., U.S. Mil. Acad., 1923; student A.O. Tactical Sch., 1934-35, Gen. Staff and Command Sch., 1935-36, Army War Coll., 1938-39; m. Gladys Rose, Dec. 26, 1923; children — Gloria Rose, Hoyt Sanford. Commnd. 2d Lt., 1923, U.S. Army Air Corps; promoted through grades to lieut. gen., Mar. 1945; instr. Fighter Tactics, A.C. Tactical Sch, 1936-38; asst. chief of staff Army Air Force, 1940-41; chief of staff, Northwest African Strategic Air Force, 1942-43; deputy chief of air staff, 1943-47; asst. chief of staff G-2 (Intelligence), War Dept. Gen. Staff, Feb. 1946; apptd. by President of the U.S. dir. of Central Intelligence, June 1946; deputy comdr. A.A.F. and chief of air staff, Army Air Forces, June 1947. Promoted to major general March 1944; appointed deputy comdr. in chief of Allied Expeditionary Air Force; comdg. gen. of U.S. 9th Air Force in France, Awarded Distinguished Service Medal..., Silver Star, Distinguished Flying Cross, Air Medal..., Legion of Merit, Bronze Star. Home: Qts. 12B, Fort Myer, Va.

Gen. Vandenberg: Successive Projects (Former Chief of the Allied Expeditionary Force and Dir. Central Intelligence)

Having noted these remarkable men, did they remain effective as a group? Lt. Col. Corso, former Chief of the Army's Foreign Technology Division at the Pentagon, notes metaphorically that like other *ad hoc* government groups,

Maj-12 became self-serving "the longer it functioned."[17] Like the Wizard of Oz, the group was absorbed by lower groups who believed blindly that they were "managed by higher-ups."[18] However, although eminently competent in the military and scientific domains, would they have a requisite background in theology and philosophy?

Λ

Recent proofs of God and evolution are harmonious! But would a radical specter of aliens causing our evolution influence political policies that discourage traditional religious belief?

Y

Suppose for the sake of argument that Judeo-Christianity, which affirms a Creator, was questioned by Maj-12. In virtue of esoteric information, as radical as it seems, its members could well think that belief in God is a superstition challenged by extraterrestrials who biogenetically caused or upgraded our evolution. Evolutionary accounts of humans, after all, remain incomplete and controversial. Could a strengthened reliance on science and an increasingly weakened belief in theology, due to the Enlightenment, be weighed against a future need to admit of aliens to the public? Despite recent scholarly inroads into proofs of God and their harmony with evolution, might a specter of alien contact incur political policies that further dissuade religious belief? Precisely, in this vein US Supreme Court Chief Justice William Rehnquist noted (*Reuters*, 20 June 2000) that a string of historical rulings bristle "with hostility to all things religious in public life." These ignored questions, as examined later in more detail, are of vital concern in virtue of a perennial role of religion in the moral, social, and political development of our culture.

Intriguing Issues

Certain issues bear on the above organizations. As they represent a sample with philosophical commentary, the issues below are also a sample, starting with appeals to balloons and rockets on the assumption that UFOs are not scientifically possible.

An A-9 Rocket Versus a Flying Saucer

An A-9 Rocket is appealed to disingenuously for rejecting a UFO crash at Roswell in July 1947. In both a phone interview on 19 June 1994 and a two-hour personal meeting with the author in 1995, the Roswell Public Relations

The Air Force has repeatedly changed its story over the years about the UFO at Roswell in 1947. In its latest reply, this purported photo of a Viking space probe after a supersonic test flight presumably shows how easily aerial debris can be confused with UFOs.

Officer Lt. Walter Haut stated that he was given direct orders by the base commander, Col. William Blanchard, to inform the news media that the US Army Air Forces had recovered a "flying disc." Reportedly, after conferring with President Truman, the Fort Worth CO Gen. Roger Ramey ordered another report be made — that the object was a military weather balloon.[19]

The balloon story conflicted with both the military's own radar detections and local citizens who saw a glowing object descend like "lightening." And civilians and military personnel at the crash scene said they and their families were warned by threats to their lives to not talk about what they saw — an odd response even if other secret military devices were the object, such as the Air Force's most recent account of a mogul balloon to detect whether the former Soviet Union was testing nuclear bombs. And this does not even consider the fact that all records at Roswell were found unaccountably missing by the GAO for the months before, during, and after the suspected UFO crash in July 1947.

Λ

USAF Gen. Arthur Exon stated that an unusual material had secretly entered the base, after the crash, and that the "boys in the lab" said it was from space.

Y

A radio station manager in Roswell said he was threatened with losing his broadcast license, for reporting a 'saucer', by a call from Washington DC — a strange threat if the military was trying to conceal *another* secret. Though the Air Force sought consistency with its first story through the years, attempts at damage control led to other speculations. US Air Force Reserve Capt. Kevin Randle, activated in 2003 during the Iraqi war, notes that NASA engineer Jerry Brown thought the debris was an A-9 rocket, not a tested Nazi V-2, because it could leave the reported sort of devastated crash site and heat-resistant debris. This debris would have consisted of the A-9's thin lightweight metal called 'Duraluminum.'

The US Government's General Accounting Office (GAO) published that all the Roswell records were *intentionally* destroyed. Noting the destruction is senior research analyst Steven Aftergood who served on the Aeronautics & Space Engineering Board of the National Research Council and is Editor of *Secrecy News* — a journal for the Federation of American Scientists founded in 1945 by Manhattan Project scientists. Their project on official secrecy is featured in the *Biodefense Quarterly* of the Johns Hopkins Center for Civilian Biodefense Strategies. He wrote on 28 July 1995: "Now it's been discovered that the CIA has destroyed the budget records for that period in Roswell also." Having been demanded by Rep. Steven Schiff, it was "found that 'some of the records concerning Roswell activities had been destroyed' and 'there was no information available regarding when or under what authority.' In fact, all of the activity records for the Roswell Army Air Force Base between 1947 and 1952 had been destroyed in direct violation of the law. No reason for the destruction of the records was ever found."

The controversy was exacerbated by Maj. Jesse Marcel, an intelligence officer at Roswell, who said he and a Counter-Intelligence (CIC) agent tested some of the wreckage. Having rare properties of strength not "from this earth," it was sent to Wright Field. His testimony was reinforced by retired Gen. Arthur E. Exon who stated that an unusual material had arrived secretly at the base shortly after the crash and that the "boys in the lab" said it was from space. The upshot of these reports is that the material was not an A-9's Duraluminum. The latter would *not be* impervious to the heat and stress to which was subject the Roswell material. Some of the tin-foilish material, as thin as that on a cigarette pack, brings to mind recent curing processes for wrinkle-free clothing insofar as when the material was crinkled, it abruptly unfolded again with no creases. This property was confirmed by Maj. Jesse Marcel's son. Now a medical doctor, he states that as a boy he played with the stuff before it was seized by his father when the balloon-story retraction was made. And on 9 July 1947, Robert Smith of the First Air Transport Unit was present when the material was put on C-54s. He testified that he saw a piece of the crated debris, a foil "that could be wadded and then... straighten itself out."[20]

These strength and crease-defying traits illustrate a metallurgical technology that, far superior to ours, implies an equally superior scientific theory. This point bears on the belief that all theories are impractical speculations, especially for pragmatic Americans who not believe in UFOs. One thinks of the dictum *I'm from Missouri, I have to see it to believe it!* One does not *see* theoretical entities in terms of which even terrestrial metals are designed. For example, Al no. 13 with atomic wt. 26.98 — a theoretical name of Aluminum — is *not seen*. That is, from technological artifacts that *are observed*, we infer theoretical entities that *are not*. Precisely, despite exceptions such as 'electric charge', this inference reveals a classical observation-theoretical distinction in the philosophy of science. Even without certifying a UFO, admittance of a super-advanced material that was manufactured, not on this earth, would be remarkable. It evokes the idea of an alien scientific theory and technology that is quantum leaps beyond ours.

Some scientists think that aliens may be visiting earth in nano-tech space probes, too tiny to be observed. Are humans themselves developing such a microscopic technology? Above is a micro-chain 'engine', each link of which could rest atop a human hair. It was fabricated at the DoE's Sandia National Laboratories. The 50-link silicon microchain transmits power somewhat like the drive belts in 19th-century sewing factories. In these factories, a central engine shaft was powered by steam-turned drive belts to power distant work stations.

Ironically, a naïve divorce of theory from technology is noted in the journal for the Phi Kappa Phi Honor Society, *The National Forum:*

> On the one hand, Americans have traditionally been characterized by an innovative industry and underlying pragmatism. On the other, a tacit pragmatism together with common sense has too frequently rendered hostility to theory.... An increased appreciation [of theory] is necessary for... a technological success endemic to America's survival.[21]

Enmity towards theory has spared most Americans from skepticism. They do not *know* that God and UFOs are not "theoretically possible." God cannot exist since He is not an observational or theoretical entity! And UFOs cannot exist since they would have to exceed the speed of light! An underestimation of science by laypersons may offset its overestimation by many scientists. Indeed, NEC physicists astonished their colleagues by announcing that they had exceeded the accepted speed. And it is bypassed also in terms of string theory and warp-drive speeds on interstellar space-time waves.

Above Top Secret

What does not seem controversial is that a status of official secrecy on the 'Extraterrestrial Matter' is so high that even most top military and elected officials are not privy to it. Perhaps, even the US President is not cleared.

What amounts to a category of Above Top Secret, that refers to some classification (Q+ or R?), was evidently not in place on 8 July 1947 when the US Army Air Forces announced its recovery of a "flying disc" at Roswell.

⅄

Prof. Robert Sarbacher was Chairman of the Washington Institute of Technology and a consultant to the U.S. government's R&D Board. In virtue of being privy to UFO files, he stated that they were classified higher than the H-bomb.

⅄

A lack of this sort of classification would hold at that time if, as the military subsequently reported, the disc had actually turned out to be a military balloon. The question of an Above 'Top Secret' status for UFOs is murky after mid-July 1947. Suppose the initial balloon report and later altered accounts were cover stories. In that case, a complete inability of the public or even Congress to obtain *any* official information on the crashed Roswell "disc" in over half a century would certainly indicate a more than ordinary secrecy.

But if an 'Above Top Secrecy' is supposed, there are also some anomalies. There is the oddity that waves of reported flying saucers after mid-July 1947 resulted in a less than top secret report by Gen. Nathan Twining, in September 1947, that the discs were real. Moreover, the report led to a military investigation of UFOs, called Project Sign, which culminated in a startling *Estimate of the Situation* in late July 1948.

Oddly, *Estimate* did not have the highest security classification even though it concluded after painstaking research that some flying discs were probably interplanetary (extraterrestrial). Yet a mere admittance of extraterrestrial vehicles was not as significant as one that crashed. And in addition to the possibility that there were either ultra-secret units such as Maj-12 or agencies composed of Men in Black (MIBs) that did not yet censor other reports, even if it did crashed ones, a more inclusive 'Above Top Secret' may have evolved with Sign. Was its report a proverbial straw that broke the camel's back? Could civilian investigators point to Project Sign to claim that most of the super secrecy *began* when the Sign document, defending a reality of interplanetary craft, was rejected and destroyed by Air Force Chief of Staff Gen. Hoyt Vandenberg? His action and the document acquired the status of rumors. Yet former government official and Professor of Physics Dr. Edward Condon was in a position to know the difference. As head of an official UFO Report at the University of Colorado, commissioned by the Air Force, he expressed agreement with Vandenberg's conclusion!

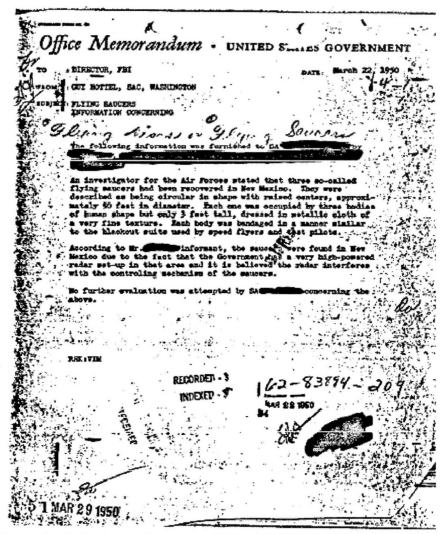

Released by the Freedom of Information Act (FOIA), this FBI memo of 22 March 1950 advises: "An investigator for the Air Forces stated that **three so-called flying saucers had been recovered** in New Mexico... circular in shape with raised centers, approximately 50 feet in diameter. **Each... occupied by three bodies of human shape but only 3 feet tall, dressed in metallic cloth of a very fine texture.**"

Agreement with his conclusion renders credible the document's existence in virtue of Condon's admitted collaboration with CIA officials and high-ranking AF officers during the Condon Report's formulation. Taken with Vandenberg's need-to-know penchant for secrecy due to his CIA Directorship from 1946 to 1947, these facts indicate some kind of formal status for Above Top Secret assigned to UFOs at the time of his action in late

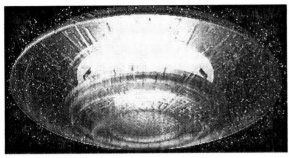

A curious image from the US Government's scientific projects in July 2003. Underneath, there is the caption: "If you watch *Star Trek*, you will note that they often do neutrino scans of objects, occasionally they find that all neutrinos are left-handed.... Reference to Papers relating to detecting unknown sources of neutrinos: Title: The Status of Neutrino Astronomy: Quest for Kilometer-Scale Instruments" in *Nuclear & Particle Physics* 22 (1997) 155-170, Professor Francis Halzen.

July 1948. The time between then and the 1947 Roswell UFO crash was evidently one when an informal supersecrecy attached only to the crashed material.

Many skeptics charge that paranoid UFOlogists exploit a classification of *Above Top Secret* for anything they cannot prove — as by analogy religious believers abuse the notion of *mystery* for whatever they cannot explain (which, of course, can be all cleared up by a rationality of modern science!). Skeptics often hold that Gen. Vandenberg's rejection of interplanetary space probes visiting earth was based, according to the rumor, on his findings that there was insufficient evidence. And even if he was incorrect, there is no evidence of a conspiracy to impose 'Above Top Secret' on information. But in addition to revelations in later decades that the Pentagon used "lies and stonewalling" to cover-up even more mundane secrets such as a high "mortality among [atomic-bomb] test veterans," there are various responses in the philosophy of science.[22]

A criterion for rational belief, that evolves from an influential twentieth-century philosophy called 'logical positivism', is that when nothing counts against a belief it is meaningless and scientifically indefensible. Called a 'liberal verification principle',[23] the criterion is how some UFO believers actually reasoned in contrast to most disbelievers. What would count against their belief includes the CIA not denying it has UFO documents, then withholding them for reasons of national security, and the Government Accounting Office (GAO) not discovering that all military records at Roswell were unaccountably taken during the Roswell UFO crash! And to admit of what counts against a belief is to admit of what counts for it. Our own history of science indicates the probability that future theories will afford interstellar travel that bypasses or exceeds the speed of light — wherein a more advanced ET could already achieve that travel, sightings by highly skilled and experienced observers, and a plethora of private and official reports.

Aside from a host of government reports already noted, private reports range from those of Dr. Robert Sarbacher to Robert Lazar. Sarbacher was a consultant to the government's R&D Board as well as Chair of the Washington Institute of Technology. In virtue of having access to some government UFO files, he stated in 1950 that they were classified higher than the H-bomb. An even higher status, while more controversial, comes from Lazar.

Λ

If contact with ET may cause political chaos, should not the UFO documents be kept secret? For what good is our right to know if it undermines the very society in which those rights obtain?

Ⅴ

Lazar's apparent disaffection with his government employers in the 1980s at Los Alamos, New Mexico, where he stated that he worked as a staff physicist, led to his allegation that there was an ultra-secret status many levels above a top-secret Q clearance (R?) Reportedly, this clearance was used by either the Department of Energy or Nuclear Regulatory Commission. A justification for Q+ clearances, wherein even the highest elected officials might not be fully informed, would obviously depend on the most formidable threats to US national security.

Above Top Secret and Ethics

From the standpoint of social and political ethics, attempts by citizens or elected officials to legally uncover top secrets on UFOs, possessed by governments, reveal a perennial concern of democracies — often more accurately called 'republics' — having representatives elected by citizens. The United States is important in perusing such states because its citizens are among the most vociferous UFO believers and since it is one of the oldest open societies in human history.

On the one hand, elected representatives have authority over the unelected military and intelligence communities. The latter should be responsive to elected officials who, in turn, are in principle answerable to the public. If the public is master over the highest elected official, who is both a civilian President and military Commander in Chief, is not the public all the more master over intelligence agencies and the armed forces? An affirmative answer is patently evident every time that civilians enter open military bases, are saluted, and called 'sir' or 'madam.' Clearly, the public has a right to government information without which political choices cannot be rationally exercised.

One may object that the exercise of legal and moral rights is no more unqualified than the right to free speech. Are we free to yell 'Fire!' in a crowded auditorium? There is a further rejoinder that the public may be properly denied information about UFOs and their occupants in as much as the public chooses, even surreptitiously, to continue to elect representatives who maintain the secrecy. And yet given that UFOs are initially and understandably classified in investigations but turn out to be exempt, say, under Title 5 of the US Code for the freedom of information, it is the case *prima facie* that the information ought to be provided to citizens.

The impact of Judeo-Christianity on the cultures of our world cannot be overestimated. One might say in terms of Christianity that ET's visit would be the most astonishing event in human history — *next* to the resurrection of Christ. This might explain why Christians and other religious believers are, or should be, more epistemologically open to the specter of UFOs than disbelievers. In the case at hand, Judeo-Christianity has perennially both influenced ideas such as 'secrecy' and illuminated their propriety. In the caption above, US President George Bush visits Pope John Paul II on May 18, 2003, at Vatican City.

On the other hand, if such information might reasonably be deemed a threat to national security in terms of causing public panic, weakening faith in social-political orders, inducing a culture shock that impedes daily functions, and providing technological aid to actual or potential enemies, then it seems reasonable to hold that public officials should not reveal UFO information. What good is the public's 'right' to know if the knowledge undermines the society in which the rights obtain? By analogy to the example of a military base, members of the public can be either detained or restricted to certain areas when they enter a base on the grounds that its security is a necessary condition for the greater public good and preservation of society.

UFO investigators may hope that their unrelenting pressure to expose "unjustified secrets" does not reveal security-threatening material that will take them down along with their nation.

The survival of some institutions *over the ages* in virtue of a psychobiological fulfillment of our nature, which is a sadly neglected role of ethics, is related to the issue of UFO secrecy. All institutions change, but

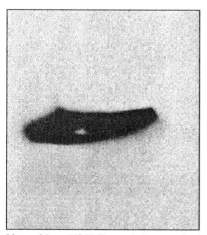

Many of these US Air Force photos were kept secret from the public for decades after Roswell. This Roswellian crescent-shaped UFO, filmed by William Rhodes with a simple Brownie camera in Phoenix on 7 July 1947 — the approximate date of the Roswell UFO crash, was 20–30 feet in diameter and traveling about 100 mph at 1,000–2,000 feet.

some more gradually than others, which also cease to endure. The survival of a select few indicates that they have accommodated or successfully resisted radical events in history as, by analogy, the laws of thermodynamics survived revolutions in physics by virtue of reflecting Nature in a primary way. Thus in the same way as those laws continue to apply historically from Newton to Einstein, the ongoing historical applicability of an institution is explicable in terms of its significantly reflecting truths about an unchanging human nature, or it would be abandoned. And thus despite faults of all organizations, one of the most abiding is the Roman Catholic Church. Its stance, whatever one's moral or spiritual convictions, illuminates a perennial view on this issue:

> *Professional secrets* — for example, those of political office holders... given under the seal of secrecy must be kept, save in exceptional cases where keeping the secret is bound to cause very grave harm to the one who confided it, to the one who received it or to a *third party*, and where the very grave harm can be avoided only by divulging the truth.[24]

In the case of UFO-secrecy, the third party is the public. Thus a knotty question, framed this way, is how the public can know if top-secret government information about UFOs could cause grave harm without possibly incurring the very harm that ought to be avoided. Undoubtedly, these considerations are an over simplification. There are questions of counterintelligence agencies that generate disinformation in a possibly illegal way. This possibly includes giving false information to the public about UFOs for the purpose of deception — but perhaps for the public's own good, if not for its survival.

Is it ethical for government agencies to lie to the public for a moral end? Does the end justify the means? A common answer among philosophers is 'no'. But the question brings to mind a law passed by Congress that forbids the assassination of political leaders when we are not at war with their nations. Many citizens might be hard pressed to find immoral the assassination of a

Hitler, Stalin or Pol Pot, even under these conditions, if their death would achieved an end of preventing incalculable suffering and a loss of millions of lives.

Moreover, for example, there are complex questions about balancing a power of the legislative, executive, and judicial branches of the US government. This government's Founding Fathers recognized inalienable rights and an inherent goodness of people. At the same time, they knew of people's corruptibility. No person or branch, however popular, was entirely trusted. But trust is a double-edged sword: No one may patronizingly decide *in toto* what the public needs to know. Yet no one has an unfettered right to national-security information, much less to divulge it. Did Jimmy Carter, an Anapolis graduate with a degree in nuclear physics, discover this fact despite his idealism? As Governor of Georgia he filed a UFO report and declared top-secret information would be released if he became President, only to evade the issue after his election.[25]

The public faces a *constructive dilemma*: If the government reveals its information, then there will be a national-security crisis and if it does not reveal the information, there will be an illegal cover-up. Either the government will or will not reveal the information. Hence, there will be either a national security crisis or an illegal cover-up. And in terms of a cover-up, mounting disinformation for over fifty years would ruin reputations and undercut faith in government. And thus the consequences of an illegal cover-up may, ironically, rival the undesirable consequences of a national-security crisis to which government officials appeal!

⋏

UFO Skeptics, influenced unwittingly by Hume's radical eighteenth-century empiricism, say sensations are confused with sensed things. But these things are ordinarily that of which we are incontrovertibly conscious. "The eye does not see itself" is the sardonic response in *The Phenomenology of Perception*.

⋎

The classic rejoinder to the dilemma is to say that it *falsely dichotomizes* two extremes. There can be a middle ground of revealing some information. Investigators who are sensitive to such nuances might continue to seek as much information as they can through various legal avenues. These avenues include both their Congressmen and Freedom of Information Act (FOIA). It seems unreasonable to expect investigators to suppress their curiosity about what would be one of the most astonishing discoveries in human history. At the same time, they may hope that their unrelenting pressure through both the courts and their elected officials to expose an "unwarranted secret" does not reveal a security threatening matter that will take them down along with their nation.

After-Images and Illusory UFOs

Also called photogenes, images of phenomena will remain in the eyes' retinas after a visual stimulus ceases. Theoretically, the after images can challenge UFO sightings because when a bright stationary object is seen, such as a star, the eyes' movements may affect the images. The images that move may cause the object to both appear as if *it* is moving and have various colors. Also, the visual stimulus ceases momentarily each time an observer blinks and may leave a variety of colored images in an eye's retina. Thus the observer of an alleged UFO may not be able to discern colored images that change due to eye movements from an unmoving external stimulus, such a either a star or the planet Venus, whose colors are quite different.

Exploitation of afterimage sensations by skeptics applies to an unsolved investigation, Case 6,

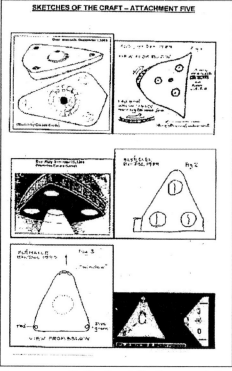

SKETCHES OF THE CRAFT – ATTACHMENT FIVE

The more specific the description the more likely a hoax and, say Air Force Officials for a *Catch-22*, the less likely a hoax the more ineffective the description. Nonetheless, these descriptions of UFOs were 'Attachment Five' for a Freedom of Information Act (FOIA) request, 5 USC § 552., submitted to the US Department of Defense by attorney Peter Gersten of Citizens Against UFO Secrecy (CAUS).

of the Condon UFO Report. Headed by Professor Edward Condon, the Report was issued under auspices of the US Air Force at the University of Colorado. Case 6 was conducted by two of Condon's staff, physicist Roy Craig and research engineer Norman Levine, in the Spring of 1966. Two policemen and three women made the spectacular claim that differently colored starlike lights were noiselessly darting like fireflies over a school at 9:00 p.m.. Suddenly, one of the lights responded to a beckoning gesture by one of the women.

The light flew directly towards her and hovered 20 to 30 feet overhead, whereupon it turned out to be a dull aluminum-like disc, about the size of a large car, with glowing lights around its top. Two other women fled in panic. But in looking back, they confirmed her general description. Having responded to a neighbor's call about the fracas, the policemen reported that by the time they arrived, an extraordinarily bright object at close range was seen speeding off. Skeptics dismissed the UFOs as natural phenomena.

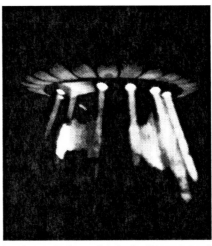

Is this an afterimage? Actually, it is a photo of the multicolored UFO that was seen by multiple witnesses. Courtesy: Maj. George Filer, US Air Force Intelligence Officer (Ret.).

The planet Jupiter, during this 30- to 45-minute incident, was brightly visible. In arguing that because some of the observations were of this planet, all of them were, some skeptics appealed to the afterimage phenomenon. A glib relegation of all the sightings at different locations to afterimages is untenable, though the afterimage phenomenon still poses a problem for the veracity of other UFO reports since those reports can be made in good faith about what *are* actually bright planets or stars. Precisely, psychologists and psychiatrists who certify that persons are being truthful about what they believe they observe, when reporting UFO phenomena, may simply beg the question of what is true about the 'things in themselves.' For this reason, the question remains of whether or not what is believed to be true *is* true in fact.

The challenge of afterimages to UFO reports is aggravated by the influence of David Hume's radical empiricism in the eighteenth century. This empiricism contrasted ideas derived from *internal* sensations of the world to the world in itself. Any ideas about phenomena which are formulated as empirical descriptions bring to mind, by analogy, an inkblot test. The test is used by either a psychologist or psychiatrist to diagnose a patient's mental state by the patient's description. But the description tells the therapist nothing about the blot in itself.

By analogy, since those who report UFOs can seek to prove only the truth of what they observe by appealing to the internal sensations in question, their descriptions may merely amount to descriptions of sensations in terms of what they *want* to see! That is, in being reminiscent of the twentieth-century physicist Thomas Kuhn, as much as Hume on whom Kuhn leans, UFO reporters might be imposing conceptual paradigms on their sensations. A peculiar metaphysical realism is suggested, namely that true ideas do not point to an external reality but to a reality of internal sensations.

There are compelling rejoinders to Hume's intellectual heirs whose skepticism about UFOs is rooted in afterimage sensations. Some responses are from existentialism and phenomenology. In emphasizing that to see is to be conscious of both what we see and our seeing, they hold we are also simultaneously aware of the sensations and things sensed. It is no more coherent to speak of sensations without being conscious of them than to speak of sensed things without also being aware of the things. We can appeal not only to an integrity of our immediate, incontrovertible and phenomenological experience for discerning one from the other but to our inability to survive if we could not.

∧

A skeptics' disbelief is either based on science or not. If it is based on science, then they cannot exclude evidence for UFOs out of hand. And if their disbelief is not based on science, then they are only expressing their own unscientific opinion.

∨

This criticism may not entirely avert afterimage objections. But the point about survival indicates that they hold as an exception and that we are patently aware, ordinarily, that what is seen is not an image. Thus a response to the Humean skepticism, based on the inkblot analogy, is that psychologists must themselves be observationally aware of what are on the blots to distinguish them from the descriptions of their patients. Why not admit that the patients, and UFO observers as well, can also be aware of a difference between the descriptions and things described? Why not be less patronizing and grant them an equal power of observation?

Nevertheless, some scientists who do *not* believe in UFOs and religious phenomena, such as the dancing sun at Fatima or Virgin Mary at Medjugorje, argue that true believers merely see what they want to believe. And although their belief might have a basis in sensations, the sensations are not external phenomena and beg for mundane interpretations. The interpretations, disbelievers argue, usually in condescending fashion, reflect the wishes and cultural conditioning of believers. But in fairness to the believers, does it not follow both that the disbelievers may be equally socialized and that they do *not see* what they want to *disbelieve?* In the end, the unqualified appeal to sensations and afterimages of such things as stars or bright planets does not merely conflict with astronomers who sometimes conclude that unidentified bright objects (UBOs) are intelligently controlled. In addition, it conflicts with the methodology of science itself.

For example, one test of a genuine scientific theory, hypothesis, or explanation is that it permits, if not outright falsification, some evidence that may count against its truth. The skeptics' disbelief is either based on a science or not. If it is, then alien UFOs cannot be excluded out of hand. And if it is not based on science, then the skeptics are merely expressing their own

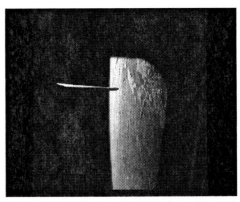

UFO outside a window of NASA's Space Shuttle? Despite unsuccessful challenges to ET-flown UFOs by explaining them away as afterimages, there is a principle in the philosophy of science called an Underdetermination-of-Theory-by-Data (UTD) Thesis. This government photo illustrates that data are often subject to inconsistent theoretical interpretations that, nonetheless, have equal power in explaining them.

unscientific opinion. The opinion, in this case, stems from relegating all UFOs to interpreted sensations or afterimages and is itself a pseudo-scientific position. The position, by their *own* understanding, would expresses only what disbelievers want to believe.

Further, a skeptic's radical distinction between images of phenomena and the phenomena themselves can be criticized by noting that an observed *success* of theories in predicting, explicating and manipulating phenomena could not occur unless the phenomena were reflected truly by the theories. The integrity of speaking theoretically about what reality is really like, which scientists skeptical of UFOs presumably do, depends precisely on their own ability to distinguish sensorial images of phenomena from phenomena. Patronizingly, however, this discernment excluded for those who observe UFOs.

Ironically, empiricist distinctions of sensations from sensedata (phenomena) would be faulted by an Anglo-American analytic philosophy that has its origin in the traditions of Hume, Kant, and Logical Positivism. Following the pioneering analysis of Ludwig Wittgenstein, it is senseless to talk of *seeing* sensations and not phenomena of which, admittedly, there are sensations.[26] And the distinctions would be criticized equally by Continental philosophy in terms, for instance, of Merleau-Ponty's phenomenology of perception. He quipped that the eye does not see itself![27]

Even coherent talk about afterimages supposes possibly objective descriptions. If there were no bright objects which were not moving, the skeptics could not even begin to doubt UFO observations in terms of the afterimages to which they appeal! By this fact itself there may also be veritable observations of real anomalous bright objects that move. And by moving in controlled ways that are physically impossible for natural phenomena and terrestrial craft, UFOs may reasonably beg for an Extraterrestrial Hypothesis.

Spider-like alien in H.G. Wells' *War of the Worlds*. Born in 1866, Wells foresaw a mind-boggling future for science by his science fiction. "Note," declare NASA officials, "the absence of a space suit for the Martian. What type of Mars environment does the... drawing suggest?" One that can be terra-formed to fit human comfort? But a press release by the University of Massachusetts on 16 January 2002 states: "Deep below the surface of the Beverhead Mountains of Idaho, a research team led by Professor Derek Lovley, head of the microbiology department at the University of Massachusetts, and Francis H. Chappelle of the U.S. Geological Survey (USGS), has found an unusual community of micro-organisms that may hold the key to understanding how life could survive on Mars. Their findings are spelled out in the Jan. 17 issue of the journal *Nature* (vol. 415)."

Age Factor and Belief in Alien UFOs

Orson Welles' 1938 radio broadcast of *The War of the Worlds*, leaving "thousands of listeners... panic-stricken,"[28] indicates that persons in both diverse age groups and a wide variety of vocations believe that extraterrestrial beings might manifest themselves, technologically, in the form of UFOs. Notwithstanding that the broadcast turned out to be a mere script, or because of it, a reality of alien UFOs was not accepted by many of middle-aged persons who were interviewed randomly in 1947 about a crashed UFO at Roswell, New Mexico. A military officer announced at the time that the US Army Air Forces had taken possession of a flying disc.

Perhaps, Welles' trickery reflects the dictum 'Fool me once, shame on you; fool me twice, shame on me.' But the adage does not apply to unfooled generations born after the broadcast. After the US Army Air Forces became the Air Force, it initiated a Project Blue Book in 1953 that publicly investigated UFOs. Since it was highly publicized, even becoming the basis for a television series, it begot some intriguing polls in later years. A Gallup Poll in 1966 showed that "younger and better educated persons are more likely to say that flying saucers are 'real' and that there are 'people somewhat like ourselves living on other planets'."[29]

Also, a scientific poll of 1,200 persons in Louisiana in 1967 by David

Deemer revealed that 61 percent of those under age twenty five, 48 percent aged twenty five to twenty nine and only 34 percent over fifty, believed that UFOs were alien visitors. The disparity of belief in age was even more dramatic for a 1973 Gallup Poll. It showed that 66 percent of those from ages eighteen to twenty four were believers, in contrast to 60 percent from twenty five to twenty nine, 49 percent from thirty to forty nine, and only 31 percent fifty and over. Educators explain this.

Whereas young people are quick to learn and accept new ideas, ideas must be related to personal experience for older persons who also have vested interests in the status quo. Given the space program since the 1960s, these facts explain why youths are more open to our visiting other planets and those planets being inhabited by aliens who visit us. Nevertheless, a poll of those eighteen and older, in January 2000, showed that 30 percent think aliens have come to earth and 43 percent that UFOs are real (Yankelovich, 2.5 percent margin of error). One may object that an *ad populum* fallacy is committed if popular belief is used to infer that UFOs are alien. Skeptics note that a belief that the earth is flat does not make it true.

Scientists open to astonishing discoveries imagine an expanding island of knowledge surrounded by an equally increasing ocean of unknown possibilities that make science fiction seem mild.

Truth is not decided by a popular vote. Some scientists are liberal and others conservative, *epistemologically*, in terms of how seriously they entertain possibilities that conflict with their theories. When Einstein's theory was challenged by UFOs which indicate that the speed of light was exceeded, conservative scientists countered with experiments, such as ones which speed up atomic particles, that seemed to pose hard evidence against it — though a possibility of UFOs going faster than the accepted speed was supported by NEC physicists whose experiment was duplicated internationally. But besides a principle of Pessimistic Induction, this challenge to liberal scientists is met by their imagining an expanding island, analogous to increasing scientific knowledge, with a watery perimeter of the *unknown* that increases in direct proportion to the known.

Popular Polls and the Politicization of Truth
Not 'knowing we know' (the K-K Thesis) and a relativistic democratization

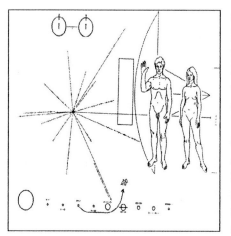

A NASA plaque sent into space informs ET about basic biological differences of men and women. It bears on ET's more informed scientific inferences to sexual behavior and social-political institutions that *fulfill* our psychobiological nature. But our nature is only a social construct, in the past by dead white men from whom we are liberated, say politically correct feminists (per Professors Daphne Patai and Noretta Koertge's *Professing Feminism*). Yet the impact of a feminist counterculture on confusing gender identities and undercutting marriage led to health warnings for women who adopt traditional male norms (*Newsweek*, 13 Aug. 2001). Unique social pathologies have ensued since the late 1960s: divorce up 200%, out-of-wed-lock childbirth 300%, violent crime 500%, plunging SAT scores, and an epidemic of sexual disease from AIDS to Herpes (*PBS McLaughlin Group Report & CDC*). Might ET be cautious about contacting a species that pathologically and immorally ignores the biology of which it informs others?

of science have led to a *political* liberalism beholden to, among others, physics professor Paul Feyerabend. He favors a scientific anarchy that was exploited by a Marxian counter culture of the New Left to politicize 'truth'. In his *Against Method: Outline of an Anarchistic Theory of Knowledge* (New Left Books), 'truth' is relative to different theories that are not known to be true about Reality since we have only subjective internal sensations of it — *à la* Hume's radical empiricism — that circularly beg for interpretations by statements whose truth is relative to inconsistent theories. Since Einstein's theory would be no more true than one that permits exceeding the speed of light, and therefore UFOs, a consensus of those in power determines which theories are true. Thus although this determination is exalted by some UFO researchers because it challenges a conventional realism in which 'truth' is objective, their own research would itself be unobjective and 'outvoted' by those who seek an equally unobjective theory that does not detract from a *this worldly* agenda of sexual, racial, and class struggle. Tellingly, the only response the author evoked from a colleague in talking to her about ET was if *he* would be sexist! Indeed, there is no shortage of articles in *The Philosopher's Index* (Bowling Green) to expose the mischievous roots of a pseudo-scientific political correctness in Feyerabend.

Professor Feyerabend, of course, suffers objections to which Hume's view on sensations was subject, as well as criticisms that his epistemology is logically incoherent. There would not be no objective truth, since theories could not be corroborated by theory-neutral observational predictions. Predictions would be understood initially in terms of sensorial sentences, to which 'truth' *is not* ascribed, that are interpreted by possibly contradictory statements of logically inconsistent theories to which 'truth' *is* ascribed. Consequently, there could be ascriptions of 'truth' to the statements S and $\sim S$ as well as to any given theories T_o and $\sim T_o$ in terms of which the statements are understood! How could this dilemma enhance a credibility of truth-claims about UFOs?

⋀

It is *rational* to infer that today's scientific impossibilities will be possible tomorrow. Thus if there exist older or superior alien intelligences, it is *irrational* to infer that our tomorrow is not in their hands today.

⋁

Feyerabend's notion that observation is theory dependent reminds one of Thomas Kuhn's *The Structure of Scientific Revolutions*. His equally liberal position, also exalted by the counter culture, is lauded tacitly by psychiatrist John E. Mack, M.D., who notes his friendship with Kuhn.[30] With a caveat that this has *not* led Professor Mack to politicize truth, in accord with Aristotle's dictum *truth before friends*, different theories of scientific paradigms in history are imposed *a priori* on observation. Accordingly, although Kuhn challenges the observations of a 'normal scientific paradigm' in which UFO research is not tolerated, tolerance is once again gained at the cost of objective truth because the truth of any observation depends circularly on the theoretical interpretation of a paradigm. And since a paradigm would undercut the objective truth-claims of abductees, who Dr. Mack defends, he does not evidently rely literally on Kuhn's formal position. Might the latter position enjoy a more figurative gloss?

This gloss might construe Kuhn as marshaling the courage of scientists to entertain UFOs as alien craft. Those who *do* and those who *do not* are like persons with opposing *Weltanschauungen* (worldviews): Both are viewing the same world but "see different things."[31] Though Kuhn is a determinist whose position rests partly on the idea that scientists are caused or conditioned by 'disciplinary matrixes' to interpret phenomena by conventional paradigms, his insight might be recast to permit free will. Here, scientists are free from the matrixes as surely as from a deterministic spatiotemporal realm in terms of which logically inconsistent truth-claims would be equally true because equally caused. In not being caused, revolutionary scientists can be praised for their heroic choices to be open-minded and 'normal scientists' can be blamed for being closed-minded — praise and blame being incoherent if there is no

While UFOs and Black Holes are not entirely understood by *normal science*, it does not follow that they do not exist. In a spectacular attempt to detect a black hole, NASA says that the Hubble Telescope peered into the core of galaxy M87 and found a jet of particles traveling near the speed of light, going 5,000 light years beyond the galaxy's borders. The jet was powered by a super-massive black hole devoid of space, time, and matter.

free will. That is, normal scientists, in feeling threatened by the unknown or having a false pride, can be blamed for neglecting the pursuit of truth in favor of a normality with which they feel comfortable!

The philosophical acrobatics of Feyerbend and Kuhn, in holding *a priori Weltanschauungen* (worldviews), less Kant's categories, permit open-mindedness at a cost of coherence. Ironically, polls show that open-minded scientists are even less imaginative than young persons who have no stake in the conventions that bias rational belief. While it would be an *ad populum* fallacy to appeal to polls for truth, objective truth can support UFO belief by way of two considerations.

Ex-NASA Chief Daniel Goldin declared that "if spectroscopic analysis revealed a blue planet with an oxygen atmosphere just 4 light years away," the demand for "a warp drive would start right away!"

First, consider the history of science. Since scientific impossibilities of yesterday are today's possibilities, it is reasonable to suppose that today's impossibilities will be possible tomorrow — if not probable. Second, the technologies and theories of tomorrow, thousands and even millions of years from now, may now be in the hands of many extraterrestrial civilizations whose inhabitants are either a much older species or more intelligent than human beings. This likelihood, given thousands of reliable sightings, brings to mind Hubble Space Mission astronaut Story Musgrave, M.D. Having six college degrees, besides being a medical doctor, he stated: "It's almost a statistical certainty there are beings out there millions of years more advanced than we.... It's one chance in a billion I'd succeed in making contact."[32]

Visits of Aliens from Alpha Centauri?

Alpha Centauri is the nearest star system about 4.4 light years away. Supposedly, this system illustrates an impossibility of ET visiting earth. Relevant to this supposition is University of Iowa astronomer Dr. J.A. Van Allen. He is an open-minded pioneering astrophysicist best known for his work in magnetospheric physics. His 1958 Explorer-1 experiment established the existence of radiation belts — named after him — that encircled the Earth, representing the opening of a broad research field. Accordingly, it is worth noting that Van Allen addressed the UFO controversy by stating that a one-way 100,000 year trip to Alpha Centuari could be lowered to 10,000 years by atomic-powered propulsion.[33] But even this is as allegedly impractical for us as it is for ET. Since Alpha Centauri is *traditionally* deemed a bad candidate for alien life, this life would be from stars many more light years away. And stars so far away would still ostensibly pose insurmountable problems even if the speed of light were approached.

Craft traveling even a significant percentage of light speed in future centuries by new propulsion systems, without other technologies, would suffer radiation hazards to electronics equipment and passengers caused by collisions with interstellar gas. The craft would be similar to a radioactive target in a cyclotron. In denying any scientific knowledge of short-circuiting the distance between stars by a "method like *Star Trek's* warp drive," Van Allen draws an implicit conclusion of many scientists that interstellar alien spacecraft visiting earth is an impossibility.[34] But since tenable arguments have only one conclusion, counter-conclusions reveal a fallacy. We need only consider the position of physicist Miguel Alcubierre in *Classical and Quantum Gravity* (1994) to see how shortsighted is such an anti-UFO reasoning:

> 1994 may be remembered as the year that warp drive was first conceived... a physical *possibility*. Long a cliche of science-fiction writing, the warp drive thought to be inconsistent with the laws of physics. But all that has changed.[35]

Reviewing this work for the *American Scientist*, physicist Dr. M. Szpir notes that a union of three dimensions of space with the dimension of time "is not an inert substrate, but rather a dynamical entity that twists and distorts under the influence of concentrations of energy."[36] This fact suggests that "it might be possible to exploit this phenomenon to travel from one star to another faster than the speed of light," and that this travel "could be done by creating a disturbance in space-time such that the region directly in front of a spaceship is expanded."[37] Therefore, Szpir adds, the "distortion of space-time would,

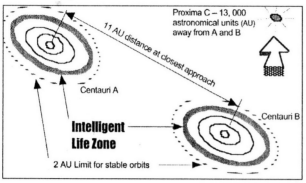

Does ET Exist at Our Closest Stellar Neighbor?

Is Alpha Centauri, once used to debunk alien contact, a promising star system for terrestrial-like planets and intelligent extraterrestrial life? Ex-NASA chief Dr. Daniel S. Goldin declared in 1992: "if spectroscopic analysis revealed a blue planet with oxygen atmosphere just 4 light years away orbiting Alpha Centauri, the "demand to build a warp drive would start right away!" The Hubble Space Telescope has been checking for their existence, and it may be no coincidence that at a recent meeting of experts in Toledo, Spain, the chief of NASA's research program Dr. Mike Kaplan said that "we will discover extraterrestrial life in the next 25 years." (See K. Croswell's "Does Alpha Centauri have intelligent life?," *Astronomy* 19 (1991), No. 4).

in effect, propel the spaceship forward [without radiation hazards] like a surfer riding the crest of a breaking wave."[38]

A physicist at the University of Wales, Alcubierre holds that warp-drive spacecraft can use exotic matter with a negative-energy density. Related to Dutch physicist H. Casimir's *Casimir effect* of 1948 and its measurement in 1958 by M. Sparnaay, usually viewed as confirming that negative-energy densities are possible, their experimental existence is further related to a possibility of either terrestrial or extraterrestrial spacecraft traveling two million light years from the Galaxy Andromeda and back the same day. This extraordinary possibility is held to be consistent with Einstein's general relativity without a causal paradox of his special relativity.[39] Exceeding the speed of light, does not lead to a causal paradox where actors in the present alter their own past. For light also travels in space-time and is carried along with the craft. The light beam would still be traveling at light speed, relative to the spaceship, that itself is not accelerating relative to the space-time in its immediate vicinity.

Further, the shortsightedness of taking current science as the 'last word' is augmented by logic. When there is not adequate knowledge about a bizarre thing, whether of alien technology or short-circuiting space-time, concluding it does not exist or cannot be achieved commits a fallacy of *appeal to ignorance* (*argumentum ad ignorantiam*). When we are ignorant about something, a conclusion cannot be inferred.

Finally, Professor Van Allen is obviously a competent and even esteemed astrophysicist. And those interested in possible visits by ET to earth are fortunate to have such a man with whom to discuss the issue. Many less notable scientists will not even address the topic. Having said this, it is

Cockpit view at warp speed? The premise of the Alcubierre warp drive is that although Special Relativity excludes objects moving faster than light in space-time, space-time itself may not be limited. It expanded faster than that speed at the early moments of the Big Bang on the inflationary model. Hence, ask NASA scientists, why not for warp drive?

unfortunate that he responds to the criticism that organized science is locked into its own way of thinking by affirming a "vested interest in... a scientific explanation."40 That is, his *identification* of the thinking of "organized science"with scientific explanations is, on the face of it, either trivial or significant but unscientific.

Without delving into the controversy over whether or not a theory is an explanation, the identification is trivial if all it means is that by definition an explanation is not scientific when

it lies outside orthodox scientific thinking. And the identification is significant but unscientific if anything inexplicable by our science, say our feasible travel to Alpha Centauri and Centauri's inhabitants traveling here, should be relegated to either science fiction or natural phenomena that are compatible with our

$$\vartheta = -\alpha \, \mathrm{Tr}(K)$$

Alcubierre Warp Drive Bubble. According to a "Breakthrough Propulsion Physics" Report by Dr. Marc Mills at the NASA Glenn Research Center in 1997, "Experiments and theories were discussed regarding the coupling of gravity and electromagnetism, vacuum fluctuation energy, warp drives and wormholes, and superluminal quantum tunneling." The Report adds that "government researchers and select innovators to jointly examine new theories and phenomena from scientific literature" have "reawakened consideration that such breakthroughs may be achievable."

limited scientific understanding.

Dr. Van Allen invites this understanding when he asserts in a *weak analogy* that "If you're... in Wyoming and you hear the thundering of hoof beats, you don't expect a herd of zebras."41 But zebras can be entertained if there are reliable witnesses. And ordinary hoof beats, in any case, are not analogous to extraordinary UFOs. Precisely, a hallmark of scientific explanations is that they are subject to the test of extraordinary phenomena which count against them. If nothing does because the phenomena are relegated to familiar hoof beats, then science is not being embraced but rather something akin to personal bias or politics.

Political ideology has already been related to an academic reluctance to deal openly with UFOs. A link to politics is further illustrated by Professor Lewontin's *Biology As Ideology*. Bringing to mind professors Paul Feyerabend and Thomas Kuhn — Kuhn lamenting in the *Scientific American* (9 May 1991) that he had bolstered political correctness in higher education, Lewontin doubts an objectivity of science since it is shaped "by social and political needs."[42] He raises a troubling question: Is there a political need for the scientific study of UFOs and, if not, does that make true an ideological denial of their existence?

In sum, problems of traveling to Alpha Centauri were exploited to the hilt by mainstream scientists, from Dr. Van Allen to rocket scientist Werner von Braun decades earlier, for disavowing the feasibility of human interstellar travel and, by implication, travel to earth by alien UFOs. And the disavowal is exacerbated by a politicized science that intimidates scientists and obfuscates truth, wherein what is empirically true can be politically incorrect. But the alleged infeasibility involves an unscientific fallacious reasoning that is apparent in logic as well as the philosophy of science.

Alternative Space Craft (ASC)

Alternative Space Craft is a name for alien UFOs, used by former NASA engineer Robert Oechsler, after he claimed that he flew in a strange high-performance, black helicopter, in January 1990, from an installation of the North American Aerospace Command (NORAD) to an oil rig near Pensacola, Florida.[43] After an astonishing trip in the craft, whose speed was about 750 mph, he stated that they landed at the oil rig which had a huge control room similar to rooms at the Johnson Space Center. Oechsler heard 'ASC' used for five blips moving on an enormous screen whose three-dimensional quality permitted seeing 'up-range' over various sectors of the United States. The blips were labeled ASC on the screen in the manner of normal radar screens. That they resembled intelligently-flown UFOs was indicated by a flight pattern of 'coming down' and 'spreading out' as well as by one of the UFOs glowing red when it landed.

Reports of alien alternative spacecraft, some of which may look like either a helicopter or conventional plane, are so sensational that even an insider of the space program is not sufficient to fully believe them. Yet Peter Gersten's Citizens Against UFO Secrecy (CAUS) states that the Freedom of Information Act revealed numerous cover-ups of state-of-the-art NORAD trackings of UFOs.[44] Thus for example, CAUS member Robert Todd was informed in February 1990 that no tracking records were kept, only to be told in May by FOIA Manager

Barbara Carmichael that they were exempt from FOIA due to national security.

A compromise was reached in June when some of the records were released. Despite most of the sentences having been blacked out, the uncensored designations ***NORAD SECRET***, NBR OF TRACKS, SECTOR, and NBR OF OBJECTS, were said to refer to alternative spacecraft. The word 'craft' implies that the objects were technological. They could not be relegated to either natural phenomena or unknown terrestrial aircraft. In short, nothing excluded — and there was reason to believe — that they were not of this world.

Science and Religion in Ancient Times?

Among ancient anomalies is a high plateau in Altiplano, at the region of Bolivia, Peru, and Argentina that some scientists link to geometrical effects of an enormous airfield. Since the airfield challenges our technological duplication, it has been suggested that either a prehistoric people developed aviation or aviators of an extraterrestrial origin made the field. These interpretations may seem too radical. But among experts who are renowned for being able to estimate what can be achieved technologically, the former British Chief of Defense Lord Hill-Norton, G.C.B., notes the "astounding geometrical effects" of "what seem like gigantic airfields laid out in the days of prehistory by means which would tax today's technology to the limit."[45]

The limits of human technology raise the specter of either our ancestors having had an advanced aeronautical science or extraterrestrials having visited earth tens of thousands of years ago — a possibility, discussed soon, that is seriously entertained by physicists at the US Air Force Academy. And while these ideas seem extreme, their mere consideration would surely strengthen the muscles of our scientific imagination. This effect would not be all bad in light of Einstein's famous dictum that imagination is more important than intelligence. Ironically, bearing on this point is the anarchical view of physics professor Paul Feyerabend.

It is an understatement to say that Feyerabend overstates theoretical interpretations of 'fact' when he denies factually objective truth-claims that are not already interpreted by theories. Since there is no theory-neutral evidence that can impartially support any theory, he favors the anarchy of a

Optical illusion: One face? Or two profiles facing each other? Despite some Paleolithic art that has an uncanny similarity to today's descriptions of ET and UFOs, this illusion reveals some dilemmas faced by archeologists who try to interpret the paratypic art.

theoretical pluralism. However, a pluralism *per se* may actually bolster objective science in our case insofar as we acknowledge that, although there are facts that beg for theoretical interpretations, no interpretation exhausts all of the empirical predicates of any phenomenon. For example, as classical mechanics ignores 'electric charge', it seems fairly easy to see how an archeological theory may ignore empirical data that would be proper to advanced propulsion systems of various craft.[46]

Debunkers of UFO interpretations in the social sciences may still respond, of course, that their investigations of the pyramids in Egypt and Stonehenge show that primitive people with ingenuity and sheer man power can fabricate what appear to be space-age structures that were erected despite archaic technologies. Today's technologies have a more pragmatic focus than the impressionistic nature of ancient art crafts. Those crafts, they will typically add, were often infused with mystical and religious symbols that did themselves constitute a form of primitive science.

Notwithstanding this anachronistic imposition of science on religion, religious constructs may still be viably interpreted as geometric coordinates for space travelers. The self-serving arrogation of ancient religion to primitive science cannot be used to glibly dismiss the prospect that ancient people could have related their religion to a science that was far superior to ours. It might not only have been a science of either their own or extraterrestrials but a mock-up of the latter. There is no shortage of modern artifacts, by analogy, that were copied by people who still lived in conditions of the Stone Age:

When the Stone Age tribespeople of Micronesia first saw planes landing and their cargoes..., they deduced that white people must have some powerful magic to lure such wondrous creatures down from the sky. They set about building mock-ups of airfields, with mud runways and control towers made of sticks and grass, in the hope of directing some of the heavenly cargo to themselves. These "cargo cults" still flourish in remote areas of Papua New Guinea.[47]

In light of the above points, there are several more to be made. First, what is so radical about a vastly superior technology in ancient times if cosmic evolutions, in light of the Big Bang Theory, render plausible the existence of alien civilizations millions of years before those of humans? Could not aliens have either visited Earth or inhabited it first? Second, the idea of ancient religion being a primitive mode of science brings to mind St. Augustine of Hippo (354-430 AD). He contrasted the Bible's depth to intellectuals about whom Ludwig Wittgenstein also had grave misgivings.

Ludwig Wittgenstein, of whom most philosophers are in a virtual 'religious awe', was impatient with those who patronized 'primitive people'. These people would never be so far from a spiritual understanding as today's scientific elite!

A cylindrical UFO, allegedly 600 meters long, with a diameter of 110 meters. Nikolay Subbotin and Emil Bachurin, of the Russian Research Station (RUFORS), have rendered famous the claim of its crash in 1991 at Impervia, in the area of the Tian-Shan mounts (Celestial mounts) east of Kirghizistan near the Chinese border. Scientists opposed to the interpretation that the Altiplano 'Airstrips' were meant for either ET or an advanced ancient people, since their spacecraft would not need runways as do airplanes, ignore cigar-shaped UFOs for which the strips might be suitable.

After Wittgenstein read scientist Sir James Frazer's *The Golden Bough*, which likened a savage people's religion to naive science, he snapped that Frazer's own scientific outlook was infinitely more shallow: "Savages" will never be "so far from any understanding of spiritual matters as an Englishman of the twentieth century."[48] This brings us back to the issue at hand. Certainly, the proverbial flying saucer is inconsistent with airstrips. But given their anomaly in the social sciences, it needs to be admitted that the constructs were suitable for torpedo-shaped mother ships which are seen by well-regarded witnesses. Taken with an analysis of physicists at the US Air Force Academy who admit that alien visits best explain some ancient writings, coupled to recent developments of logic in philosophical theology which reveal that theology and science are harmonious, social science does not undercut an extraterrestrial hypothesis.

Admittedly, hypotheses should first utilize mundane explanations. It is well to remember, however, that explanations in the social sciences do not coordinate observation and theoretical terms in a context of higher mathematics, as do formalized theories of physics, by which there are rigorous predictions, explications, and manipulations of phenomena.[49] It is not trivial to distinguish phenomena in physics from persons as understood in the human sciences. The distinction reveals not only their inexactness, permitting broad disagreement about UFOs, but their anthropocentric nature. In short, the human-centered nature of the social sciences makes them inadequate for many of the cosmic issues raised by military projects. The latter are now addressed.

Notes

[1] Dr. B. O'Brien, Ad Hoc Committee Report, USAF Scientific Advisory Board, from *Final Report of the Scientific Study of Unidentified Flying Objects*, University of Colorado Under Contract to USAF, ed. by S. Gillmor (NY: Bantam Book, 1969), p. 7.

[2] *Ibid.*, p. 7.

[3] Dr. Allen Hynek, *The UFO Experience: A Scientific Inquiry* (Chicago: Henry Regnery, 1972) pp. 192-213.

[4] W. Sullivan, "Introduction", *Final Report of the Scientific Study of Unidentified Flying Objects*, Ed. by Gillmor, p. vi.

[5] See Sir Karl Popper, *Realism and the Aim of Science* (Totowa, NJ: Roman and Littlefield, 1983), p. 244.

[6] Dr. F. L. Whipple, "Introduction," in Drs. D. Menzel and E. Taves, *The UFO Enigma: The Definitive Explanation of the UFO Phenomenon* (New York: Doubleday & Co., 1977), p. xiv.

[7] The excitement expressed to my colleague was over LTC Philip J. Corso's just published *The Day After Roswell* (1997). And although there were subsequent challenges to its veracity, an initial excitement was most appropriate. Moreover, the later challenges were often met with rejoinders which make that book still worthwhile to ponder.

[8] From an official letter issued by the Pentagon. See scientist and government consultant Dr. Allen Hynek, *The UFO Experience: A Scientific Inquiry*, p. 189.

[9] *Ibid.*, p. 262.

[10] *Ibid.*, p. 262.

[11] Dr. Edward Condon, "UFOs: 1947-1968." From Gillmor, *Scientific Study of Unidentified Flying Objects*, p. 548. Typically, saying that scientists had taken a side on UFOs was a euphemism for saying they had reacted with either angry contempt or sarcasm.

[12] S. E. Ambrose, *Eisenhower: The President, Vol. Two* (NY: Simon & Schuster, 1984), pp. 395-96.

[13] Timothy Good, *Alien Contact: Top-Secret UFO Files Revealed*, With Commentary by

Admiral Lord Hill-Norton GCB, Chief of Defense Staff (NY: William-Morrow, 1993), p. 123.

[14] See Timothy Good, *Above Top Secret: The Worldwide UFO Cover-Up*, With a Foreword by the Former Chief of Defense Staff, Lord Hill-Norton, GCB. (NY: William-Morrow, 1988), p. 545, for a copy of this document.

[15] Stanton T. Friedman, *Top Secret/MAJIC*, With a Foreword by Whitley Strieber (NY: Marlow & Company, 1996), p. 166. Interestingly, *Filer's Files* (#3, 1999), pp. 5, 6, quotes Kevin Randle as asserting that Dr. Richard Haines seriously erred in his remarks on new Mj-12 documents: "Of course the burden of proof lies with the debunkers, since the documents are claimed to be authentic and not vice versa. No. No. No... The burden of proof is not on us to prove them [inauthentic]. How can a respected scientist like Dr. Haines twist this around?... This twisting of scientific principles is one of the major problems with all Ufology." Randle has made enormous contributions to UFO research, but it is highly dubious that Dr. Haines concludes they are authentic because they are simply claimed to be — much less that a burden of proof is on scientists who disagree with a hypothesis. Is it not more reasonable to suppose, whatever was said as opposed to *meant*, that he supports a burden for both those who affirm and deny hypotheses? Otherwise, we could not make sense of the sheer volume of his scientific contributions which suppose a thorough knowledge of elementary scientific methodology. See Haines' enviable scholarship in the last several volumes of *Marquis Who's Who in America*.

[16] Good, *Alien Contact*, p. 121.

[17] Corso, *The Day After Roswell*, p. 82.

[18] *Ibid.*, p. 84.

[19] Lt. Walter Haut told the author, in a personal interview, that he received "direct orders from Col. William Blanchard" to publicly report the crashed flying disc. The *Roswell Daily Record's* headline "RAAF Captures Flying Saucer On Ranch in Roswell Region," and subtitle "No Details of Flying Disk Are Revealed," is retained in copies of that edition (V. 47, No. 99, 8 July 1947). Though the military later claimed it was a weather balloon, this paper quotes Roswell citizens who stated that they saw "a large glowing object [that] zoomed out of the sky from the southeast, going in a northwesterly direction at a high rate of speed."

[20] Kevin D. Randle and Donald R. Schmitt, *UFO Crash at Roswell* (NY: Avon Books, The Hearst Corporation, 1991), p. 213.

[21] See Robert Trundle, "Philosophical Reflections at the Bicentennial," *National Forum* 57 (1975) 3-9, from the abstract in *The Philosopher's Index*, 1989, Vol. 23, p. 622.

[22] See C. Hanley, "Spotlight: Pacific Bombing Fallout," *The Associated Press*, July 14, 1996, in *The Cincinnati Enquirer*, p. A.

[23] A post-positivist liberal verification principle specifies that sentences are meaningful *if and only if* something counts for or against their truth. While applicable to sentences expressed as laws or their conjunctions (theories), it becomes self-refuting when used without caveat for other extra-scientific sentences since nothing counts for or against the *principle's* truth. A form of its dogmatic application to UFO experience is analogous to Near-Death-Experience (NDE). Ironically, a positivist defender of the conservative interpretation of the principle

was the famous English philosopher A. J. Ayer. Specifying that sentences are meaningful *if and only if* they are empirically or logically true (false), this version is even more rigid. Ironically, years after using it to dismiss 'life after death' as meaningless, Ayer confessed to an incontrovertible experience of conscious life after death in "What I Saw When I Was Dead" in London's *Sunday Telegraph* (Oct. 1988) !

[24] Joseph Cardinal Ratzinger, Ed., Interdicasterial Commission, *Catechism of the Catholic Church*, (Liguori, MO: Liguori Publications, English translation for the USA, 1994, US Catholic Conference, Inc., Libreria Editrice Vaticana), p. 597, #2491, emphasis added.

[25] Timothy Good noted in correspondence to me, 22 June 1999, that "[former President Jimmy] Carter never said he would 'release top-secret UFO data'!" Notwithstanding Good's extraordinary knowledge of UFO history, however, there is the following quote in his own *Above Top Secret* (p. 368): " 'If I become President,' Carter vowed, 'I'll make every piece of information this country has about UFO sightings available to the public'." Reference for the quote is made by him as well as by others to the *Daily Telegraph* (London, 2 June 1976).

[26] See Ludwig Wittgenstein, *On Certainty* (San Francisco: Arion Press, 1991), #41, where doubting if one has two hands is *not* ameliorated by looking to see them because the doubt may as well be extended to testing one's sight. I would add: What is the point of testing one's sight if one's eyes 'see' merely sensations? And since sensations are something internal and may be private, truth-claims about them would be subjective in contrast to an admitted objectivity about the objects sensed.

[27] See Maurice Merleau-Ponty's *Phenomenology of Perception* (London: Routledge & Kegan Paul, 1978), p. 4: "to see is to have colors.... But red and green are not sensations, they are the sensed (*sensibles*), and quality... a property of the object".

[28] See Adora Lee's, "Public Attitudes Toward UFO Phenomena," *Final Report of the Study of Unidentified Flying Objects*, Ed. Gillmor, p. 238, for these and the following statistics.

[29] See *The New Encyclopedia Britannica* (V. 12, 1998). It is also noted, in *The Encyclopedia Americana* (V. 28, 1993), that the broadcast "caused panic in the New York area because of its realism."

[30] For various references to Thomas Kuhn by Dr. John Mack, see Mack's *Abduction: Human Encounters with Aliens* (NY: Charles Scribner, 1994). Also, see Kuhn's futile reply to the dilemmas of incoherence in his Postscript to the 2nd edition of *The Structure of Scientific Theories*, ed. by Frederick Suppe, and Tim McGrew's distinction of global and local epistemological relativisms that "still leaves him [Kuhn] vulnerable to the charge of self-refuting relativism" (in "Scientific Progress, Relativism, and Self-Refutation," *The Electronic Journal of Analytic Philosophy* 2:2 (1994). And while others sought to defend Kuhn, he did himself admit that his defenders were often sympathetic with a radical new left politics, *e.g.* political correctness. See J. Horgan's interview of Kuhn in "Profile: Reluctant Revolutionary," *Scientific American*, 9 May 1991, p. 49. Finally, in regard to how Kuhn's philosophy of science relates to UFOs and is exploited politically, see Paul E. McCarthy's Ph.D. dissertation *Politicking and Paradigm Shifting: James E. McDonald and the UFO Case Study*, the University of Hawaii, at http://912a-87.umd.edu/mccarthy/intro.html: "The primary focus of this study is the personal politics of science which tends in this instance to cast light upon the reception of potentially new observational data in science, Thomas Kuhn's notion of

paradigm shift, and the history of the UFO controversy itself. Moreover, through a decidedly different look at the scientific process than that which is traditionally put forth, *an effort is made to demonstrate that the scientific process can be profitably regarded as a political process* [emphasis added]."

[31] Thomas Kuhn, *The Structure of Scientific Revolutions* (Chicago: The University of Chicago Press, 1969), p. 149.

[32] From an interview of Story Musgrave, M.D., by F. Kuznik, "The Must-Win Mission," *USA Weekend*, 26-28 November 1994, p. 6. Interestingly, Dr. Musgrave inspired many students when he spoke at N. Kentucky University in the summer of 1994.

[33] Randle, *The UFO Case Book*, pp. 193-4.

[34] *Ibid.*, p. 194.

[35] See a critique of physicist's M. Alcubierre's work in M. Szpir's "Space-time Hypersurfing?," *American Scientist — Sigma Xi, Scientific Research Society* 82 (1994) 422-23.

[36] *Ibid.*, pp. 422-23.

[37] *Ibid.*, pp. 422-23.

[38] *Ibid.*, pp. 422-23.

[39] *Ibid.*, pp. 422-23.

[40] Randle, *The UFO Casebook*, pp. 195-6.

[41] *Ibid.*, p. 196.

[42] See Dr. J. Horgan's interview of Thomas Kuhn in "Profile: Reluctant Revolutionary" in the *Scientific American* (May 9, 1991, p. 49) and Professor R. Lewontin's *Biology As Ideology* (NY: Harper Collins, 1998), from the author's own abstract.

[43] Good, *Alien Contact*, pp. 233-4.

[44] *Ibid.*, p. 234.

[45] See Adm. Lord Hill-Norton in Good's *Above Top Secret*, p. 8.

[46] See R. F. Kitchener's "Toward a Critical Philosophy of Science," *New Horizons in the Philosophy of Science*, Avebury Series in Philosophy of Science, Ed. D. Lamb (London: Ashgate Publishing, 1992), p. 19.

[47] See "The Week," *The National Review*, Vol. 52, April 17, 2000, p. 8.

[48] Quote of Wittgenstein by Ray Monk, *Ludwig Wittgenstein: The Duty of Genius* (London: Penguin Books Ltd., 1990), p. 310.

[49] Predictive reasoning in any formalized natural science, from physics to astronomy, might be specified in the following simplified way:

$$[e_1 \wedge (e_1 \approx t_1) \wedge (t_1 \wedge T_o \vdash t_2) \wedge (t_2 \approx e_2)] \rightarrow e_2$$

Where a physical system might be any phenomenon from a solar system to a steel structure, an empirical state description of the system e_1 is treated *as if* ($\approx$) it is identical to a theoretical state description t_1 of the *system at a past or present time* ($e_1 \approx t_1$). The 'as if' identification of theoretical and empirical state descriptions is a 'big if'—given an Underdetermination-of-Theory-by-Data (UTD) Thesis that should render humility in regard to UFOs or any other phenomena (since t_1 and $\sim t_1$ can be inferred from e_1). In any case, t_1 is taken with a theory T_o

and rules of logical deducibility ($\vdash$) to yield t_2, and t_2 is identified with e_2 ($t_2 \approx e_2$). In conjunction with the other symbols in the antecedent $[e_1 \wedge (e_1 \approx t_1)...(t_2 \approx e_2)]$, there is the implication ($\rightarrow$) of the empirical description e_2 of the *physical system at a future time*. In terms of a liberal verification principle, if the physical system at that time is predicted truly by e_2, there is evidence counting for a truth of T_o. And if e_2 does not obtain ($\sim e_2$), there is evidence against T_o. The famous philosopher of science Stephen Körner did not address a UTD Thesis when he formulated this schema, simplified here, since the Thesis was introduced later. See Körner's *Experience and Theory: An Essay in the Philosophy of Science* (London: Routledge & Kegan Paul, 1966), pp. 182-190.

7

Military Projects and Moon Missions

ILITARY PROJECTS ARE rooted in a crashed airborne device in 1947 in New Mexico. Given that the projects arose afterwards to study UFOs and that the military initially reported that a flying disc had been recovered, it is reasonable to suppose that the device was not an object of this world. Thus there should be serious attention to select secret projects, the Apollo missions, and a *startling* Presidential mandate for manned missions to Mars in 2004.

There should be attention not only to an admission by one of the most admired US Presidents, who established a top secret status for UFOs, but a possibly sinister intention of their pilots. If alien pilots are visiting earth, a question ensues of why the academic community has been silent about the topic since the time of the moon missions in the 1960s. While traditionally jealous of its academic freedom and adverse to any government censorship, this community has studiously ignored mind-boggling implications — from the sciences to religion and general culture — of the possible visits.

Begrudging Projects that Signify Contact?

Having noted that the projects arose after a reported UFO crash at Roswell, attention is limited to the most significant ones. Among them are Projects Sign and Grudge. Previously, it was observed that there is some evidence for an authenticity of the document, 18 November 1952, designated OPERATION MAJESTIC 12 PREPARED FOR PRESIDENT-ELECT DWIGHT D. EISENHOWER.[1] Ostensibly prepared by Adm. Roscoe Hillenkoetter, it specified that "A need for as much additional information as possible about these [UFO] craft... led to... U.S. Air Force Project Sign in December 1947."[2] And in adding that security liaisons between Sign and Majestic-12 be limited to two persons in the Intelligence Division of the Air Materiel Command, it stated that "Sign evolved into Project Grudge in December, 1948."[3] While reference to Grudge is odd since it apparently denied extraterrestrial UFOs and the Maj-12 document affirms them, Sign also affirmed them. And although there is no copy of Sign, reference is made to its 2A security classification as well as to the Air Force which, after first denying that it existed, later admitted that it did.[4]

'Must' is a strong word for reasonable belief. In the philosophy of logic, the word would not be used to hold that descriptions of a UFO *must* be false since they are inconsistent with our accepted scientific theories.

Project Sign Affirms the Impossible

There was the report to Sign of Capt. C. Chiles and his copilot. While flying a DC-3 at 5000 feet on a moonlit night on 24 July 1948, they saw an oncoming torpedo-like object with blinding lights shoot by them at about 700 mph. After its sudden reversal of direction, deemed impossible, they noted that the lights came from windows and a blue glow underneath! An Air Force official admitted that there was no aircraft to fit its description. Later assessments indicated a sighting of the same object by passenger C.L. McKelvie and Air Force maintenance crewman W. Massey, several hours before at Robins AFB in Georgia.[5] Noting that the UFO was about the size of a B-29, Massey added that it was more cylindrical with an odd phosphorescent glow.

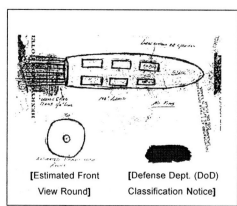

[Estimated Front View Round]   [Defense Dept. (DoD) Classification Notice]

With the typed words "DOWNGRADED AT 2 YEAR INTERVALS DECLASSIFIED AFTER 12 YEARS DoD DIR 620010" (*indicated by arrow at the right*), the above is Copilot Capt. John Whitted's sketch for Project Sign officials. It depicts six large rectangular lighted panels on a 100-foot torpedo-like craft that left a long trail of orange-red flames.

Finally, despite other witnesses who saw a similar craft with two rows of lights, reported from The Hague, The Netherlands, a specter of the 'impossible' led Sign officials to seek an acceptable explanation for the remarkable phenomenon. Their consulting scientist, Dr. Allen Hynek, later held that some UFOs were extraterrestrial spacecraft, resulting in his plea for philosophers to

**'Sky Devil-Ship' Scares Pilots;
Air Chief Wishes He Had One**

Plane Makers
Dubious About
Alabama 'Thing'

Capt. John Whitted and Capt. C.S. Chiles
Atlanta Journal (July 25, 1948)

In a memo on 3 Nov 1948, Col. McCoy of the AMC's Technical Intelligence Division informed Gen. Vandenberg that Project Sign interviewed Nobel Laureate Dr. Irving Langmuir on the origin of Green Fireballs (*above*). Col. McCoy stated that another interview was planned about "all the data now available, and it is hoped that he will be able to present some opinion as to the nature of the objects...." Being either tiny remote- or internally-controlled, the balls needed to be urgently studied after October 1st when, over Fargo, North Dakota, an Air National Guard pilot had a long dogfight at night with a small fast-moving light that blinked and that was obviously under intelligent control.

help scientists explain them.[6] However, his inexperience at the time led him to say that a "sheer improbability of the facts as stated," means that it "*must* have been an extraordinary meteor."

The word *must* is a strong modal term that begs for caution in regard to reasonable belief. In the philosophies of logic and science, the word would not be used to say that descriptions of UFOs are *necessarily* false when they are inconsistent with our scientific theories. Since the theories undergo virtual revolutions, the word is typically applied only to laws that have not undergone them such as those of thermodynamics. And even interpretations of the first law, say either that a periodically functioning machine *must* be limited to work by given energy sources or these sources *cannot* be exceeded, may be modified gradually over time.[7]

⋀

Government knowledge of UFOs is increasingly revealed by the Freedom of Information Act. Not since Professors Hynek and McDonald have scholars been so provoked by what is purportedly impossible.

⋁

The gradual modification is unlike major theories that are dramatically superseded by others. While the laws of thermodynamics were adapted to a revolution in physics from Newton to Einstein, the laws were not replaced by others. Hence, '*must*' was not used viably by Hynek. Rather, we can say that the sheer unlikeliness that a machine violates our theories means that if it did, it was extraordinary. But with a caveat that we are not talking about logical

CONFIDENTIAL   15

UNCLASSIFIED

FORT MONMOUTH, NEW JERSEY - 10-11 September 1951

On 10 and 11 September 1951, a series of incidents occurred in the area of Fort Monmouth, N. J. An initial sighting of an unidentified object was made on a radar set. Soon after the radar sighting, two Air Force officers in a T-33 aircraft unsuccessfully attempted to intercept an unidentified object. Later several more radar sightings were reported.

Status of Investigation

A complete investigation of this incident was carried out and will be reported in Project Grudge Special Report No. 1. It has been tentatively determined that the T-33 pilots probably observed a balloon that had been launched a few minutes prior to their arrival in the area. Two of the radar sightings were returns from balloons and the others were probably due to weather phenomena and excitement of the student operators due to previous sightings. Only one radar return cannot be explained. The operator who observed this incident assumed the object was traveling over 700 mph because the radar set's automatic tracking would not follow the target. It is possible that the inability to track the object was due to his inability to properly operate the set under mental stress.

Bringing to mind the Chiles-Whitted description of an oncoming UFO that passed them at an estimated 700 mph and then abruptly reversed its direction to pace their plane, this Project Grudge Memo on 10 September 1951 states that a UFO "was traveling over 700 mph because the radar set's automatic tracking would not follow the target."

necessity, we can viably say a machine *must* be consistent with the laws of thermodynamics. And thus accepting the explanation of a meteor over one for a UFO, to preserve an integrity of a scientific theory, was unwarranted. In contrast to Hynek, Sign investigator US Air Force Capt. R. Sneider was impressed by the aerodynamic feasibility of the object as well as by the caliber and number of witnesses. He advised that an alien origin of the flying object was certainly "a logical premise."[8]

The premise of an alien origin was shared by physicist Dr. G.E. Valley, of the Massachusetts Institute of Technology, in February 1949. A consultant to the Air Force Scientific Advisory Board, he proposed a secret Sign report that had an amazing import. "If there is an alien civilization which can make such objects, it is most probable that its development is far in advance of ours."[9] And in rejecting appeals to an unqualified truth of our scientific theories about which doubt is rational because their supersession by future theories is likely, he added: "This argument can be supported on probability arguments alone without recourse to astronomical hypotheses."[10]

The hypotheses beg for criticism in light of an Inductive Pessimism. It is rational to be pessimistic about an untarnished truth of any present-day theory, much more hypotheses. For an inductive reasoning about the theory in light of past theories renders probable its falsification in certain domains. A truth about invariant mass in Newtonian theory, for instance, does not hold in a relativistic domain where bodies approach the speed of light. Another sort of pessimism was suggested by Dr. Valley when he stated in the same report that "In view of the past history of mankind, they [ETs] should be alarmed."[11] For we now possess atomic bombs and will soon have spacecraft, and "We should, therefore, expect at this time above all to behold such visitations."[12]

Despite this, did Valley take back with the left hand what he had given with his right? CIA official H.L. Bowers wrote to Dr. Machle on a CIA memorandum, 31 March 1949, that while it is possible the UFOs are from outer space, a group consisting of MIT's Dr. Valley, GE's Dr. Langmuir, RAND's Dr. Lipp, and Aero Medical Lab's Dr. Hynek "have concluded that it is highly improbable."[13]

Opposing views did not stop Sign officials from submitting an *Estimate of the Situation*. Its conclusion that UFOs are interplanetary vehicles echoed Gen. Twining's AMC report that the discs "are real." But the document went up channels to Air Force Chief Staff Gen. Vandenberg who destroyed it since there was no "proof." Witnesses to the document, such as Capt. E.J. Ruppert of Air Technical Intelligence at Wright Patterson, said they saw the document before its destruction. Later heading Project Blue Book, he said that some copies were "kept as mementos of the golden days of UFOs."[14] Yet Vandenberg shot a message over the bow of the military rank-and-file that such appraisals were to be denied. Even Ruppert began to deny they were alien, at least in public.

In Denial of the Impossible? Project Grudge

In short, rejection of UFOs as extraterrestrial craft led to an end of Sign and start of Grudge, in early 1949, whose personnel *grudgingly* pursued UFO reports because they were told to believe that they were all mundane phenomena. Nevertheless, various documents indicate that any special report, that might confirm a reality of alien UFOs, were kept from the public and sent up channels for top-secret analysis. Outside reference to the analysis ranges from plausible to bizarre. The latter was typified by Navy petty officer William Cooper on 18 December 1988. He alleged over a computer network that when working with an intelligence team under Adm. Bernard Clarey, The Pacific Fleet's Commander in Chief, he saw Project Grudge Special Report 13 as well as a Majority briefing.[15]

The letters 'Maj' in Maj-12 denoted *Majority*, not Majestic, said Cooper. And he stated not only that he had seen photos of the big-nosed gray aliens, similar to the ones that had landed at Hollomon Air Force Base in either 1964 or 1971, but that he was privy to the fact that the aliens were both in contact with the US Government and cooperating with it in advanced projects at its top-secret base at Area 51 in Nevada. In going beyond investigators who include John Lear, Robert Lazar, and William Moore who had reported parts of the same story, a 25-page report *The Secret Government: The Origin, Identity and Purpose of MJ-12* was displayed on 23 May 1989.

This document specified that a top-secret government, that actually ran the country, had been initially composed of a special group of CIA and other intelligence agents. Allegedly, they had murdered Truman's former Secretary of State James Forrestal, who was thought to have committed suicide, to prevent his exposure of the UFO cover-up. The agents eventually formed a

Hotel of Bilderberg, The Netherlands: Former Navy Petty Officer William Cooper states that the Bilderbergers are a supranational group (founded in May 1954 in this Hotel for NATO members to harmonize the Allies). Officially said to be composed of the world's political and business leaders who are selected by merit, its members are prohibited from speaking openly about decisions. Oddly, Laurence Rockefeller — brother of a founding member David Rockefeller — commissioned a major report by renowned scientists to debate the physical evidence in puzzling cases for UFOs that was presented by investigators. This led to publication of Stanford University Physics Professor Peter Sturrock's *The UFO Enigma*, supported by Laurence Rockefeller and implemented by the esoteric Society for Scientific Exploration, with a Foreword by Laurence Rockefeller.

secret international society of elite officials called the 'Bilderbergers' for marshaling worldwide money and power as well as for coordinating contact with these aliens.

Cooper's reports, affirming a conspiracy to spread AIDS, rival in bizarreness the TV series *X-Files*.[16] Admittedly, his writing in *Behold a Pale Horse* is so curiously inspired that it has the ring of authenticity. But a general question arises: Are some fantastic claims either hoaxes to make money or military intelligence disinformation? In bringing to mind Moore's confession that he had worked with the military, it may be suspected that true information is mixed — knowingly or unknowingly — with false claims for 'raising the public's consciousness'. At the same time, the false claims would discredit UFOs in the eyes of the academic community lest public awareness come too quickly.

𝄃

Reportedly, an end to the security risk of UFOs was due to developments of the Strategic Defense Initiative (SDI). This led to one Army officer's remarks about Sign as well as Grudge

𝄂

Government information on UFOs has been increasingly uncovered by The Freedom of Information Act. Not since the time of Professors Allen Hynek and James McDonald has there been such an interest by elite scholars, such as Harvard psychiatrist John E. Mack and retired NASA scientist Dr. Richard Haines, in what is purportedly impossible. One would expect public denials of UFOs from Grudge personnel, given the Air Force's repeated denials of a

CIA recruiting Ad. Established the same month and year as the Roswell UFO crash in July 1947, the CIA is clearly warranted in its attention to foreign terrorists who exploit universities as safe havens. Vigilance may bear properly on UFOs as well. Over 900 pages of UFO documents have been released by the CIA due to the Freedom of Information Act. Obviously, secrecy might be endangered by tenured professors who pursue truth about UFOs. An article in the academic journal *Lingua Franca* notes a cloak-and-dagger link of professors to the CIA. Since 1996, it has made public outreach a "top priority and targets academia in particular. According to experts on US Intelligence, the strategy has worked." Esteemed academics are quoted, including a Harvard professor and Columbia University's former President-Elect of the American Political Science Association. Both men admit of having worked for the CIA. And Yale's H. Bradford Westerfield is quoted as stating that "There's a great deal of actually open consultation and there's a lot more semi-open, broadly acknowledged consultation." *(David Gibbs, "Academics & Spies: The Silence that Roars," LA Times, 28 Jan 2001)*

Roswell UFO crash — with explanations that border on the absurd — and an audacious stonewalling of elected members of Congress who include Rep. Steven Schiff and Sen. Nancy Kassebaum.

However, caution about the possibility of disinformation should not blind one to possible 'inside' truths. As former Chief of the Army's Foreign Technology Division and a certifiably decorated career soldier, retired Army Lt. Col. Philip Corso is not likely to have fabricated his report on a Roswell UFO cover-up. His evidently credible claims may not merely hold in virtue of being devoid of any outlandish government conspiracy. In addition, he properly acknowledges a previous need for secrecy, reveals more modest but plausible events — even in a matter-of-fact way, and explained his ability to release top-secret information without any longer risking national security. Purportedly, an end to the security risk was due to developments of the Strategic Defense Initiative (SDI) that afforded protection from alien craft. This led to his remarks about Sign and Grudge.

Precisely, a restraint of Corso's remarks, and their accord with other research, makes them of great importance. He notes as a backdrop to his brief comments: "To those of us inside the military/ government machine the government is dynamic... when it comes to devising ways to protect its most closely held secrets."[17] After the Roswell UFO crash, the 'machine' was not merely one step ahead of the news media and UFO researchers but a thousand or more steps.

The truth was never hidden but only camouflaged. The curious "just didn't... recognize it for what it was when they found it. And they found it over and over again."[18] Thus although Blue Book was to satisfy the public's need for a response to UFO sightings, Grudge and Sign "were of a higher security to allow the military to process sightings... that couldn't easily be explained away."[19] Contrary to most UFO literature, Grudge took UFOs as seriously as Sign. Despite Gen. Vandenberg's rejection of interplanetary UFOs, he directed "the air force to evaluate and track UFO sightings... in response to the working group [Maj-12]" after December 1949.[20] Yet, it was not Sign that submitted the *Estimate of the Situation* which held that UFOs were of alien origin, after 243 sightings, but the Air Technical Intelligence Center (ATC).

A

Being unable to will a moral law that harms one's self by causing panic or revealing murderous technologies, in Kant's *secular philosophy*, converges with a *philosophical theology* in which secrets should be kept except when greater harm is avoidable only by divulging the truth.

Y

Rather than viewing UFOs as threats to airborne operations and military installations, the ATC concluded that aliens were only observing humans. Worse, for some in the 'loop', a few officials sought to use *Estimate* for making UFO data public. Why didn't Gen. Vandenberg just agree with Gen. Twining and Adm. Hillenkoetter to ask Truman to begin releasing the data to the public?[21] Due to a fear of public panic and the threat to security by foreign nations if the Roswell technology were revealed, Vandenberg "had a cow, and not a mutilated one."[22] His destruction of the report on ET's presence that led to Grudge — having more to do with *begrudging* a release of information than not taking UFOs seriously, was one of the "last official government assessments of the UFO situation ever to get even close to being distributed before the real cover-up clamped down."[23] Contrary to most UFO investigators, however, there is the possibility that a cover-up may have been justified. The justification might be made by an ethics of secular philosophy that converges with one of philosophical theology.

Kant held one should always act as if a moral maxim was a universal law.[24] Can one rationally will a law that permits others to divulge secrets harmful to one's self? This holds for causing public panic or revealing technologies that could kill millions, and converges with a principle of theology that professional secrets must be kept except when greater harm is avoidable only by divulging the truth.[25] After hundreds of sightings examined by Grudge,

Unusual UFO photo from Blue Book taken at Tulsa, OK, on 2 August 1965. On that night, states French researcher Patrick Gross, "thousands of people in four Midwestern states witnessed spectacular aerial displays by large formations of UFOs." A "multicolored disc was photographed... while several persons watched it perform low altitude maneuvers." This photo was much analyzed, pronounced authentic, and later published by *Life* magazine and many newspapers."

there was a 1949 CIA memo and 1952 complaint from the Office of Scientific Investigations about a lack of UFO data. In order to maintain a compartmentalized secrecy and satisfy both public and intelligence demands, Project Blue Book was established.

Project Blue Book and Its Report 14

Becoming a basis for the popular television series in the 1970s, the Air Force's Project Blue Book was created to salve civilian alarm and even concerns in the government. This fact reveals that most government officials are as *in the dark* as most civilians who may suppose that there is some sort of monolithic conspiracy. After some of the project's salient features are discussed, another dimension of this project is examined: a 'Report 14' that has been studiously disregarded by most UFO debunkers.

Λ

Since this intelligence officer assumes the public could handle the traumatic cultural implications, from science to philosophy to religion, the only fathomable secrets to keep might be specifics of the technology.

Υ

The Bluff of Blue Book

Blue Book was begun in 1952 because whatever Maj-12 was supposed to be doing, "it wasn't satisfying the National Security Council."[26] Thus the Council, says Col. Corso, had the CIA determine if UFOs posed a danger to the United States.[27] These claims are important since the CIA had compromised US security, he adds, as much as the USSR's security had been breached by the KGB. In being privy to this fact, certain inner intelligence circles retained all UFO secrets and only cooperated with the CIA on the surface.

Superficial cooperation included a meeting of CIA and Air Force officials on 14 January 1953, prior to the inauguration of President-elect Eisenhower.

Author's computer-enhanced blowup of the Blue Book UFO photographed on 2 August 1965 over Tulsa, Oklahoma. A lighter scratchy oval or saucer shape is now relatively visible.

FLYING SAUCERS:
AN ANALYSIS OF THE
AIR FORCE
PROJECT BLUE BOOK
SPECIAL REPORT No. 14

Aside from the fact that more clear-cut cases of alien UFOs were almost certainly sent up to Maj-12 members, Blue Book Report 14 found that the *lower* an ability to identify UFOs — that is, the more exotic they were, the *higher* were the quality of the photographed sightings!

Within two months after Adm. Hillenkoetter presented his Maj-12 BRIEFING DOCUMENT to Eisenhower, Maj-12's Dr. Loyd Berkner and Air Force officers reviewed UFO case histories under direction of the CIA physicist Dr. H.P. Robertson, Director of the Weapons Evaluation Group. Becoming known as the Robertson Panel of Blue Book, it advised that all official parties concluded that UFOs posed no threat and should be officially debunked by the government; this, after watching two films of purported UFOs and hearing superficial testimony about them!

In not sitting well with more serious external officials who knew it was patently absurd to debunk all UFOs, the conclusion led to an outcry that was not lessened by Blue Book's relegation of virtually all UFOs to mundane phenomena or by its being made up of overworked enlisted staff headed by skeptical Air Force officers. So, controversy over how to handle UFOs led in 1966 to a third-party mediator, namely the University of Colorado. Headed by physics professor Dr. Edward Condon and friend of Harvard astronomer and UFO debunker Donald Menzel, the inquiry got off to a bad start.

Condon not only ridiculed UFOs before an examination of the evidence had even started by publicly making jokes about them but, at about the same time, university Dean Dr. Robert Low made the notorious remark that the "trick" would be to *appear* to be impartial. Notwithstanding the partiality, however, the scientific staff objectively examined a host of compelling case studies. The study's fruit was the now famous *Final Report of the Scientific Study of Unidentified Flying Objects* (1969). Ironically, Condon's disingenuous concluding chapter that UFOs were not worth any further investigation, colliding head-on with most of the contents, satisfied both naive government skeptics and those who knew better such as the elite members of Maj-12.

Blue Book ended and the government was supposedly no longer in the UFO business. In an interview with Jaime Shandera, a former intelligence officer stated that Maj-12 was afraid "that if they continued with Project Blue Book, some of the information... would get out into the public, so they decided to end [it]."[28] He added: "they did that with an investigation by the University of Colorado... the government had decided to go underground...."[29]

$$\Lambda$$

Give-'em-hell President Harry S. Truman may have tacitly apologized to posterity for initiating the UFO cover-up in his *Memoirs*. Does this open diary suggest a reason?

$$\Upsilon$$

Again, a troubling question of ethics is raised. On the one hand, the same intelligence officer stated that "sometimes when you deceive the public, you're protecting the public."[30] On the other, he noted that Blue Book was a charade since there *was* a general knowledge of aliens coming to earth "back in the fifties, late forties, and... communication with them... [that] shouldn't be kept."[31] If this is true, there is a question about what secrets would be left to keep. Since the officer evidently assumes that the public could handle the traumatic cultural implications, from religion to philosophy to science, the only fathomable secrets to keep might be specifics of the technology and whether ETs are friend or foe.

Either ETs are friendly or not. If *they* are, the public's ability to handle traumatic cultural implications are still in question. And if *they* are not, the question is exacerbated by that of our physical security — indeed, survival. In either case limited revelations of *their* technology would eventuate in public pressure for more information, with a slippery slope to other specifics. And specifics of advanced technologies, being more exposed to foreign intelligence, would deepen security concerns even further. There could now be hyper-advanced weapons of hostile nations to contend with as well as with ET.

These points may not entirely justify official deception of the public. But they surely illustrate why UFO enthusiasts who probe secrets can worsen government worry over control and, thus, actually impede official forthrightness.

In any event, this intelligence officer noted that since the time of the 1950's film *The Day the Earth Stood Still*, in which the inside of a flying saucer actually resembled a crashed UFO, until the 1980's movie *E.T.*, the government has been conditioning the public to accept the information so that it would not have to be revealed abruptly at some future time.[32] Reportedly,

NASA photo of President Truman presenting the Collier Trophy for aircraft design to Lew Rodert, of the National Advisory Committee for Aeronautics (NACA), in Dec. 1947 – five months after the alleged Roswell UFO crash. In a 1955 review in the *New York Times*, Allan Nevins states of Truman's *Memoirs*: "Though we have seldom had a leader of more unphilosophical mind – he never once in this book discusses an abstract idea – his insights were very often unusual." What was the unusual secret he would not talk about, due to "national security and out of consideration for some people still alive"?

Project Blue Book played a central role in this conditioning process, and UFO enthusiasts may have been part of the procedure. There would, of course, be official concern that not either too many shocking secrets or alarming information be adduced in too short a time.

Intriguingly, no one has noticed since the time of Blue Book that "give-'em-hell" President Harry Truman, acclaimed for his integrity and plain speaking, wrote a few lines that might have an astonishing import. He reportedly set up Maj-12 and a top-secrecy of the Roswell UFO crash. In the same year Gen. Douglas MacArthur warned of an attack "by people... from other planets" in *The New York Times* (1955),[33] did Truman tacitly apologize for a cover-up by revealing its reasons in his *Memoirs*? What might be some of the reasons?

> For reasons of national security and out of consideration for some people still alive, I have omitted certain material. Some of this material cannot be made available for many years, perhaps for many generations.[34]

Given that a generation is about twenty years, from the birth of parents to their offspring, *many* generations would likely be at least sixty years. Since Truman wrote his memoirs in 1955, this means that the material might be released, in terms of his limited foresight then, in 2015. Surely, material omitted for this span of time does not refer to the atomic bomb. Everyone knew about it by 1945, two years before a reported Roswell UFO crash, and details of its construction would not be in his memoirs anyway.

Aside from the UFO crash, the bomb is the only other evident item bearing enormously on national security and would not, in any case, be omitted in deference to people still alive. Yet there were living people from military and government icons to Truman himself whose reputations would be damaged

Rectangular hole in a rock, evidently artificial, filmed in 2004 by the Spirit Lander, per NASA/ JPL/ MSSS. In 1997, the Pathfinder spacecraft also sent down a lander to Mars at an area called the "Sagan Memorial Station," in honor of astronomer Carl Sagan, and filmed some curious phenomena. While Sagan debunked UFOs and publicly pooh-poohed the possibility that ET had been to Mars, a little known fact is that there are many peer-reviewed scientific articles which affirm precisely that finding. With revealing titles, they include:

1. M. Stein & M. Carlotto, "A Method for Searching for Artificial Objects on Planetary Surfaces," *J. Brit. Interplan. Soc.* 43 (1990)
2. B.T. O'Leary, "The 'Face' on Mars & Possible Intelligent Origin," *J. Brit. Interplan. Soc.* 32 (1990)
3. V. DiPietro, G. Molenaar, & J. Brandenburg, "The Cydonia Hypothesis," *J. Sci. Expl.* 5 (1991)
5. T. Van Flandern,"New Evidence of Artificiality at Cydonia on Mars," *MetaRes Bull.* 6 (1997)
4. M.J. Carlotto, "Imagery Analysis of Unusual Martian Surface Features," *Applied Optics* 27 (1998)
6. H. Crater & S. McDaniel, "Mound Configurations on the Martian Cydonia Plain," *J. Sci. Expl* 13 (1999)

Among the findings were that scientific analysis of the Face on Mars strongly evidenced the artificial origin of it and other objects nearby. Despite these astonishing findings which are the sort of thing that spark revolutions in science, other scientific journals such as *Nature* refused to even entertain the subject since there was a "risk to the reputation of a commercial publisher" and "a group of 'big name' scientists and their institutions would have to assume the risks involved in such a controversial action." (Quoted from Physics *MetaResearch.org*)

by a UFO cover-up — a cover-up of cosmic proportions that pales a 'Watergate' or 'Whitewater', even if it was justified. As a self-proclaimed watchdog of our republic, the embarrassed news media would surely engage in a frenzy of exposing officials since it had virtually aided a conspiracy by allowing itself to be deceived and manipulated.

There is an intellectual incest of university debunkers who conform to political ideology in the name of 'science' and share an anti-science dogma against UFOs.

Though these points about Truman's *Memoirs* hold independently of Lt. Col. Corso's book, the latter bears on this issue. In the presence of trusted inside officials, Truman reportedly asked about the Roswell UFO: "Do we ever tell the American people?" He added that "Winchell [a scandal-mongering newsman] would crucify me with this...." And although Secretary of State Forrestal and Adm. Hillenkoetter "had a knee-jerk reflex answer, 'no'," Secretary Forrestal "quickly saw that it wasn't that easy."[35] Did an uneasiness of the question, coupled to one about ET's intentions to use far superior weapons on humans, play a role in Forrestal's suicide?

Finally, researchers are beholden to physicist Stanton Friedman for a less known side of Project Blue Book.[36] During the time of this project in the early 1960s, Friedman had a security clearance for work on intermediate reactors for space vehicle applications. This background was incidental to his discovery, but relevant to his analysis, of a privately published version of the *US Air Force Project Blue Book Special Report 14* at the UC library at Berkeley. The novel document had more than twenty charts, tables, graphs, maps and references to over 3,000 UFO sightings of which more than 600 were unidentified by professional investigators whose work was not publicly known.

Well Kept Secret of Special Report 14

What struck Friedman was that the famed Batelle Memorial Institute at Columbus, Ohio, had evaluated the sightings. Its Investigators found, contrary to skeptics who assume exactly the opposite, that there was an inverse relation between the quality of sightings and scientific ability to identify them: The lower an ability to identify the objects, the higher a quality of the sightings. The sightings were rigorously categorized into insufficient information, knowns as aircraft, astronomical objects and so on, and unknowns. The unknowns were cross-compared to knowns by features that included size, shape, speed, and color. It was found that there was a less than a one percent chance that unknowns were missed knowns.

Friedman's assertion that he was "certain that the government's own information strongly indicated that a significant number of UFO sightings were probably of alien vehicles"[37] brings to mind the previously noted intelligence officer who stated that members of Maj-12 were afraid that "some of the information... would get out into the public, so they decided to end Project Blue Book."[38] The effectiveness of this stratagem and the significance of Report 14 is clear by the highly publicized remarks of UFO debunkers who include the late astronomer Carl Sagan. The debunking gambits are illustrated by Friedman when he notes Sagan's distortion of the evidence for UFOs in *The Demon Haunted World*. Sagan states with no caveat that there "are reliably reported cases that are un-exotic, and exotic cases that are unreliable."[39] But aside from the fact that more clear-cut cases were almost certainly forwarded to Maj-12, Blue Book Report 14 found exactly the opposite: The lower an ability to identify UFOs — the more exotic they were, the higher was a quality of the sightings.

Consequently, it is odd that Frederick Crews, after fondly recalling his decades in a "congenial university setting," debunks a new book sympathetic to UFO research in *The New York Review of Books*. He says of the book, which criticizes Sagan, that it expresses "contempt for critical reasoning."[40] But given the facts provided by Friedman and others, the reasoning of Crews

In praising Thomas Paine for being far ahead of his time because he opposed religious superstition, astronomer Carl Sagan's *Demon Haunted World* took its place in a long line of literature that upheld an absolutist view of science. "Science is the true theology," Paine was quoted as stating in Emerson's *The Mind on Fire*. Paine's *Age of Reason* (1794) touted a *critical* deism that presaged modern efforts to render scientific our modern religion. But Paine's religious rationality exalted physics and *uncritically* ignored a presupposed metaphysics of determinism that undercut coherent scientific 'truth'. Hence, it was one of the naive steps that led to a dogmatic denial of both traditional religion and a natural theology of Nature's God — despite this God's role in the *Declaration of Independence* — in the thought of many professors who the public looks to for enlightenment. Indeed, the Enlightenment evolved into a humanistic worldview that, shaping most universities, paradoxically viewed changing theories of science as unchanging cosmic knowledge. And thus a humility enjoined by traditional religion was replaced pridefully by an oxymoronic "pre-Copernican anthropocentrism" that denied a more intelligent ET.

is itself uncritical. It belies nothing less than an intellectual incest of like-minded university debunkers who conform to a liberal-left agenda in the name of 'science' and share an anti-science dogma against UFOs. And the reasoning that Crews praises, of Cornell University's Carl Sagan, collides head-on with a fundamental principle of scientific rationality. Its quintessential expression is to acknowledge facts that count against a belief. Understandably, Friedman sighs, "These people know about these works [Report 14]... but choose not to pass this knowledge on...."[41]

Unhappily, due to Friedman's long friendship with Sagan, in virtue of their being classmates at the University of Chicago from 1953 to 1956, Friedman misses the point about a culpability of UFO debunkers. He laudably brings to mind Aristotle's dictum *truth before friends*. However, he misguidedly asks "why denigrate a scientist who has done so much to bring science to the masses?" Here, he contrasts Professor Sagan to both UFO debunker Philip Klass and a researcher, with whom he disagrees about some matters, "who do not have Sagan's scientific credentials."[42]

Precisely, the point is that prominent scientists such as Professor Sagan should be far more sensitive to the facts, because of their education, than less educated laypersons who are more influenced by their biases and self-interests. Sagan's greater responsibility proceeds *pari passu* with his greater blame, the consequences of which were exacerbated

In contrast to the heirs of Paine, Voltaire's *Micromégas* (1752) imagined aliens coming to earth. While Voltaire was as devoted to the Enlightenment as Paine, his similar rejection of religious faith and acceptance of a human-centered faith in science could not quell a genuinely rational specter of ET's visit.

by the influence of his popular books and television appearances. Friedman's otherwise courageous stance of valuing truth over friends may, in this case, have been skewed by his friendship. A larger point about this fallacy of *missing the point* (*ignoratio elenchi*) bears on an increased blame of the academic community that has largely either ignored or ridiculed the scholarly study of UFO phenomena.

Project Snowbird

Project Snowbird is a post-Blue Book project that some researchers, such as Timothy Good, hold to be bogus — although he concedes that the project may contain some truth. Snowbird was allegedly established in 1972 to test fly a nuclear-powered device as part of an effort to implement some of the recovered extraterrestrial technology. This allegation brings to mind a project called 'Aquarius' which was reportedly begun in the early 1980s for purposes that ranged from accumulating information about alien life to reverse engineering extraterrestrial craft. Having noted the supposed nature of the project, we turn to some brief details of its history.

Reportedly, the Snowbird craft was an atomic device, escorted by over 20 unmarked helicopters, whose radiation burned three witnesses who got out of their car.

The bizarre history of Snowbird includes its link to a rash of cattle mutilations, largely in the Western United States from the 1960s to 1980s. After 130 mutilations in two years, confirmed by the Colorado Bureau of Investigation (CBI) in August 1975, Senator Floyd Haskell requested FBI assistance. Still, residents continued to see unmarked helicopters in the vicinity of UFOs and the enigma was not lessened by Haskell's later reassurance in the Colorado Springs *Gazette Telegraph* (September 24, 1975): "I am hopeful... that the CBI officers will be able to learn who or what is behind the mutilations..." in which, in many cases, blood was completely drained from

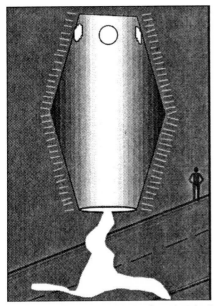

While Snowbird has been an alleged project since 1972 for developing extraterrestrial technology and test flying recovered UFOs, it is linked to virtually every aspect of the UFO enigma. In addition to a link between cattle mutilations and the government, it was related to the encounter of a glowing craft hovering over the road — with flames melting the asphalt (*above*) amid nearby military helicopters. And although Snowbird was reportedly also revealed in a so-called Briefing Document of Project Aquarius, another 'Project Snowbird' was said to be a "joint Army/Air Force military exercise in the subarctic region in 1955," according to Gale Research's *Names Directory* (1963).

the animals and various organs removed by an advanced laser process. These anomalies were exacerbated by government officials who certified that in some cases there were odd inhuman 'footprints' and cadavers that were evidently dropped from airborne craft.

These accounts were acknowledged in reports of the New Mexico State Police in October 1975 and May 1976, in documents by the state medical pathologist, in letters by Senator Schmitt, and by US attorney G. Bell and K. Rommel who was head of the DA's Office in New Mexico.[43] In view of the laser process, unduplicable surgeries, and UFOs sighted with military helicopters, there understandably arose the idea of a collusion between the UFOs and top-secret elements of the US government. Belief in a liaison with the government was bolstered by the research of Paul Bennewitz. Indeed, his inquiry raised a specter of technology being provided by aliens in exchange for their right to carry out the mutilations.

This story began in 1980, when Bennewitz took 8mm film of UFOs near Kirtland Air Force Base in New Mexico. This fact was acknowledged in documents of the Air Force Office of Special Investigations (OSI). Reportedly, the Air Force was alarmed by his interception of some peculiar low-frequency radio signals that he believed were communications from aliens.[44] That this possibility was not altogether a howler, was supported by William Moore. He confessed to having worked with military counterintelligence officers who sought to discredit and harass Bennewitz, he said, because they were disturbed by his detection of the signals. Approached precisely in virtue of his affiliation with UFO investigators, Moore later apologized for having cooperated with the officers in order to uncover government information.

The information never materialized. But Bennewitz stated to Timothy

Bringing to mind the nuclear-powered UFO, that radiated Betty Cash and two other persons, is this nuclear rocket engine (*left*) being transported to a test stand. The XECF — Experimental Engine Cold Flow — experimental nuclear rocket engine was part of project Rover/NERVA (Nuclear Engine for Rocket Vehicle Application). Its main objective was to develop a flight-rated thermodynamic nuclear rocket engine. NASA says that this testing was at the AEC's Nevada Test Site in Jackass Flats, Nevada, about 100 miles west of Las Vegas, and that "cutbacks were made in response to a lack of public interest in space flight, the end of the space race after the Apollo Moon landing, and growing use of low-cost unmanned, robotic space probes.

Good, with Good's proviso of possible disinformation, that research and computer communication indicated that the "humanoids are made from specific cattle parts."[45] Suggesting that the humanoids were apparently either green or wore green outfits, Bennewitz said that the aliens and government had made a deal wherein mutilations were evidently allowed by the authorities in trade for an alien atomic craft and technology. Government unmarked helicopters were to safeguard the alien operations.

"The 'green humanoids' are supposedly manufactured by the aliens," notes Good, "one of whose bases is said to be under the Archuleta Mesa near Dulce, New Mexico [near where cattle mutilations also occurred]."[46] He was told that the Snowbird craft was the atomic device, escorted by over twenty unmarked helicopters, whose radiation severely burned Betty Cash and two others when they saw the glowing device hover in front of their car in 1980. Despite Moore's confession and possible disinformation, an effectiveness of the latter would depend on some degree of truthful data. Does this data bear on humanoids seen near the mutilations?

Accordingly, before turning to the case of Betty Cash, it is relevant to note the complaint of a rancher to investigative journalist Linda Howe in April 1980 near Waco, Texas.[47] While searching for a cow ready to give birth, he saw two 4-foot tall creatures in a clearing among mesquite trees about a hundred yards distant. They wore tight-fitting green clothes, the color of spring mesquite leaves, and had the same color of green egg-shaped hands, reminiscent of cattle hooves, with pointy ends turned toward the ground. They were carrying the newborn calf with their free arms swinging uniformly. While no noses or hair were noticed, the beings looked like slightly muscular small adults who were neither fat nor thin. When the they became aware of the rancher and turned toward him, he noticed their big 'sloe' dark almond eyes — similar to ones of cows — angled to pointed ends. He ran because of the sight and fear of being abducted. He was so scared that he did not tell his wife or son about the event for two days. Then, they returned to the site and saw the 'calf.' In a tidy setting there was a hide folded inside out pulled over the skull, a backbone without ribs, and no blood. Related to this report are others. Whereas Colorado State University pathologist Dr. Albert McChesney stated that it was "impossible to withdraw not even all of the blood... without leaving some telltale mark" by any known technology,[48] a security guard testified that mutilations were found in an area over which he saw hovering lights with "no beam, no sound — nothing."[49]

Historically, these reports are part of a project Snowbird that includes the government's allowance of mutilations by aliens for their technology, some of which includes UFOs themselves. The reports may seem fanciful, but considered in view of the philosophy of science, the issue of a pessimistic induction arises wherein our theories that cannot now utilize genetic technologies to fabricate creatures may be superseded by future theories that can. Even now, our ability to clone animals and to at least imagine reproducing dinosaurs out of DNA in the blood of fossilized mosquitoes, inconceivable even a decade ago, makes it difficult to glibly dismiss some of the reports. A more sober report includes the well established case of Betty Cash.

Betty, Vickie Landrum, and Landrum's seven year-old grandson Colby witnessed a fiery object in the sky at about 9:00 p.m. on 29 December 1980 as they were driving near Huffman, a suburb of Houston, towards Dayton, Texas.[50] The object descended to treetop level and hovered about 135 feet from the front of their car. They stopped, as flames shot down beneath the object, after which the two women got out to get a better look and Colby pleaded with them to get back inside. Understandably, Vickie was the first to get back inside the car in response to her grandson's screams, whereupon she

FBI Joins Investigation of Animal Mutilations Linked to UFOs

By WILLIAM BARNHILL, BOB PRATT and DAVID WRIGHT

The FBI has joined in the investigation of the bizarre mutilation of thousands of grazing horses and cattle over an 18-state area — attacks which have been linked to UFOs.

Disclosure of the FBI role was made at a recent conference of officials from seven states where the attacks have reached an alarming level.

Sen. Harrison Schmitt (R.-N. Mex.), the ex-astronaut and scientist who organized the conference, declared: "Either we've got a UFO situation or we've got a massive, massive conspiracy which is enormously well funded."

At least 8,000 cattle and horses have been butchered with surgical precision over an estimated 1.25 million square mile area stretching from Tennessee to Oregon since the mutilations began around 1970. The 1.25 million square miles is more than a third of the total land area in the continental United States.

In many cases the attacks have coincided with UFO sightings. Baffled investigators say the strange pattern of the mutilations includes these startling facts:

● No tire marks, footprints or other signs of human activity are found near the mutilated carcasses.

● Only the blood and certain parts of the animals — usually the reproductive organs — are removed.

● Trace elements found on and in some carcasses are the same as those collected after a UFO sighting in New Mexico.

BAFFLING incidents have occurred in 18 states.

EXPERTS: New Mexico state trooper Gabe Valdez takes tissue samples from a mutilated cow found at Dulce, N. Mex. Assisting him is retired scientist Howard Burgess, who's investigated several similar incidents with Valdez.

know why they're doing it, so therefore we should leave it alone.'

"Those are their exact words ... The 'star people' know what they're doing and should be trusted."

Dr. Monteith said he has no doubt that aliens from outer space are responsible for the

UFO researcher: "What few clues we have concerning those responsible for the mutilations suggest that we are dealing with well-equipped, highly capable airborne entities ... We are forced, I feel, to the hypothesis that unidentified aircraft are the means — UFOs."

To aid in solving the mystery,

This news item was found in an FBI File. Declassified by the Freedom of Information Act, the File had comments by agents who acknowledged the extraordinary: Organs were removed and blood completely drained from animals in an inexplicable manner.

began to pray. Her less amount of time outside the car, as soon noted, would bear on various controversies. Whereas Betty described the object as being an "extremely bright light" with an indistinct shape, possibly since she stared at it more intensely for a longer time, Vickie stated that it seemed "oblong with a rounded top and a pointed lower half" and Colby described it as being "diamond-shaped."[51]

Flames continued to shoot down from the hovering object with both bursts that sounded like a flame thrower and a beeping noise which was heard during the whole ordeal. Betty noticed that the car got so hot that she was even unable to touch the door. But this encounter abruptly changed as the object began to move upward. Seeking to follow its flight in their car, the three became aware of about two dozen twin-rotor helicopters, later said to be Chinooks, that were accompanying the device at a distance of about three-quarters of a mile. After trying to follow it with several stops to get a better view, Betty drove the others home and then returned to her own house at about 9:50 that evening. Though Vickie's inflamed eyes were followed by a loss of her hair and Colby developed a 'sunburn' on his face as well as damage to his eyes, it was Betty who suffered the most.

Afterward, Betty was admitted to Houston's Parkway General Hospital after suffering swollen shut eyes, diarrhea, nausea, headaches, neck pains, and scalp nodules that seeped clear fluid when they burst. Doctors were unable to diagnose a disease, but there was no doubt that she was exposed to radiation. She received medical bills totaling $10,000 in two months, developed breast

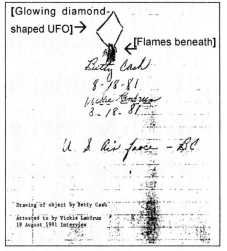

[Glowing diamond-shaped UFO]→

←[Flames beneath]

Betty Cash

8-18-81

Vickie Landrum

3-18-81

U. S. Air force - DC

Drawing of object by Betty Cash
Attested to by Vickie Landrum
18 August 1981 Interview

Betty Cash's sketch of UFO that radiated her, from declassified Air Force documents *via* the FOIA.

cancer, and had a mastectomy. As military veterans and victims of nuclear fallout discover, causal relations of cancer to a source—from radiation to agent orange used in Vietnam—are difficult to prove. In this case, the problem of proof was exacerbated by the nature of the UFO and government denials of any involvement.

After more than six years in August 1986, Betty and Vickie lost their suit against the US government for $20 million when Judge Ross Sterling, upon hearing testimony from the Air Force, Army, Navy, and NASA that the UFO was not operated by them, stated: "he will not hear the evidence they wanted their attorneys to present."[52] Some reports note that the Judge was made privy to a top-secret document, *not* about details of Snowbird, but, about other government documents bearing on the law suit that could not be revealed without a grave risk to national security. The suffering of Vickie, Colby, and especially of Betty, not to mention her medical expenses, seem so outrageous that one can scarcely imagine any judge or jury dismissing the evidence.

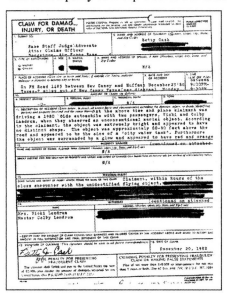

US Air Force Form 95 – CLAIM FOR DAMAGE, INJURY, OR DEATH, declassified by the FOIA (but heavily blacked out), which describes in detail the diamond-shaped UFO that inflicted severe harm on both Betty Cash and Vickie Landrum.

The evidence included other witnesses, at distances far apart, who testified that they saw the UFO and odd helicopters. Also, there were medical experts who certified that the peculiar radiation injuries of the closest witnesses were related to an unconventional cause. The cause rendered highly improbable that the vehicle was any civilian aircraft. What security concern and government authority would be so great that an egregious injustice to American citizens was not only tolerated but covered up? The question bears on a 1984 meeting of Congressmen, including Ohio Rep. John Seiberling, to demand information from top ranking Air Force officials.

The officials were enjoined to explain why the Air Force had violated the laws of the land by its seizure of 90,000 acres of private land near Groom Lake, Nevada, for a military facility that the government did not even admit to exist, namely the notorious Area 51 or Dreamland. When asked by what or whose authority the land had been expropriated, in violation of the fundamental laws of the US, the officials responded obscurely that it was a "*much much higher authority*" than that of the US Air Force. They added that the authority could be disclosed only in a secret "closed briefing."[53]

The mounting questions about Project Snowbird, intensified by the judge who suspiciously dismissed a law suit against the government without hearing any of the considerable evidence for the UFO, assured an ensuing host of unearthly explanations. A former intelligence officer code-named 'Falcon' stated during a videotape, with the two main witnesses in attendance, that the object "was an alien craft piloted by military aircraft pilots."[54] He said they could not control the strange machine and that the military, fearing a crash, sent out search-and-rescue helicopters.[55] And Good noted that "alleged briefing papers seen by Linda Howe and Bill Moore" contained brief references to Project Snowbird, "established to 'test-fly a recovered Alien aircraft'."[56]

In light of the Air Force officials who were challenged by Congressmen for seizing land around Area 51, it is interesting that Snowbird was said by Falcon to be continuing at a base in Nevada. The aliens were not merely said to be in control of the base. Another former intelligence officer, code-named Condor, added that there was a signed agreement between the aliens and the US Government. The agreement specified, in effect, that "we won't disclose your existence if you do not interfere in our society, and we allow you to operate from... Area 51, or 'Dreamland'...."[57]

Obviously, given that these assertions are true, the agreement was violated if operating from Nevada's Dreamland meant that test flights of the craft were to be confined there. The Betty Cash incident occurred in Texas. Moreover, if the claim about an agreement between aliens and humans is true, Lt. Col. Philip Corso comes to mind. As former Chief of the Army's Foreign Technology at the Pentagon, he noted that a cover-up of the crashed UFO at Roswell placed government authorities at the mercy of aliens who could use blackmail to get what they want by threatening to publicly disclose the secret.

Furthermore, the cover-up hindered our understanding of a technology,

Project Snowbird allegedly includes the government's operation of recovered UFOs. Skeptics say UFOs are often confused with human devices. NASA stated, for instance, that this photo is of a domed tank top at the Lewis Research Center's Electric Propulsion Lab, from which technicians emerged after an ion-engine had been tested.

hundreds or even hundreds of thousands of years more advanced than ours, by excluding it from our greatest scientists. And the cover-up squandered government time and public taxpayer money in order to needlessly sustain the secret. Finally, a tantalizing alleged secret, before and after Project Snowbird, is the cover-up of alien UFOs which interfaced with America's manned missions to the moon in the late 1960s that, if true, raises intriguing questions about the startling announcement by President George Bush of manned missions to both the moon and Mars in 2004.

The Apollo Moon Missions

In Greek mythology, Apollo was the god of the sun. In inspiring the modern prospect of space travel, the name was adopted in response to President John F. Kennedy's goal in 1962 of achieving a lunar landing by the end of the 1960s. Given reports of a Roswell UFO crash in 1947 and five-star General Douglas MacArthur's warning in the *New York Times* of a future "attack by people from other planets"[58] — said at the same time as President Harry Truman's *Memoirs*, one may well wonder if Apollo was as much of a race against other planetary people as against the Soviet Union.

Precisely, this point is made by former Army Technology Chief Lt. Col. Philip Corso. He reports that there were highly classified concerns about ET: "An extraterrestrial presence on the moon, whether it was true or not in the 1950s, was an issue of... military importance."[59] Before examining alleged encounters on the Apollo moon missions, there is a brief history of Apollo in regard to the UFO issue.

Historical Outline of Apollo and UFO Issue
Three Apollo modules included a command module, for carrying astronauts

Prophetically, a day before his assassination John F. Kennedy quoted the Bible in Houston: "Where there is no vision [God's Wisdom], the people perish." Above, addressing 35,000 faculty and students at Rice University, he stated: "We choose to go to the moon in this decade and do the other things, not because they are easy, but because they are hard, because that goal will serve to organize and measure the best of our energies and skills...."

to and from the moon, a service module for housing the propulsion and environmental systems, and a lunar module for landing on the moon. After a setback in 1967 when a flash fire in the Apollo AS-204 killed three astronauts before takeoff, three successful unmanned missions were made. These included Apollo 4 for testing the command-module engines, Apollo 5 to test the lunar module's descent and ascent propulsion systems in space, and Apollo 6 for testing the entire vehicle. Subsequently, Apollo 7 was the first manned mission in the Fall 1968, with Walter Schirra, Donn Eisele, and Walter Cunningham, to test the vehicle's systems together in preparation for a lunar flight. The astronauts also practiced such things as navigational techniques and focusing a telescope on bright objects.

However, Apollo 8 with Frank Borman, James Lovell, Jr., and William Anders, was the first manned flight to the moon in December 1968. The craft did not land but, in preparation for it, underwent lunar orbit. Though Lovell later stated "I don't believe any of us in the space program believe that there are such things as UFOs,"[60] a relevant tape recording of astronaut Gordon Cooper in 1973 was noted. To an admittance that he has lived for "many years... in a secrecy imposed on all specialists in astronautics," Cooper added "I can now reveal that every day, in the USA, our radar captures objects of form and composition unknown to us."[61]

Indeed, Army Lt. Col. Philip Corso states that while NASA cooperated with military intelligence on reporting the UFO buzzing of our moon-mission spacecraft, NASA "had an even more classified back channel to the Hillenkoetter working group [Maj-12] and were keeping them updated on every single alien craft appearance... especially during the early series of Apollo flights...."[62] Moreover, the controversy increased with Apollo 11. Preceded by Apollo 9 that had tested the lunar module in earth orbit and Apollo 10 that again orbited the moon, Apollo 11 left for a lunar landing with astronauts

Neil Armstrong, Michael Collins, and Edwin Aldrin on 16 July 1969. Interestingly, the Soviet Union put the unmanned probe Luna 15 in lunar orbit, of whose progress the astronauts were informed, for reasons evidently unknown to the Apollo mission or US government.

Extraterrestrials on the Moon?
After six hours in the module, Armstrong stepped on the lunar surface and uttered the famous words "That's one small step for (a) man, one giant leap for mankind." And most intriguingly in terms of the UFO issue, an official plaque was posted that read "HERE MEN FROM THE PLANET EARTH FIRST SET FOOT UPON THE MOON JULY 1969, A.D."

In response to a mission control question "What's there?," Apollo control stammered dismayingly: "These babies are huge, sir... Oh, God, you wouldn't believe it! I'm telling you there are other spacecraft out there...."

"MEN FROM... *PLANET EARTH*?" A euphoria and international generosity over the achievement may explain the words. But a fierce race with the Soviets to get to the moon first during the Cold War, coupled with denials of UFOs by NASA, would lead one to expect "MEN FROM *THE UNITED STATES*." Without wishing to exaggerate a significance of the actual words, they *did* allow for other planetary beings to have previously set foot on the moon. Did official sensitivity to a virtual immortality of the words evoke caution and an unaccustomed honesty? That the honesty about alleged extraterrestrials might have been the case is suggested by reports that describe alien interference with the moon missions. There were even shocking admissions by former officials at NASA.

Former NASA employee Otto Binder stated that radio hams, with VHF reception, evaded NASA's broadcast censorship. Apollo 11 was heard, in response to a mission control question "What's there?", to say, "These babies are huge, sir.... Oh, God, you wouldn't believe it! I'm telling you there are other spacecraft out there...."[63] Whereas Good notes that the report was dismissed as science fiction until 1979, when it was confirmed by former Chief of NASA Communications Systems Maurice Chatelain, Corso claims that the extraterrestrials "tried to scare us away from the moon... more times than even I know."[64] And although Corso guessed that the aliens were from other solar systems, Chatelain believed that some UFOs may have come from

NASA depiction of a giant space station in 1959 for up to 50 people. Why did plans for such stations, the US presence on the moon, a greatly enhanced ability to have manned visits to Mars, the advent of novel scientific discoveries, and a vastly increased national security end so abruptly? Why did *not* the Soviet Union, at the height of the Cold War, exploit this abrupt termination for a sake of its own international prestige and military-scientific advantage? That a sudden about face by both superpowers would have bewildered the early astronauts is patently clear in *First on the Moon* (1970) by Neil Armstrong, Michael Collins, and Edwin Aldrin who stated: "From now onwards, the moon is irrevocably bound up with the future of mankind. And well before the end of this [20th] century, the first human child will be born there." Was the answer to why Apollo ended provided by these same authors when they added that, aside from a possibility that ET intends harm, contact will "have a shattering impact upon human philosophy, religion and... politics"?

Saturn's moon Titan.

Though Apollo 12, 13, and 14 followed, there were mishaps that seemed curious in light of reports that alien craft did not merely surveil our spacecraft. By electrical and magnetic field interference, they were said to have disrupted "launch vehicles... by buzzing them, jamming radio transmissions, causing electrical problems with the spacecrafts' systems, or causing mechanical malfunctions."[65] Malfunctions included a lightning bolt that crippled Apollo 12's power system during its launch on 14 November 1969 and, after another attempt, a failure of its television. And in route to the moon on 11 April 1970 there was a failure of the electrical and life-supply systems of Apollo 13. Blamed on a faulty installation, this led to the requirement that the lunar module operate as a lifeboat for the rest of the trip, causing grave concern among NASA officials. The Apollo 14's lunar launch on 31 January 1971 had failures to complete docking maneuvers of the command and lunar modules, a possible failure of the ascent-stage batteries, and a false abort signal of the lunar module even before the latter's descent. And while Apollo 15 made a safe trip to the moon and back without problems from late July to early August, Apollo 16 was launched on 16 April 1972 with problems of the command module's backup control system.

Only after deliberating four hours, was it decided not to abort the mission. The lunar module landed six hours late on a lunar plateau called 'Descartes' in honor of the Father of modern philosophy. Finally, a safe return of the mission was followed by launching Apollo 17 on 7 December 1972 and its successful return a week later. Curiously, despite the decades of planning for a permanent lunar base and a potential for major discoveries, the Apollo missions came to an abrupt end.

It scarcely needs to be said that the foregoing mishaps on the Apollo missions are explicable by our own fallible technology. And the perceived threat of aliens can be accepted without accepting their hostility. If they were hostile and had a vastly superior technology as manifest by the anomalous maneuvers of their UFOs, then presumably not either the astronauts or humanity itself would still be here to discuss the problem! In this regard, astronaut Gordon Cooper stated to a Special Committee at the United Nations: "I believe that these extraterrestrial vehicles and their crews are visiting this lanet from other planets, which are obviously a little more advanced than we are on earth... we need to determine how best to interface with these visitors n a friendly fashion."[66]

NASA depicts the Aerover Blimp for exploring Saturn's moon Titan. Former NASA Communications Chief Maurice Chatelain thought that aliens may well occupy Titan, possibly because—as he confirmed—astronauts reported UFOs on the moon. Oddly, a reporter investigating reports of UFOs at Edward's Air Force Base, stated that on his unauthorized visit there he observed a Blimp-like UFO (*see below*). A DARPA intelligence document, obtained by the Freedom of Information Act, referred to a top secret blimp named "Fat Man." The obvious question is whether it was a recovered UFO or an experimental craft for space exploration. In any case, Titan resembles Earth before life emerged. While there seems to be little oxygen, water vapor was detected in its atmosphere! If it turn out to resemble the above image, this fact is to be confirmed in 2004 by the Huygens Probe — named after Christian Huygens, father of the wave theory of light. Ironically, he inferred in 1690 from the Copernican revolution that there would be intelligent beings on other celestial bodies!

Did ET End Apollo?

On the one hand, Kennedy was assassinated in 1963. Even if his enthusiasm for Apollo was shared by Presidents Johnson and Nixon, there was not the same personal stake in it. Also, economic difficulties skyrocketed with Johnson's guns-and-butter policy of expanding the Vietnam War in 1965 and, concurrently, achieving the Great-Society goal of eradicating poverty. Dwindling tax money for Apollo was worsened by unrest over the further assassinations of Robert Kennedy and Martin Luther King, Nixon's political dilemmas, a nation torn by dissent over Vietnam, and countercultural agenda from a sex revolution to a women's liberation, the black struggle for civil rights, and riots in major cities.

Drains on political and monetary resources were compounded by the Cold War. Having resulted in mounting taxes for an arms race over three decades,

Top Left: Bringing to mind the proposed NASA blimp to explore Titan, a UFO photo was taken over Edwards AFB in 1997 by Coleman Fluhr. Looking like the cross between a blimp and rocket, though it was neither since its operations were anomalous, he states that "When I tried to make this photo public, three men came to my home [notorious Men in Black?]. They identified themselves as military police and threatened to have me imprisoned for violating national security if I didn't hand over all photos of that airship." He refused and the men said they would be back. "They left me their card and said I was in deep trouble." His house was broken into that night and his computer equipment, files and cameras were stolen. "The next morning I called the number on the card given to me by the men, it was a dead phone number. The army had no... record of these men!"

Bottom Left: A highly censored document on the UFO by the Defense Advanced Research Projects Agency (DARPA), released by the Freedom of Information Act. It refers to the object as "Fatman," name of an atomic bomb dropped on Japan in WWII. The words "classified top secret" are legible on the second line from the bottom.

resuming Apollo was out of the question when President Reagan proposed a Strategic Defense Initiative (SDI). A specter of the Soviets losing the War raised fear of their launching a nuclear strike. And SDI led to even more revenues for a missile-launched kinetic energy beam weapon. Locking onto incoming warheads or low-orbit space vehicles, the weapon would neutralize their electronics. In short, the Apollo mission was viewed as a dream America could ill afford.

Astronaut Edgar Mitchell, whose Ph.D. from MIT makes him a credible voice among scientists, stated that Washington should "tell the truth about aliens."

On the other hand, many of the developments were entangled with the UFO issue. Corso notes that Army Intelligence suspected that the Apollo mission was "abandoned because there was no way to protect the astronauts from possible alien threats."[67] And astronaut Edgar Mitchell, whose Ph.D. from MIT makes him a creditable voice in science, stated on 11 October 1998 that Washington should "tell the truth about aliens" and suggested that "there are humanoids manning craft... not in the arsenal of any nation on Earth."[68] In view of his remarks and Roswell, there may be as many reasons to suppose that Apollo's end was due to aliens as to politics. But again, these points do not imply an aggression of the alleged aliens.

Apollo Revitalized?

What then are we to make of President Bush's announcement on 14 January 2004 that he plans a renewed lunar colony and manned-missions to Mars? Given that ET brought about the end of such plans before, at least partly, there are several considerations.

Assuming that Bush's plans are not a pre-election gambit, and there is no indication that they are, Washington may feel forced to rejuvenate missions in light of other nations probing our solar system. There is no doubt about the political prestige as well as military and scientific advantages at stake. *Nature News Service* stated on 23 August 2003 stated that "A flotilla of small, cheap landers — Europe's Mars Express, NASA's twin Mars Exploration Rovers, and Japan's Nozomi — left Earth earlier this year

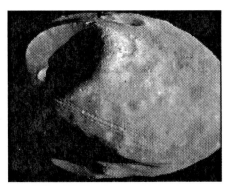

A question of ET on the moon bears on Mars and its moon Phobos (*above*). There is an intriguing timeline of controversy: In 1984 scientists Tom Rautenberg and C.W. Churchman began the Mars Investigation Group at the University of California at Berkeley. But within a year they went from Berkeley to Brown University to the Geo-Space Research Center when their hosts became "sensitive to the unusual nature of the investigation." In 1991, after the Cold War, a photo was smuggled out of Russia by Air Force Col. Marina Popovich that showed a 15 1/2 mile long tubular UFO near Phobos. Soon afterwards, the Soviet Probe that took the film mysteriously lost contact and disappeared. I.S. Shklovsky and Carl Sagan featured the chapter "Are the Moons of Mars Artificial Satellites?" in *Intelligent Life in the Universe*. The Soviets had earlier adduced evidence that the density of Phobos was so low that it might be hollow. Troubled in 1993 by official disregard of possible alien artifacts, S.V. McDaniel, Professor Emeritus at Sonoma State University, wrote *Setting Mission Priorities for NASA's Mars Observer: Failure of Executive, Congressional & Scientific Responsibility?* In 1993 scientists formed the Society for Planetary SETI Research (SPSR) for research on extraterritorial intelligence in our solar system. In 1996 geologists J.L. Erjavec and R.R. Nicks stated that some landforms on Mars may "have significant features for humanity." In 2003 the Chief of Celestial Mechanics at the US Naval Observatory T. Van Flandern (Ret), stated in *Meta Research* that *a priori* denials of Martian artifacts "fly in the face of observed evidence." (Sirag's "Face on Mars," www.stardrive.org.)

to race to the red planet." Moreover, Washington would be concerned about sightings and even about possible contact with ET. Second, the mere specter of ET, in this regard, would be a sufficient incentive. All nations on Earth would seek a first contact that is vastly enhanced by the missions. Third, aside from missions of other nations, the execution of Bush's plan would stir up UFO sightings in a way analogous to other major developments such as the atomic bomb tests in the 1940s. Washington may be seeking to more quickly 'raise public awareness' for the inevitable day of both admitting of the Roswell UFO crash and a previous contact with ET. Indeed, there has been a series of shocking announcements by former NASA employees and eminent scientists

A NASA photo analyst stated that he and another employee were told by a top Administrator to destroy the negative of a blue sky on Mars, in earlier missions, in order to make it appear that there was no life.

In the article "Life Once Existed on Mars" in *ABC News On Line* (January 29, 2004), Australian scientists reported that there is evidence of life having once existed on Mars. Their discovery, published in *Journal of Microscopy*, found microscopic fossils of primitive bacteria-like organisms in a Mars meteorite that matched characteristics of bacteria found in mud in Queensland. Dr. Tony Taylor from the Australian Nuclear Science and Technology Organisation stated: "We've now got enough evidence to warrant a sample retrieval mission and I believe it's going to be a matter of developing the technology to get authentic pieces of Mars back here on Earth where we can subject them to the really high-tech high resolution instruments that you can't possibly put on Mars." Even more immediate are reports of apparently artificial objects detected in photos of recent landings. They include some evidence of Martian microbes that may secrete concentrated organic acids that bore into rock. Rock-boring organisms on Earth are common, and this *or other* mechanisms are occurring on Mars. Indeed, the new Mars Rovers, noted Maj. George Filer (*Filer's Files* 4 February 2004), are picking up artifact-like square holes in rocks. Further, the recently revealed blue coloring of the Martian sky indicates a much greater biosphere than has been admitted by NASA.

The book *Mars the Living Planet* by Barry Digregorio, Dr. Gilbert Levin and Dr. Pat Straat, notes Filer, reveals disturbing facts on a disparity of colors in photos that are seen today and in previous landings on Mars. Jurrie Van der Woude worked thirteen years at the California Institute of Technology which operates the Jet Propulsion Laboratory (JPL) for NASA. He reportedly said:

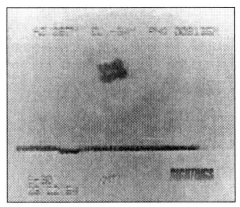

This cross-shaped UFO was inadvertently captured on film by a government controlled state-of-the-art radar tracking system at a remote tracking location known as Area S30 on the Nellis Test Range in 1994. The radar-controlled cameras monitor the 7,700 square miles above the range. Though a plausible explanation for this UFO may be that it is a secret project, as the Blimp-like UFO at Edwards AFB may have been, this explanation is not by any means settled. Might UFO sightings dramatically increase if either NASA or the European Space Agency reestablish manned missions to the moon and Mars?

"Both Ron Wichelman and I were responsible for the color quality control of the Viking Lander photographs, and Dr. Tomas Mutch, the Viking Imaging Team Leader told us he got a call from a NASA Administrator asking that we destroy the Mars blue sky negative created from the digital data." The images were then *falsely* reddened to make it appear there was no life, no green algae or lichen on the planet." Filer asks: Why is NASA withholding blue-colored photos from the public?

These matters bring us back to a relationship of UFOs to American culture. The connection spans from comedian Jackie Gleason and popular films to a political turnabout, in terms of a new interest, of higher education.

Afterword on American Culture

At the apex of President Richard Nixon's political problems back in 1973, the wife of comedian Jackie Gleason stated that Gleason had a bizarre experience because of his long-standing friendship with the President. Returning home one night in that year, Gleason was visibly shaken after visiting Homestead Air Force Base in Florida because he saw "a top secret repository where the bodies of aliens were stored. The visit was conducted under extremely tight security."[69] Is it farfetched to suppose that the invitation by Nixon reflected his need for nonpartisan friendship with a renowned man of humor at a time when he suffered singular political attacks and immanent impeachment? Gleason's refusal to deny the story made it all the more intriguing. Army News Service Editor, L. Bryant, noted that his attempt to solicit testimony in the AFB newspaper, from anyone who could confirm the story, was strictly forbidden by Air Force officials. These cultural considerations are augmented by first-rate commercial films. For example, *Close Encounters of the Third Kind*, *E.T.*, and *Signs* have led skeptics to think that depictions of ET and UFOs in the films caused later sightings and abductee reports. But astronomer and Air Force consultant Dr. Allen Hynek was an advisor for the production of, and even made a cameo appearance in, *Close Encounters*. Spielberg's *E.T.* was inspired as well by both official reports and reported encounters. And the film *Signs* echoed scientific research on peculiarities of crop circles by biophysicists such as those at BLT Research. Hence, the debunkers may actually illustrate a fallacy of reasoning that is called the *false cause*. That is, depictions of the aliens may have been caused by the credible reports rather than the reports by the films!

Interestingly, when the film *E.T.* was shown at the White House for President Reagan in 1982, he reportedly whispered to Spielberg "There are probably only six people in this room who know how true this is."[70] And a former counter-intelligence agent with the US Air Force Office of Special Investigations (AFOSI) stated in an interview that since the early 1950s, from when *The Day the Earth Stood Still* was filmed until *E.T.*, the US Government has been trying to gradually condition the public. "The public would readily accept it more than they would if it was all shoved out at one time."[71]

Professor Jodi Dean now concedes that "Those of us attracted to left-wing causes, to critical positions against political... authorities, or maybe just to underdogs in general may feel at home in UFOlogy."

Finally, charges of communist subversion in the McCarthy era of the early 1950s, which led to an *externally* coerced conservatism in universities, gave way to an *internally* coerced New-Left liberalism from the late 1960s into the next century. This fact bears heavily on the UFO issue because universities are the only institutions that even make a pretense of pursuing objective truth. But what is empirically true too often became politically incorrect. While there was only a sizable minority of professors who politicized the curricula, a majority of faculty and administrators passively acquiesced to the trendy politics. The controversy over 'PC' evidenced a noxious influence on society.

The influence later dominated public issues in America such as the Thomas-Anita Hill Supreme Court hearings, gays and women in the military, and a military purge of sexist soldiers. Though in ideal positions of job tenure to challenge a UFO cover-up, professors were not promising candidates for taking on government censors, who they traditionally eschew, or for exposing facts irrelevant to their agenda. Academic enmity to UFO interest, ostensibly fixed in practical concerns for making society better, was itself a cover-up. In

Official saucer photo from the US Government? Information on this widely published photo is murky at best. But the photo was reportedly taken around 1951 by the US Marine Air Group over the North China Sea during the Korean war. That year precedes the date that most popular sci-fi films were made, such as the classic *The Day the Earth Stood Still*, revealing that the films were as likely to be 'caused' by the reports as they by the films.

embracing a terrestrial eradication of sexism and racism while pushing a multicultural awareness of ethnic *differences*, utopian aims for a revolutionary new Earth collided head-on with any interest in extraterrestrials that would stress the *same* human race.

Taking their cue from academia, the corporate world was as worried about law suits as for politically correct workplaces. And the news media was as concerned with reporting workplace progress as with world events. But academic ex-feminists such as Professor Camille Paglia have been influential against this trend. She won a full page in *Time* for arguing that "An antiseptically sex-free workplace is impossible and unnatural."[72] In addition to women such as philosophy professors Christina Sommers and Noretta Koertge who scathingly criticize feminist epistemology,[73] does Paglia signal a trend that bodes well for an openness to UFOs by a liberal to left news media and academia? Many in academia will remain hostile to the world's great religions that, unless they are themselves politicized, warn against the arrogance of political utopias. But Professor Jodi Dean now also concedes: "Those of us attracted to left-wing causes, to critical positions against political... authorities, or maybe just to underdogs in general may feel at home in UFOlogy."[74] This new found interest in UFOlogy brings us to some recent developments.

Notes

[1] See "The Enigmatic Group: Majestic-12" in the previous chapter.

[2] For a copy, see the Appendix in Timothy Good's *Above Top Secret*, With a Foreword by the former Chief of Defense Staff, Lord Hill-Norton, G.C.B. (NY: William Morrow, 1988), p. 549.

[3] *Ibid.*, p. 5.

[4] See Jerome Clark, *The UFO Book* (NY: Visible Ink, 1998), pp., 486, 517.

[5] See "Columbus Man Tells of Seeing Brilliant Flash," *Columbus Dispatch*, July 25, 1948, *Ibid.*, p. 77.

[6] Allen Hynek, *Comments on Project Sign Case 144, Chiles-Whitted Case 7/24/48*, N.d. *Ibid.*, p. 78, emphasis added.

[7] See B. Yavorsky and Yu. Seleznev, *Physics* (Moscow: MIR Publishers, 1979), p. 165. This text was sold at the University of Colorado at Boulder.

[8] Clark, *The UFO Book*, p. 78.

[9] Project Sign Report No. F-TR-2274-1A, February 1949. From T. Good, *Above Top Secret*, pp. 268.

[10] *Ibid.*, p. 268.

[11] *Ibid.*, p. 268.

[12] *Ibid.*, p. 268.

[13] *Ibid.*, P. 330.

[14] *Cf.* Clark, *The UFO Book*, p. 177.

[15] See William Cooper, *Behold a Pale Horse* (Sedona, AZ: Light Technology Publications, 1991). In his correspondence to me June 22, 1999, Timothy Good noted that Cooper's claims were false.

[16] Clark, *The UFO Book*, p.161.

[17] Lt. Col. Philip J. Corso (Ret.), *The Day After Roswell* (NY: Simon & Schuster, Pocket Books, 1997), p. 78.

[18] *Ibid.*, p. 78.

[19] *Ibid.*, p. 78.

[20] *Ibid.*, p. 261.

[21] *Ibid.*, p. 261.

[22] *Ibid.*, p. 261.

[23] *Ibid.*, p. 262.

[24] *Cf.* Immanuel Kant, *Groundwork of the Metaphysics of Morals* (NY: Harper & Row, 1948), pp. 82-92.

[25] Joseph Cardinal Ratzinger, Ed., Interdicasterial Commission, *Catechism of the Catholic Church* (MO: Liguori Publications, 1994), p. 597.

[26] *Ibid.*, p. 262.

[27] *Ibid.*, p. 262.

[28] Timothy Good, *Alien Contact*, With Commentary by Admiral Lord Hill-Norton CGB, Chief of Defense Staff (NY: William Morrow, 1993), p. 129.

[29] *Ibid.*, p. 129.

[30] *Ibid.*, p. 130.

[31] *Ibid.*, p. 131.

[32] *Ibid.*, p. 131.

[33] See Gen. Douglas MacArthur's remarks in "MAC'ARTHUR GREETS MAYOR OF NAPLES," *The New York Times*, Oct. 8, 1955. The page numbers were blurred in my university microfilm.

[34] Harry S. Truman, "Preface," *Memoirs by Harry S. Truman: Years of Decisions*, V. I (NY: Doubleday, 1955), p. X. As Truman declared repeatedly in this very volume, it pays to read about history — especially that history which is written by those who helped make it!

[35] See Corso, *The Day After Roswell*, pp. 69, 70.

[36] See Stanton Friedman, *Top Secret/MAJIC* (NY: Marlow & Co., 1996).

[37] *Ibid.*, p. 5.

[38] See Good's interview, *Alien Contact*, p. 129.

[39] From Friedman, *Top Secret/MAJIC*, p. 140.

[40] See Fred Crews, "The Mindsnatchers," *The New York Review of Books*, Vol. XLV, No. 11, June 25, 1998, pp. 18, 19.

[41] Friedman, *Top Secret/MAJIC*, p. 119.

[42] *Ibid.*, p. 140.

[43] For copies, see the Appendix in Good's *Alien Contact*.

[44] Good, *Alien Contact*, p. 86.

[45] *Ibid.*, p. 87.

[46] *Ibid.*, p. 87.

[47] See Linda M. Howe, *An Alien Harvest* (Huntingdon Valley, PA: LMH Productions, 1989), pp. 83, 84. From Good, *Alien Contact*, p. 58.

[48] Quoted on videotape by Linda Howe in *A Strange Harvest* (Huntingdon Valley, PA: LMH Productions, 1980). From Good, *Alien Contact*, p. 59.

[49] Linda M. Howe, "1989: The Harvest Continues," *The UFO Report 1991*, ed. by Timothy Good (London: Sidgwick and Jackson, 1990).

[50] *Cf.* Good, *Above Top Secret*, pp. 303-305.

[51] *Ibid.*, p. 304.

[52] J. Schuessler, "Cash-Landrum Case Closed," *MUFON Journal*, No. 222, October 1986, pp. 12-17. From Good, *Above Top Secret*, p. 305.

[53] For a videotape of Congressman Seiberling and Air Force officials, see *UFOs: The Best Evidence*, George Knapp, KLAS TV, Channel 8, March 1989.

[54] This and the following quotes are from interviews in the video "UFO Cover-Up Live", from Good, *Alien Contact*, p. 136.

[55] *Ibid.*, p. 137, emphasis added.

[56] *Ibid.*, p. 137.

[57] *Ibid.*, p. 137.

[58] See Col. Corso, *The Day After Roswell*, p. 126.

[59] *New York Times*, October 8, 1955. For further details, see fn. 33.

[60] From Good, *Above Top Secret*, pp. 377.

[61] *Ibid.*, p.380.

[62] Corso, *The Day After Roswell*, pp. 128, 129.

[63] Good, *Above Top Secret*, pp. 383, 384.

[64] See Good's *Above Top Secret*, p. 384, and Corso's *The Day After Roswell*, pp. 156, 157.

[65] Corso, *The Day After Roswell*, pp. 124, 125

[66] Reported in *The Gazette*, Montreal, 29 November 1978. From Good, *Above Top Secret*,

pp. 379, 380.

[67] *Ibid.*, p. 243.

[68] Maj. George Filer (Ret.), *Filer's Files #41*, October 14, 1998, p. 1, in reference to a report by Tom Rhodes in the *Times of London*.

[69] Timothy Good, *Alien Contact* (NY: William Morrow, 1993), pp. 104.

[70] *Ibid.*, p. 81.

[71] *Ibid.*, p. 131.

[72] Camille Paglia, "A Call for Lustiness," *Time*, March 23, 1998, p. 54.

[73] Christina Sommers, Professor of Philosophy at Clark University, is the W.H. Brady Fellow at the American Enterprise Institute. She has authored *The War Against Boys*, *Who Stole Feminism?*, and *How Women Have Betrayed Women*. Feminists, she says, are obsessed with practical agenda, devoid of fact and intellectual credibility, that seeks to mold the public into believing that girls are academically "shortchanged" and drained of self-esteem by a society that favors boys. But boys, not girls, have been systematically shortchanged. Noretta Koertge is Professor Emerita of the History & Philosophy of Science and Adjunct Professor of Women's Studies at Indiana University. She has a Ph.D. in Philosophy of Science from the University of London where she studied under Sir Karl Popper. She received the prestigious Dewey Foundation Fellowship and is Chief Editor of *Philosophy of Science* (official journal of the Philosophy of Science Association). Her latest book with Women's Studies Professor Daphne Patai, *Professing Feminism: Cautionary Tales from the Strange World of Women's Studies*, criticizes feminist pedagogy and epistemology.

[74] See Jodi Dean's *Aliens in America* (Cornell University Press, 1998), reviewed in F. Crews' "The Mindsnatchers," *The NY Review of Books*, p.18. Crews notes his university background with evident pride. Though there is nothing wrong with this, all things equal, she derides a possibility of alien UFOs in a predictable group-think cadence of university received views. Since I am a professor out of step with the drum beat, I expect especially virulent *ad hominem* and *straw man* attacks by politically correct professors and their academic toadies who are in step — and even by UFO researchers sympathetic to some of their agenda. Sadly, the abusive attacks are defended to the extreme of asserting that even criticism of their typically self-refuting theories of truth is a patriarchal criticism rooted in the dead-white male Aristotle! For examples of these attacks on Aristotle, see the articles of feminists Marti Kheel and Val Plumwood in *Covenant for a New Creation*, Ed. by C.S. Robb et al (1991), and *Environmental Philosophy*, Ed. by M.E. Zimmerman et al (1993). The point is, if Aristotle is subject to these appalling attacks amid a deafening silence of academic pharisees, the attacks will surely happen to me and others who cross these academic martinets — as by further analogy the disciples of Christ could expect brutal attacks if they happened to their Master. No delusion of grandeur here, only sober expectation in light of a depressing history of the human condition. Is there import here for what ET can expect by contact? Lest my remarks seem to too pithy, see those of Harvard Professor Harvey Mansfield at the outset of this book in "Emerging Awareness of the Problem." He suggests that political correctness is a classic case of tyranny and that professors need the courage to start raising a "little hell."

8

Since the Missions:
Controversy over Religion
and Science

S INCE THE TIME of the moon missions, new developments
include an international panel of scientists headed by Stanford
University physicist Peter Sturrock. This panel notes physical evidence
of UFOs and endorses their study. But the proposed study is paled in boldness
by one of physicists at a US military academy. They admit the failure of our
science to fully fathom, what they take to be, advanced extraterrestrial craft
that may have been visiting earth for tens of thousands of years. Finally, the
issue of science raises some engaging questions. Given some alleged alien
references to God, is belief in God consistent with science? If they are —
indeed, if science is related logically to theology, then aliens and humans
may share a religio-scientific belief system. This system would bear on political
mishaps that the human species has perennially suffered throughout its history.

U.S. Air Force Academy: Physics 370

The US Air Force Academy's *Air Force Cadet
Manual for Physics 370* rejects the SETI Institute's
traditional dismissal of UFOs as alien spacecraft
because the speed of light would have to be
exceeded to avoid millions of years of travel. And
traveling at that speed is not exceeded according
to relativistic physics. But aside from NEC physicists who announced that
they have exceeded the accepted light speed, this Air Force Manual questions
any staid reliance on our physics by questioning so-called well established
laws. And although unclassified government material, the Manual takes
seriously a possibility of extraterrestrials visiting earth on the basis of general
science and common sense. Excerpts in the material contain the author's

223

commentary. Though the commentary acknowledges a salient openness to UFOs of the US Air Force Academy for its future pilots, it challenges some assumptions and raises other singular points:*[1]

AIR FORCE CADET MANUAL
PHYSICS 370
UNITED STATES AIR FORCE ACADEMY, INTRODUCTORY SPACE SCIENCE
TABLE OF CONTENTS

BETWEEN EARTH AND THE STARS   Page

MAN'S EFFORTS TO UNDERSTAND

Major Donald G. Carpenter is the principal author of each chapter, with co-authors and exceptions as indicated below:

Chapter   **Co-authors [Principle*]**

* Note: The illustrations displayed below were not in the USAFA Manual.

CHAPTER XXXIII: UNIDENTIFIED FLYING OBJECTS

What is an Unidentified Flying Object (UFO)? Well, according to United States Air Force Regulation 80-17 (dated 19 September 1966), a UFO is "Any aerial phenomenon or object which is unknown or appears to be out of the ordinary to the observer." This is a very broad definition which applies equally well to one individual seeing his first noctilucent cloud at twilight as it does to another individual seeing his first helicopter. However, at present most people consider the term UFO to mean an object which behaves in a strange or erratic manner while moving through Earth's atmosphere. That strange phenomenon has evoked strong emotions and great curiosity

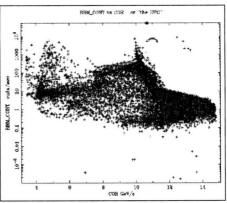

UFO: "Any aerial phenomenon or object," states US Air Force Regulation 80-17," "which is unknown or appears to be out of the ordinary...." This holds for the above image from the ASCA (Advanced Satellite for Cosmology and Astrophysics). But the object is seen by US scientists as a product of "dark earth data" and not a UFO. Thus, what is a UFO for one observer may not be for another.

among a large segment of our world's population. The average person is interested because he loves a mystery, the professional military man is involved because of the possible threat to national security, and some scientists are interested because of the basic curiosity that led them into becoming researchers.

The literature on UFOs is so vast, and the stories so many and varied, that we can only present a sketchy outline of the subject in this chapter. That outline includes description classifications, operational domains (temporal and spatial), some theories as to the nature of the UFO phenomenon, human reactions, attempts to attack the problem scientifically, and some tentative conclusions. If you wish to read further in this area, the references provide an excellent starting point.

33.1 DESCRIPTIONS

One of the greatest problems you encounter when attempting to catalog UFO sightings, is selection of a system for cataloging. No effective system has yet been devised.... The net result is that almost all UFO data are either treated in the form of individual cases, or in the forms of inadequate classification systems. However, these systems do tend to have some common factors, and a collection of these factors is as follows:

[System for Cataloging]

a. Size
b. Shape (disc, ellipse, football etc.)
c. Luminosity
d. Color

e.   Number of UFOs

Behavior
a.   Location (altitude, direction etc.)
b.   Patterns of paths (straight line, climbing, zig-zagging etc.)
c.   Flight characteristics (wobbling, fluttering etc.)
d.   Periodicity of sighting
e.   Time duration
f.   Curiosity or inquisitiveness
g.   Avoidance
h.   Hostility

Associated Effects
a.   Electromagnetic (compass, radio, ignition systems etc.)
b.   Radiation (burns, induced radioactivity etc.)
c.   Ground disturbance (dust stirred-up, leaves moved, standing wave peaks on surface of water etc.)
d.   Sound (none, hissing, humming, roaring, thunderclaps etc.)
e.   Vibration (weak, strong, slow, fast)
f.   Smell (ozone or other odor)
g.   Flame (how much, where, when, color)
h.   Smoke or cloud (amount, color, persistence)
i.   Debris (type, amount, color, persistence)
j.   Inhibition of voluntary motion by observers
k.   Sighting of "creatures" or "beings"

After Effects
a.   Burned areas or animals
b.   Depressed or flattened areas
c.   Dead or missing animals
d.   Mentally disturbed people
e.   Missing items
     We make no attempt here to present available data in terms of the foregoing descriptions.

33.2 OPERATIONAL DOMAINS — TEMPORAL AND SPATIAL
What we will do here is to present evidence that UFOs are a global phenomenon which may have persisted for many thousands of years. During this discussion, please remember that the more ancient the reports the less sophisticated the observer. Not only were the ancient observers lacking the terminology necessary to describe complex devices (such as present day helicopters) but they were also lacking concepts necessary to understand the true nature of such things as television, spaceships, rockets, nuclear weapons and radiation effects. To some, the most advanced technological concept was a war chariot with knife blades attached to the wheels.

The AF Academy Manual refers to this book published by Chaosium and Edited with an Introduction by Tim Maroney. He states in his Foreword that the "enormous two-volume work The Secret Doctrine, considered the cornerstone of Theosophy [Theology and Philosophy combined], is an extended commentary on the *Book of Dzyan*...." Purportedly, it is an "ancient scripture describing the history of the universe from the beginning of the current cosmic cycle until the early evolution of humanity." An occult teaching is praised, said to be echoed by the ancient Greek Atomistic philosophers Leucippus and Democritus who held: "Space was filled eternally with atoms actuated by a ceaseless motion... [p. 74]." The AF Academy Manual relates this book to ET: "One of the stories is of a small group of beings who supposedly came to Earth many thousands of years ago in a metal craft which orbited Earth several times before landing."

By the same token, the very lack of accurate terminology and descriptions leaves the more ancient reports open to considerable misinterpretation, and it may well be that present evaluations of individual reports are completely wrong. Nevertheless, let us start with an intriguing story in one of the oldest chronicles of India... the Book of Dzyan.

This book is a group of "storyteller" legends which were finally gathered in manuscript form when man learned to write. One of the stories is of a small group of beings who supposedly came to Earth many thousands of years ago in a metal craft which orbited Earth several times before landing. As told in the Book "These beings lived to themselves and were revered by the humans among whom they had settled. But eventually differences arose among them and they divided their numbers, several of the men and women and some children settling in another city, where they were promptly installed as rulers by the awestricken populace."

"Separation did not bring peace to these people and finally their anger reached a point where the ruler of the original city took with him a small number of his warriors and they rose into the air in a huge shining metal vessel. While they were many leagues from the city of their enemies, they launched a great shining lance that rode on a beam of light. It burst apart in the city of their enemies with a great ball of flame that shot up to the heavens, almost to the stars. All those who were in the city were horribly burned and even those who were not in the city — but nearby — were burned also. Those who looked upon the lance and the ball of fire were blinded forever afterward. Those who entered the city on foot became ill and died. Even the dust of the city was poisoned, as were the rivers that flowed through it. Men dared not go near it, and it gradually crumbled to dust and was forgotten by men."

"When the leader saw what he had done to his own people he retired to his palace and refused to see anyone. Then he gathered about him those of his warriors

227

who remained, and their wives and children, and they entered into their vessels and rose one by one into the sky and sailed away...."

Could this foregoing legend really be an account of an extraterrestrial colonization, complete with guided missile, nuclear warhead and radiation effects? It is difficult to assess the validity of that explanation... just as it is difficult to explain why Greek, Roman and Nordic Mythology all discuss wars and conflicts among their "Gods." (Even the Bible records conflict between the legions of God and Satan.) Could it be that each group recorded their parochial view of what was actually a global conflict among alien

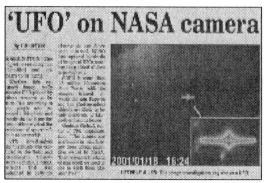

'UFO' on NASA camera

This news in the *Perth Sunday Times* (Australia) on 19 January 2003 states that this photo of a UFO — enlarged bottom right — "is certainly unidentified and appears to be flying." The article adds that "UFO investigators say the image was captured by the Solar and Heliospheric Observatory (SOHO), a NASA Satellite... launched in 1966 to observe the sun." NASA admits that the photo is from its archives but says it was produced by processing and interpolation techniques applied to a cosmic ray hit. A SOHO scientist says reassuringly that "most common sources" of UFO claims are planets and cosmic rays, although he *unreassuringly* leaves open the *uncommon sources* in which there is no evident tinkering with photos.

colonists or visitors? Or is it that man has led such a violent existence that he tends to expect conflict and violence among even his gods?

[Author's Commentary: Are *supernatural* conflicts between legions of God and Satan misidentified in the Air Force Academy Manual as scientifically advanced battles between extraterrestrials? If the battles occurred, why must they be extraterrestrial and not spiritual, unless it is assumed that the spiritual does not exist? But this reductively imposes *naturalistic* explanations on religion. Ironically, in accord with this Air Force suggestion, some atheistic UFO researchers would say that military intelligence was disclosed *to the Air Force* by secret contacts with ET — the information being intended to "raise the consciousness" of Academy cadets: Our religions were inspired by ET whose technologies were confused with miracles. Aliens, the researchers would say, are comparable to divine creators since they biogenetically created the human species. Yet this view, besides also being reductive, does not viably address various considerations of either science or theology. These include redemption — not miracles — being the essential mission of Christ, and revitalized modes of reasoning in philosophy for a First Cause and Creator that (i) bears on a common source of evolution for humans and ET, (ii) counters claims of ET's divinity, (iii) logically relates theology to science, and (iv) poses a religio-scientific belief system that may be common to extraterrestrials and terrestrials.]

Is ET shown facetiously by a Lawrence Berkeley National Laboratory astrophysicist & Einstein medal winner? In addition to arguing for God as a First Cause, this distnguished scientist has an *Introduction* where the image was shown. It states: "Due to its great penetrating power neutrino astronomy may detect sources and classes of sources that have yet to be observed since they are opaque to photons." The section *Scientific Motivation Unknown: Astrophysical and Cosmological Sources* states: "It is always exciting and good to open a new window of observation. Each time this has been done in the past, new discoveries have been made." Finally, the section *UFOs and Aliens – UFO Neutrinos* states the following: "If you watch StarTrek, you will note that they often do neutrino scans of objects, ocassionally they find that all the neutrinos are left-handed. If they do, then we will notice."

...In the Speculum Repali in Konungs Skuggsa (and other accounts of the era about 956 A.D.) are numerous stories of "demon ships" in the skies. In one case a rope from one such ship became entangled with part of a church. A man from the ship climbed down the rope to free it, but was seized by the townspeople. The bishop made the people release the man, who climbed back to the ship, where the crew cut the rope and the ship rose and sailed out of sight. In all of his actions, the climbing man appeared as if he were swimming in water. Stories such as this makes one wonder if the legends of the "little people" of Ireland were based upon imagination alone.

...In a sworn statement dated 21 April 1897, a prosperous and prominent farmer named Alexander Hamilton (Le Roy, Kansas, USA) told of an attack upon his cattle at about 10:30 p.m. the previous Monday. He, his son, and his tenant grabbed axes and ran some 700 feet from the house to the cow lot where a great cigar shaped ship about 300 feet long floated some 30 feet above the cattle. It had a carriage underneath which was brightly lighted within (dirigible and gondola?) and which had numerous windows. Inside were six strange looking beings jabbering in a foreign language... [who] suddenly became aware of Hamilton and the others. They immediately turned a searchlight on the farmer, and also turned on some power which sped up a turbine wheel (about 30 ft diameter) located under the craft. The ship rose, taking with it a two-year old heifer which was roped about the neck by a cable of one-half inch thick, red material. The next day a neighbor, Link Thomas, found the animal's hide, legs and head in his field. He was mystified at how the remains got to where they were because of the lack of tracks in the soft soil.... Hamilton's sworn statement was accompanied by an affidavit as to his veracity...

[and] signed by ten of the local leading citizens.

On the evening of 4 November 1957 at Fort Itaipu, Brazil, two sentries noted a new "star" in the sky. The "star" grew in size and... stopped over the fort. It drifted slowly downward, was as large as a big aircraft, and was surrounded by a strong orange glow. A distinct humming was heard, and then the heat struck. One sentry collapsed almost immediately, the other managed to slide to shelter under the heavy cannons where his loud cries awoke the garrison. While the troops were scrambling towards their battle stations, complete electrical failure occurred. There was panic until the lights came back on but a number of men still managed to see an orange glow leaving the area at high speed. Both sentries were found badly burned... one unconscious and the other incoherent, suffering from deep shock.

Despite openness to UFOs by the Air Force Academy Manual, the official Air Force response to the Roswell UFO is denial. Above is a body bag in which dummies were stored, said by officials to have been confused with alien bodies — though these bodies were reportedly seen by witnesses several years prior to the use of dummies.

Thus UFO sightings not only appear to extend back 47, 000 years... but also are global in nature. One has the feeling that this phenomenon deserves some sort of valid scientific investigation, even if it is a low level effort.

33.3 SOME THEORIES AS TO THE NATURE OF THE UFO PHENOMENON

There are very few cohesive theories as to the nature of UFOs. Those theories that have been advanced can be collected in five groups: a. Mysticism, b. Hoaxes, and rantings due to unstable personalities, c. Secret weapons, d. Natural phenomena, e. Alien visitors

Mysticism

It is believed by some cults that the mission of UFOs... is a spiritual one, and that all materialistic efforts to determine the UFO's nature are doomed to failure.... Some have suggested that all UFO reports were... hoaxes, or were made by people with unstable personalities. This attitude was particularly prevalent... when the Air Force investigation was being operated under the code name of Project Grudge. A few airlines even went so far as to ground every pilot who reported seeing a "flying saucer." The only way for the pilot to regain flight status was to undergo a psychiatric examination.

[Author's Commentary: This criticism of psychiatric exams for pilots who see UFOs brings to mind the former Soviet Union which put dissidents into insane asylums because they did not have a "correct" grasp of reality, according to a Marxist politics, and a political correctness in open societies which arose in the 1960s. A conservative status quo of the 1950s was worsened by a liberal to left "correctness" of the 1960s that began in American universities. They impeded attention to UFOs and, in effect, gave aid and comfort to a government suppression of investigations. The academic impediment is due to a preoccupation with terrestrial political agenda such as "raising a consciousness" of ethnic identities and a practical progress that would be sidetracked by "theoretical" concerns about extraterrestrials. An exception is the Left's hostility to alleged reactionary elements of government. There may be some support for UFO inquiries when claims of a cover-up help erode trust in government! This exception was suggested in Prof. Jodi Dean's *Aliens in America* (Ithaca, NY: Cornell University Press, 1998).]

The Air Force Academy Manual refers to the Zamora incident. The latter began at 5:45 p.m. on 14 April 1964 in Socorro, New Mexico, when thirty-one year old policeman Lonnie Zamora saw a shining object 100–200 yards away. His first thought was that a car had overturned and its gas tank had exploded. Looking closer, officer Zamora discovered that the object was an oval-shaped craft without either windows or doors. He saw an unusual red insignia on the UFO's side and then noticed two small beings, he first thought to be children, who were dressed in white overalls. The scene was investigated by Air Force consultant and Northwestern University astronomer Dr. Allen Hynek who supported Zamora's story and reported that anomalous *material evidence* was left by the craft.

There was a noticeable decline in pilot reports during this time interval, and a few people interpreted this decline to prove that UFOs were either hoaxes or the result of unstable personalities. It is of interest that NICAP (The National Investigations Committee on Aerial Phenomena) even today still receives reports from commercial pilots who neglect to notify either the Air Force or their own airline.

There are a number of cases which indicate that not all reports fall in the hoax category. We will examine one such case now. It is the Socorro, New Mexico sighting made by police Sergeant Lonnie Zamora. Sergeant Zamora was patrolling the streets of Socorro on 24 April 1964 when he saw a shiny object drift down into an area of gullies on the edge of town. He also heard a loud roaring noise which sounded as if an old dynamite shed located out that way had exploded. He immediately radioed police headquarters, and drove out toward the shed. Zamora was forced to stop about 150 yards away from a deep gully in which there appeared to be an overturned car. He radioed that he was investigating a possible wreck, and then worked his car up onto the mesa and over toward the edge of the gully. He... was amazed to see that it was not a car but instead was a weird egg shaped object about fifteen feet long, white in color and resting on short, metal legs.

Beside it... two humanoids dressed in silvery coveralls. They seemed to be working on a portion of the underside of the object. Zamora was still standing there, surprised, when they suddenly noticed him and dove out of sight around the object. Zamora also headed the other way, back toward his car. He glanced back at the object just as a bright blue flame shot down from the underside. Within seconds the egg shaped thing rose out of the gully with "an earsplitting roar."

The object was out of sight over the nearby mountains almost immediately, and Sergeant Zamora was moving the opposite direction almost as fast when he met Sergeant Sam Chavez who was responding to Zamora's earlier radio calls. Together they investigated the gully and found the bushes charred and still smoking where the blue flame had jetted down on them. About the charred area were four deep marks where the metal legs had been. Each mark was three and one half inches deep, and was circular in shape. The sand in the gully was very hard-packed so no sign of the humanoids' footprints could be found. An official investigation was launched that same day, and all data obtained supported the stories of Zamora and Chavez. It is rather difficult to label this episode a hoax, and it is also doubtful that both Zamora and Chavez shared portions of the same hallucination.

Secret Weapons

A few individuals have proposed that UFOs are actually advanced weapon systems, and that their natures must be revealed. Very few people accept this as a credible suggestion.

 [Author's Commentary: A denial by the USAF Academy Manual that UFOs are secret government weapons has merit with a caveat that, in the form of experimental aircraft, they are often alleged to be recovered alien craft or copies of them. It is claimed that these UFOs are tested at government bases such as Nevada's S4 Groom Lake area. In being an especially sensitive topic for the military, we could scarcely expect a candid allowance for this prospect.]

Natural Phenomena

It has also been suggested that at least some, and possibly all, UFO cases were just misinterpreted manifestations of natural phenomena. Undoubtedly this suggestion has some merit. People have reported, as UFOs, objects which were conclusively proven to be balloons (weather and skyhook), the planet Venus, man-made artificial satellites, normal aircraft, unusual cloud formations and lights from ceilometers (equipment projecting light beams on cloud bases to determine the height of the aircraft visual ceiling). It is also suspected that people have reported mirages, optical illusions, swamp gas and ball lightning (a poorly-understood discharge of electrical energy in a spheroidal or ellipsoidal shape...). But it is difficult to tell a swamp dweller that the strange, fast moving light he saw in the sky was swamp gas; and it is just as difficult to tell a farmer that a bright UFO in the sky is the same ball lightning that he has seen rolling along his fence wires in dry weather. Thus accidental misidentification of what might - well be natural phenomena breeds mistrust and disbelief; it leads to the hasty conclusion that the truth is deliberately not being told. One last suggestion of interest has been made, that the UFOs were plasmoids from space... concentrated blobs of solar wind that succeeded in reaching the surface of Earth. Somehow this last suggestion does not seem to be very plausible; perhaps because it ignores such things as penetration of Earth's magnetic field.

Alien Visitors

The most stimulating theory for us is that the UFOs are material objects which are either "Manned" or remote-controlled by beings who are alien to this planet. There

The Air Force Academy Manual notes that skeptics relegate virtually all UFOs to hoaxes or natural phenomena. At 9:35 am on 16 July 1952, Shell Alpert took this photo of four elliptical blobs of light in formation through the window of his photographic laboratory. He was a Coast Guard seaman assigned to a base in nearby Salem, MA. Whereas the first analysis by the US Air Force's Project Blue Book held that it was "probably" a double exposure hoax, the second analysis said it was "probably" reflections of street lamps on a window. The third analysis concluded it to be "unexplained"!

is some evidence supporting this viewpoint. In addition to police Sergeant Lonnie Zamora's experience, let us consider the case of Barney and Betty Hill. On a trip through New England they lost two hours on the night of 19 September 1961 without even realizing it. However, after that night both Barney and Betty began developing psychological problems which eventually grew sufficiently severe that they submitted themselves to psychiatric examination and treatment. During the course of treatment hypnotherapy was used, and it yielded remarkably detailed and similar stories.... Essentially they had been hypnotically kidnapped, taken aboard a UFO, submitted to two-hour physicals, and released with posthypnotic suggestions to forget the entire incident. The evidence is rather strong that this is what the Hills, even in their subconscious, believed happened to them. And it is of particular importance that after the "posthypnotic block" was removed, both of the Hills ceased having their psychological problems.

The Hills' description of the aliens was similar to descriptions provided in other cases, but this particular type of alien appears to be the minority. The most commonly described alien is about three and one-half feet tall, has a round head (helmet?), arms reaching to or below his knees, and is wearing a silvery space suit or coveralls. Other aliens appear to be essentially the same as Earthmen, while still others have particularly wide (wraparound) eyes and mouths with very thin lips. And there is a rare group reported as about four feet tall, weight around [Page 461] 35 pounds, and covered with thick hair or fur (clothing?).

Members of this last group are described as being extremely strong. If such beings are visiting Earth, two questions arise: 1) why haven't there been accidents which have revealed their presence, and 2) why haven't they attempted to contact us officially?

 [Author's Commentary: Extremely strong aliens could be from a gravity heavy planet, with a larger mass than earth, where gravitational effects on their evolution incur a greater muscular development. Question 1 in the above paragraph of the Air Force Academy Manual tacitly admits of UFO visits and likely accidents. Indeed, there seems to be as much evidence for a Roswell UFO crash, among others, as for the Socorro accident that are acknowledged by the Air Force Academy Manual. It is not farfetched to suppose that these crashes would no more be admitted by the air Force — or its institutions such as the Academy — than its possible development of alien technology. However, to accept this Manual's support for a rationality of believing in UFO accidents is to accept a rationality of believing that crashed UFOs may sometimes have been recovered.]

The Air Force Academy Manual notes controversy over whether the Tunguska explosion was caused by a UFO or natural phenomenon such as a meteor. Though witnesses reported that the object changed direction and traveled too slowly to be a meteor, NASA states emphatically: "In 1908, a 200 to 300 foot-diameter meteorite exploded 5 miles above the ground of the Tunguska region of Siberia. Trees were incinerated in a 9–mile radius from ground zero and were knocked over in a 25 mile radius. If this had occurred over a heavily populated area, the effect would have been catastrophic for the people living there."

In that chapter it was suggested that the Tunguska meteor was actually a comet.... However, it has also been suggested that... [it] was actually an alien space craft that entered the atmosphere too rapidly, suffered mechanical failure, and lost its power supply and/or weapons in a nuclear explosion. While that hypothesis may seem far fetched, samples of tree rings from around the world reveal that, immediately after the Tunguska meteor explosion, the level of radioactivity in the world rose sharply for a short period of time. It is difficult to find a natural explanation for that increase in radioactivity, although the suggestion has been advanced that enough of the meteors great kinetic energy was converted into heat (by atmospheric friction) that a fusion reaction occurred.

 [Author's Commentary: Different explanations for radioactivity left by the Tunguska explosion in Siberia, noted above, reflect a principle in the philosophy of science, discussed in my book, called an "Underdetermination-of-Theory-by-Data (UTD) Thesis." It is *as* important to give names to principles in the philosophy of science *as* in both science and logic because the names underscore well established modes of reasoning about repeating problems that might otherwise be ignored. The UTD Thesis holds that any datum or phenomenon is subject to logically inconsistent theories that, nevertheless, are empirically equivalent in their explications or predictions. This dilemma means that, since the success of predictions or explanations count for 'truth', it is not possible in these cases to rationally decide which theory or hypothesis is true, or more likely to be true.]

This still leaves us with no answer to... why no contact? That question is very easy to answer in any of several ways: 1) we may be the object of intensive sociological and psychological study. In such studies you usually avoid disturbing the test subjects' environment; 2) you usually do not "contact" a colony of ants, and humans may seem that way to any aliens (variation: a zoo is fun to visit, but you don't "contact" the lizards); 3) such contact may have already taken place secretly; and 4) such contact may have already taken place on a different plane of awareness and we are not yet sensitive to communications on such a plane. These are just a few of the reasons you may add to the list as you desire.

33.4 HUMAN FEAR AND HOSTILITY

Besides the foregoing reasons, contacting humans is downright dangerous.... On the microscopic level our bodies reject and fight (through producing antibodies) any alien material; this process helps us fight off disease but it also sometimes results in allergenic reactions to innocuous materials. On the macroscopic (psychological and sociological) level we are antagonistic to beings that are "different."

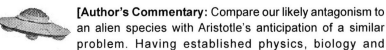

[Author's Commentary: Compare our likely antagonism to an alien species with Aristotle's anticipation of a similar problem. Having established physics, biology and psychology with Darwin's admittance that the modern founders of physiology were nothing but schoolboys compared to "good ol' Aristotle," Aristotle noted that the sciences overlap. Thus a biological aggression of male higher primates necessary for courting would spill over into their having a greater psychic aggressiveness than females. By the same token, females as well as males may have a natural psychic ambivalence towards other similar species that is rooted biologically in such things as propagation and territorial needs.]

For proof of that, just watch how an odd child is treated by other children, or how a minority group is socially deprived, or how the Arabs feel about the Israelis (Chinese vs. Japanese, Turks vs. Greeks etc.). In case you are hesitant to extend that concept to... aliens let me point out that in very ancient times, possible extraterrestrials may have been treated as Gods but in the last two thousand years, the evidence is that any possible aliens have been... assaulted (in South America there is a well-documented case [of an attack on a humanoid being])....

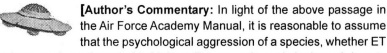

[Author's Commentary: In light of the above passage in the Air Force Academy Manual, it is reasonable to assume that the psychological aggression of a species, whether ET or humans, is partly rooted in biology. For example, male jealousy brings to mind the fact that sperm of a higher primate will attack foreign sperm in the vagina of a female mate. Biological aggression begets a psychic one. However, might there not be a self-conscious reflexive intelligence of both aliens and humans wherein they can be educated to tolerate beings different

MONUMENT TO RACIAL HATRED: victim of Nazis rests where he died, trying to escape his horrible death. (Photo by USA Sgt. E.R. Allen in Germany, 16 April 1945).

MONUMENT TO CLASS HATRED: Guard tower among hundreds of gulags across the Soviet Union, is a Marxist symbol of profound human suffering. First used by Lenin to imprison priests and political opponents, Stalin was then responsible for sending 12 to 15 million people to these camps. The prisoners were used as forced labor to work on massive industrial projects. As more laborers were needed for bigger projects and to replace those falling behind schedule, the arrest of more "class enemies" was justified. Millions were executed or perished horribly, in the name of utopia, as they labored for the glory of the state.

from themselves? This tolerance is usually held to be a hallmark of advanced cultures. At the same time, an altruistic dictum that "everyone loves everyone" of the Left, *e.g.* Marxism, is as contrary to our nature as "every race hates every other race" of the Right, *e.g.* fascism. The point is not merely that intelligence permits broad parameters of cultural fulfillment of one's psychobiological nature. By the findings of science itself, this nature may be invigorated by things such as an interracial propagation, as by analogy vegetation is strengthened by cross breeding. This point may not necessarily support a breeding between humans and aliens whose physiology could exclude it. But the point at least supports their mutual respect. At the same time there is a danger that our respect for some aliens may be transposed into an awe that could lead to a neglect of our cultural institutions which we would come to regret.]

More recently, on 24 July 1957 Russian antiaircraft batteries on the Kouril Islands opened fire on UFOs. Although all Soviet Antiaircraft batteries on the islands were in action, no hits were made. The UFOs were luminous and moved very fast. We too have fired on UFOs. About ten o'clock one morning, a radar site near a fighter base picked up a UFO doing 700 mph. The UFO then slowed to 100 mph, and two F-86's were scrambled to intercept. Eventually one F-86 closed on the UFO at about 3,000 feet altitude. The UFO began to accelerate away but the pilot still managed to get to within 500 yards of the target for a short period of time. It was definitely saucer shaped. As the pilot pushed the F-86 at top speed, the UFO began to pull away. When the range reached 1000 yards, the pilot armed his guns and fired in an attempt to down the saucer. He failed, and the UFO pulled away rapidly, vanishing in the distance....

33.5 ATTEMPTS AT SCIENTIFIC APPROACHES

In any scientific endeavor, the first step is to acquire data, the second step to classify the data, and the third step to form hypotheses. The hypotheses are tested by repeating the entire process, with each cycle resulting in an increase in understanding (we

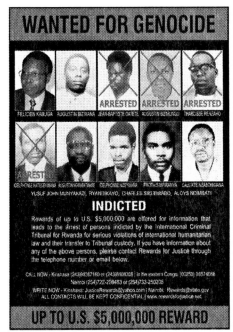

MONUMENT TO ETHNIC AND CULTURAL HATRED: The class warfare decreed by Marxism mutated into ethnic cleansing, terrorism, and culture wars from Africa and Afghanistan to Yugoslavia and even America. In "Has Global Ethnic Conflict Superseded Cold War Ideology?," *Studies in Conflict & Terror*, RAND (1996), the author notes that mass murder in the name of Marxism was not exposed by a War Crimes Trial as it was for the Nazis. Thus many intellectuals were deluded into supposing that historical conflicts were a *progressive* phenomenon of politics. In Western societies since the Cold War, the Left succeeded in "raising a consciousness" of our race, gender and class — providing a global apologetics for ethno-cultural liberations. Though the liberations were rationalized by a political correctness in universities, the virulent mutations of ideology bode ill for a kindly interest in ET and for what ET can expect by contact with the human race.

hope). The UFO phenomenon does not yield rapidly to this approach because the data taken so far exhibits both excessive variety and vagueness. The vagueness is caused in part by the lack of preparation of the observer... very few people leave their house knowing that they are going to see a UFO that evening.

Photographs are overexposed or underexposed, and rarely in color. Hardly anyone carries around a radiation counter or a magnetometer. And... there is a very high level of "noise" in the data. The noise consists of mistaken reports of known natural phenomena, hoaxes, reports by unstable individuals and mistaken removal of data regarding possible unnatural or unknown natural phenomena by [Page 463-Figure 33-1 omitted] overzealous individuals who are trying to eliminate all data due to known natural phenomena). In addition, those data, which do appear to be valid, exhibit an excessive amount of variety relative to the statistical samples available. This has led to very clumsy classification systems, which in turn provide quite infertile ground for the formulation of hypotheses.

One hypothesis which looked promising for a time was that of ORTHOTENY (i.e., UFO sightings fall on "great circle" routes). At first, plots of sightings seemed to *verify* the concept of orthoteny but recent use of computers has revealed that even random numbers yield "great circle" plots as neatly as do UFO sightings [emphasis].

[Author's Commentary: The above bears on *verifying* the truth of a theory about UFOs. But philosophers of science no longer speak of verifying theories as if their truth is unqualifiedly known. Today, a standard mode of scientific reasoning is:

$T_o \rightarrow P$ where T_o specifies a theory's interpretation of phenomena, $\rightarrow$ reads "if... then" and P denotes predictions that are true. By this truth-functional material implication, however, it is logically possible for P to be true when T_o is false. Hence, scientists can only falsify T_o since it is *logically impossible* for T_o to be true when P is false. And thus it is all the more unreasonable, as the Air Force Academy Manual indicates, to reject UFOs in terms of theories that are not either strictly known to be true or may be falsified in the future. There are even problems for a new reasoning of modal logic: It is impossible for T_o to be false when its predictions are systematically true *(P): Necessarily (P $\rightarrow T_o$)*. For T_o's truth may be restricted to a domain, as the truth of Newton's theory is limited *inter alia* to a domain in which phenomena do not approach the speed of light. In accord with the Manual, these points reaffirm that it is either naïve or dogmatic to reject out of hand the possibility of alien UFOs.]

There is one solid advance that has been made though. [Physicist] Jacques and Janine Vallée have taken a particular type of UFO — namely those that are lower than treetop level when sighted — and plotted the UFO's estimated diameter versus the estimated distance from the observer. The result is shown in Figure 33-1, and it yields an average diameter of 5 meters with a very characteristic drop for short viewing distances, and rise for long viewing distances. This behavior at the extremes of the curve is well-known to astronomers and psychologists as the "moon illusion." The illusion only occurs when the object being viewed is a real, physical object. Because this implies that the observers have viewed a real object, it permits us to accept also their statement that these particular UFOs had a rotational axis of symmetry.

(c) 2001 Alien-mbc

Above: A computer enhanced saucer photo given to Air Force Intelligence Officer Maj. George Filer (Ret.). The AF Academy Manual admits that such UFOs may well violate our scientific laws. But it does not refer to a secret government *group* that allegedly oversees a reverse engineering of crashed UFOs, as does the letter from an FBI file *overleaf.*

[Reference to Jacques Vallée on page 464 is omitted. Page 465 continues.] Further data correlation is quite difficult. There are a large number of different saucer shapes but this may mean little. For example, look at the number of different types of aircraft which are in use in the US Air Force alone.

It is obvious that intensive scientific study is needed in this area; no such study has yet been undertaken at the necessary levels of intense and support. One thing that must be guarded against in any such study is the trap of implicitly assuming that our knowledge of physics (or any other branch of science) is complete.

[Author's Commentary: The foregoing admits of the plausibility that some UFOs may be extraterrestrial, even if they are impossible in terms of our science, when it cautions future Air Force pilots against "The trap of implicitly assuming that our knowledge of physics... is complete." The incompleteness is braced philosophically by a principle of Pessimistic Induction that was formulated by Oxford Professor W.H. Newton-Smith in *The Rationality of Science*. This principle is applied to specific cases throughout our discussion of rational UFO belief. The philosophy of science and logic have developed principles that, in naming cogent modes of reasoning, enhance our ability to detect dogmatic positions. These positions include one that appeals exhaustively to present-day theories for specifying an impossibility of ET's visits since their travel would have to exceed the speed of light in violation of Einstein's physics. We are as warranted in believing that this physics is not entirely true as previous generations would have been in believing that the truth was not reflected *in toto* by the physics of Newton or Aristotle.]

An example of one such trap is selecting a group of physical laws which we now accept as valid, and assume that they will never be superseded. Five such laws might be:

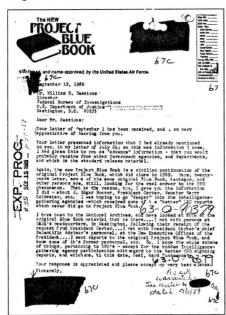

With the letterhead *THE NEW PROJECT BLUE BOOK,* the writer states that he met with President Jimmy Carter's Chief Scientific Advisor on UFOs and encourages the FBI to seek information about this elite *group* for the public's sake (Maj-12 ?).

1) Every action must have an opposite and equal reaction.
2) Every particle in the universe attracts every other particle with a force proportional to the product of the masses and inversely as the square of the distance.
3) Energy, mass and momentum are conserved.
4) No material body can have a speed as great as c, the speed of light in free space.
5) The maximum energy, E, which can be obtained from a body at rest is $E=mc^2$, where m is the rest mass of the body.

Laws numbered 1 and 3 seem fairly safe, but let us hesitate and take another look. Actually, law number 3 is only valid (now) from a relativistic viewpoint; and for that matter so are laws 4 and 5. But relativity completely revised these physical concepts after 1915, before then Newtonian mechanics was supreme! We should also note that general relativity has not yet been fully verified. Thus we have the peculiar situation of five laws which appear to deny the possibility of intelligent alien control of UFOs, yet three of the laws are recent in concept and may not even be valid. Also, law number 2 has not been tested under conditions [of] large relative speed or accelerations. We should not deny the *possibility* of alien control of UFOs on the basis of preconceived notions not established as related or relevant to the UFOs [emphasis].

 [Author's Commentary: A questionable truth of our scientific laws reveals that although strong modal words may be used, such as "must" or "cannot " in regard to how phenomena behave, the laws have been superseded historically in a way in which, say, the laws of thermodynamics and a causal principle have not been replaced. That is, by a primordial and perennial understanding of the nature of phenomena, once held to be a mere speculative metaphysics, some of the metaphysics may have been confused with modalities; a modal status of propositions that are true in an unusually strong epistemological sense. Developed by both Aristotle and medieval thinkers, modal logic was rediscovered in the twentieth century and is now evoking much excitement. For example, if there were a conflict between relativistic physics and the laws of thermodynamics, the thermodynamic laws would be less in question than the physics in terms of this modal logic. In other words, in considering an analogous case, the claim that it's impossible that Sally was born when there were not prior biological processes is more certain — has a stronger epistemological status — than ordinary laws in biology the denials of which, all things equal, we might say are false but not physically impossible. In short, these points support the Air Force Academy Manual in a significant way. The points counter cavalier dismissals of a possibility of alien UFOs on the ground that certain laws, in somehow being necessarily true, render impossible interstellar visits.]

A causal proof of God reveals a religio-scientific belief that may be common to terrestrials and extraterrestrials. The belief would bode well for how a more advanced ET might treat us in virtue of its more mature coordination of scientific truth to ethics and politics.

Does ET Share Our Religio-Scientific Belief?

The case for ET visiting Earth by a physics text of the US Air Force Academy suggests that some extraterrestrials have a rational nature, not only in reasoning

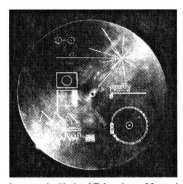

In accord with the AF Academy Manual that ET's science could be far beyond our theories, NASA's Voyager 1 and 2 spacecraft were launched into space with a gold anodized record locating Earth for ET's visit and providing other data. In the 25th year after their launch in 1977, the probes were still sending data on their surroundings through the Deep Space Network (DSN). We may hope, for reasons noted soon, that ET has a naturalistic ethics based on Nature's God!

about science but about religion. Is religious belief illogical? Suppose that logic could be used to prove a God of Nature, common to a natural theology of Judaism, Islam and Christianity, who created ET and humans as well as the world. This prospect is no more farfetched than using physics, as did physicist Dr. Frank Tipler in *The Physics of Immortality*, to prove that a resurrection of the dead is physically possible.[2]

Surely, there is no more reason to dismiss proofs of logic in a work for the general public than proofs of physics in Tipler's popular book. Both books bear on the public's attitude as much as its understanding in which theology is often viewed naïvely *as* a mere matter of irrational faith. Given that these proofs are not based on faith and are rooted in the experienced world, it is consistent with natural — if not a supernatural — theology to hold that the proofs pertain to a world of aliens. Astonishingly, aliens and humans may believe in a common Creator. Alleged references by aliens to a Creator, understood as either a First Cause or Nature's God, range from "There is only one"[3] and "God is universal" to "a Superior Intelligence governs the universe" and "Disregarding the laws of Nature is disregarding the laws of God."[4]

The proof of God, outlined in chapter 2, bears on a system of religio-scientific *belief* that may be shared with ET. One proof leans on Thomas Aquinas' reasoning to a First Cause of the world.[5] Where the world is a universe of second causes, the proof deals with possibilities in modal logic. This logic is used with evident success in terms of the *physically impossibility* that there is a universe ($\sim S$) when there is no First Cause ($\sim F$).

A First Cause in the semantics is a purely voluntary cause that, including will, purpose, and intelligence, is not itself caused as are second causes that are both voluntary and natural. The classic mistake that a First Cause must itself be caused is made by Theodore Schick who, beyond his error of holding that UFO belief is irrational in *How to Think About Weird Things*, commits a *straw man* fallacy when he states: "If everything has a cause other than itself, then god must have a cause other than himself."[6] Those who skew the proof in this way usually also espouse merely a truth-functional logic that permits only a *logical possibility* that S when $\sim F$. Modal logic accepts this possibility

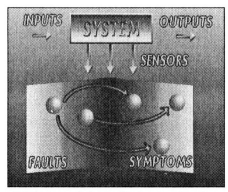

NASA's Johnson Space Center depicts a fuzzy logic for fault detection, isolation and recovery in engine and Shuttle S-Band Diagnoses on space flights. Though fuzzy and modal logic evolved in the 20th century to handle inadequacies of classical and mathematical logics, modal logic was discovered by Aristotle and developed in the medieval period up to Thomas Aquinas. While mathematical logic was used to show that Thomas' proof of a First Cause was invalid, the claims of invalidity committed a fallacy of *appeal to ignorance (argumentum ad ignorantiam)* since that logic could not formally capture causal relationships — a point made by Ludwig Wittgenstein's former student Professor Peter Geach, who was awarded the order *Pro Ecclesia et Pontifice* by the Holy See, as early as 1969 in *God & Soul*.

but not the *physical* one. Acceptance of the latter would violate a basic nature of second cause phenomena as would, by analogy, denials of the laws of thermodynamics and a causal principle presupposed by scientific inquiry. And in noting that this understanding renders coherent a phenomenon of the purposive inquiry at hand, the physical impossibility can be recast as a necessity: 'Necessarily if ~*F*, then ~*S*.'

Thomas' other reasoning in the proof may comprise a *reductio ad absurdum* for this conditional in virtue of a denial that leads to absurdity: *S* when ~*F* begets a regress of second causes that, itself being a phenomenal totality, either causes itself or comes into existence from nothing and nowhere. In taking the conditional 'Necessarily if ~*F*, then ~*S*' with *S*, we infer *F*. This reasoning is valid and evidently sound by way of a modal analogue for the inference rule *modus tollens* and finds more formal expression as $\Box\,(\sim F \rightarrow \sim S) =_{df} (\sim F \Rightarrow \sim S)$, where $\Box$ reads 'Necessarily', $\Rightarrow$ denotes 'Necessarily if...then', and $\vdash$ means 'therefore':

$$\sim F \Rightarrow \sim S,\ S \vdash F$$

Having made these points, there is a version of the Lewis conditional in modal logic that offers another approach to the proof.[7] Its formulation, a reader uninterested in further technicalities can skip, is as follows: There are truth conditions that exclude vacuous truth and that, among other things, do not require specifications of the sorts of possibilities. The antecedent of the conditional provides them since not all practically, physically, or logically possible universes are considered but only those most like our own where the antecedent is true. The main point is that the conditional premise 'If there is no first cause (~*F*), then there are no second causes (~*S*)' is true if and only if the following conditions hold, where $\Box \rightarrow$ reads "if... then" for $\sim F\ \Box \rightarrow \sim S$:

1. There must be some universe u where there is no first cause, and
2. There is no universe at least as possible relative to our universe as u is where there is no first cause but there is a second cause.

If these conditions do hold and there are second causes in our universe, the universe must also have a first cause:

$$\sim F \, \square \rightarrow \sim S, \ S \vdash F$$

The conditions hold and, therefore, Thomas' argument is sound.

On the one hand, he would presumably reject a universe u in clause 1 where there is no first cause. A lack of this cause stems from truth conditions that exclude vacuous truth. The conditions favored by Lewis that permit it, eliminate the clause by permitting the truth of Thomas' conditional even if there is no possible universe without a first cause. But in terms of Lewis' semantics his proof becomes a kind of trivial inference from the necessary existence of God as a first cause of all possible universes.[8]

On the other hand, nevertheless, this version of the conditional results in the soundness of his proof for our universe: There is a God who created our universe and any planet in it on which intelligent life evolved.

Also, although $S \, \square \rightarrow F$ is not inferable by contraposition from $\sim F \, \square \rightarrow \sim S$ in this logic, it can be used independently for the conditional 'If there are second causes (S), there is a first cause (F)': 1. there is a universe u where there are second causes, and 2. there is no universe, as possible relative to our world as u, where there are second causes but no first cause. Clause 1 is already true of our world and it is unchanged in 2 where there is a first cause. There is this cause since, as for the physical necessity that events are caused, there is no physical possibility that second causes cause themselves. And if a first cause was caused for a causal series ad infinitum, a second-cause totality would be uncaused. Thus there is an evidently sound proof with no remote universe u without a first cause. Indeed, sequent '$S \, \square \rightarrow F, \ S \vdash F$' better captures a Thomistic *theology of science* since any uncaused universe u could come from nothing, would be unlike all the phenomena that compose it, and would not be subject to scientific inquiry.

While this modal logic is subject to questions, say whether a 'universe' is primitive with a modal operator 'possible' since there are implicit 'possible universes', not either this logic or Thomas' modality — that does not have this problem— seem as problematic as a truth-functional logic that ignores impossibilities other than logical ones and cannot allow for the truth of either

Did a First Cause *cause* the Big Bang from a singularity or series of them, in which physical systems evolve into biophysical systems and they into psycho-biophysical systems of extraterrestrial and terrestrial life? In "Can Everything Come to Be Without a Cause?" (*Dialogue* 1994), Quentin Smith states: "Since definitions of causality often make explicit or implicit reference to laws, it is natural to suppose that, if there is only one completely lawless thing [a singularity], this thing will also be the only thing exempt from causality." But laws *refer* to a causal principle and not it to the laws. Laws both arise by a scientific inquiry that presupposes the principle, since otherwise the inquiry would usurp itself by admitting of events that occur for no reason, and presuppose causal regularities for coordinating predicted states of a physical system to its present state. However, the principle does not imply a given law. And while there are no lawful 'states' of a singularity that are known since it is devoid of anything subject to current theories, it is a phenomenon of energy (E) correlated to the universe's mass (m) by Einstein's physics ($E = mc^2$). This fact in itself renders relevant scientific inquiry to a study of the singularity. Thus although it is logically possible that the singularity is not subject to the inquiry, this possibility is not reasonable to suppose. And future inquiries, that are not reasonable to abandon, presuppose a physical impossibility that the singularity caused itself or came from nothing. The presupposition is not a metaphysical truth, defended intuitively as Professor Smith suggests. Rather it is a 'modal truth'.

the thermodynamic laws or causal principle. In short, the foregoing reflects millennial developments of modal reasoning from Aristotle to Aquinas and its rediscovery in the twentieth century. And it relates to issues that are disparaged naïvely as *weird* under a tedious veil of "critical thinking" in modern philosophy.[9] Our thinking situates a proof of God in a novel context. It gives expression to a system of religio-scientific belief that is common to both terrestrials and extraterrestrials. Moreover, the thinking may bode well for how more advanced aliens might treat us in virtue of their better coordination of science to ethics.

That is, ethics is rooted in a psychobiological nature of intelligent beings. Given a First Cause, Creator, or God of Nature, there is no *naturalistic fallacy* in which an 'ought' is inferred from an 'is' since their natures *are* as they *ought* to be. Thus science bears logically on how our natures ought to be fulfilled. But the fulfillment may be more fully inferred by ET from more mature sciences over much longer spans of time. The time could be tens of thousands or even millions of years and, to utter an understatement, ET's understanding of society could be light years in advance of ours. And thus its inferences from scientific truth to truths of ethics and politics, that would be analogous to the method of a Greek and medieval naturalistic ethics, would be much more eruditely

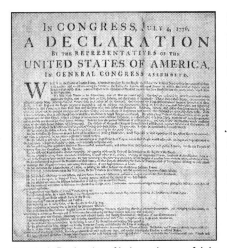

America's Declaration of Independence, of July 4, 1776, refers to Nature's God as a Creator for certifying rights that would hold independently of ET. *Pari passu* ET would have rights that hold independently of humans, despite reference to humans and Earth in the document: "WHEN in the Course of human Events, it becomes necessary for one People to dissolve the Political Bands which have connected them with another, and to assume among the Powers of the Earth, the separate and equal Station to which the Laws of Nature and of Nature's God entitle them, a decent Respect to the Opinions of Mankind requires that they should declare the causes which impel them to the Separation. WE hold these Truths to be self-evident, that all Men are created equal, that they are endowed by their Creator with certain unalienable Rights, that among these are Life, Liberty, and the Pursuit of Happiness...."

fathomed and better applied than our primitive inferences to social-political institutions. In point of fact, these institutions were skewed by fascism, communism, and an unfettered capitalism. And due to a *naturalistic fallacy* that ignores both a Creator and the modal reasoning for relating our nature to a politically institutionalized ethics, those pathological views have also politicized ethics and science.

There may be less to fear from ET than from our own political ideologies. Race, gender, and class strife has made the last century the most murderous era in human history; the ideologies provide an apologetics for ethnic cleansings and terrorism that an otherwordly intelligence may ameliorate.

The coordination of science to ethics by an alien species does not rule out, anthropocentrically, a species that is anomalous to us biologically and culturally.[10] But why rule out, dogmatically, an alien race with a similar biology and contact with them? Even if we contact a dissimilar advanced race, there is reason to suppose that this race would both expect and respect a naturalistic ethics of other intelligent beings whose nature evolved from the same origin of Nature and its laws. Again, one alien species allegedly warned that "Disregarding the laws of Nature is disregarding the laws of God."[11]

Politically, these words are not *alien* to Americans. Their own Founding Fathers wrote a Declaration of Independence that rooted moral rights in "Laws of Nature and of Nature's God." And given the above proof of God — despite contrary assumptions conditioned by our modern culture, it would be reasonable to suppose that both humans and a more advanced species have a rational nature *in common* which evolved from Nature's God. Presumably, if a morally relevant scientific reasoning about the cosmos is sound, the soundness would hold for rational species of different cultures even if the cultures had species who evolved in either diverse parts of this universe or universes that are similar. Clearly, the universal nature of the foregoing reasoning and its novel integration of theology, ethics, and science complements a Cosmological Principle and provides a novel justification for optimism in regard to future contacts with extraterrestrial civilizations.

A Final Sign of Hope?

In addition to aspirations for a religio-scientific belief system that ET and humans may share, one hopeful sign in the civilian world was the *Washington Post* article "Report Urges Study of UFO Sightings" on June 29 1998.[12] While this article was studiously ignored by all major television stations and began with the disclaimer "supposed UFO sightings," it was the first recognition by elite international scientists that the sightings are of real airborne phenomena — if not of extraterrestrial craft. Also, it is the first major news report to take them seriously, although in this case there was little choice.

"Some supposed UFO sightings have unexplained physical evidence," states the article, "that deserves serious scientific study, an international panel of scientists has concluded." Predictably, it adds that "In the first independent scientific review of the controversial topic in almost 30 years, directed by physicist Peter Sturrock of Stanford University, the panel emphasized it had found no convincing evidence of extraterrestrial intelligence or any violation of natural laws."

15 witnesses, including police, observed a 200-foot long luminous object that blocked roads and interfered with car ignitions and electrical systems before departing. Being one of the most important cases of electromagnetic effects, the effects were noted in 1998 by Stanford University Professor of Physics Dr. Peter Sturrock who said that UFOs warrant urgent funding for scientific study. The sighting was at Scarborough, NY, 10 Oct 1966, Blue Book Case 11073.

But are laws of theories known to be absolutely true? By reasoning from past theories that were falsified in certain domains, as was classical physics in a domain of bodies approaching the speed of light, a pessimistic induction holds wherein it is rational to doubt a theory's unqualified truth. This truth is tacitly doubted in the case at hand when the panel cites *"inexplicable* details, such as burns to witnesses, radar detections of *mysterious* objects, *strange* lights appearing repeatedly in the skies over certain locales, *aberrations* in the workings of automobiles... and other damage in vegetation [emphasis added]."

⅄

In holding that "the scientific community might learn something if it can overcome the fear of ridicule and get funding for research to explain these occurrences," the report has a double message: Governments need not be alarmed that scientists seek their secrets and there are no secrets that will explode our science.

⅄

Accordingly, how can the panel conclude both that there is no evidence of alien intelligence and that there is no violation of any scientific laws? The laws, known or unknown, afford technology. Technology either does or does not cause the evidence. If the evidence *is not* caused by technology, there are natural phenomena anomalous to any known natural laws. And if the evidence *is* caused by technology, then the panel virtually admits that there are craft which do not originate on earth.

Since the panel confirms inexplicable details that include strange lights, mysterious objects, aberrations in car engines, radiation detection, and damage to witnesses as well as to vegetation, a conclusion cannot be drawn one way or the other by the admitted evidence. Also, the panel ignores a host of reliably reported technological artifacts that cannot be explained by our science, not to mention well regarded witnesses who claim contact with aliens themselves. Former Air Force Intelligence Officer Maj. George Filer, has witnessed the phenomena firsthand and stresses the paratypic technology.

Maj. Filer states that UFOs "defy the laws of physics, make turns on a dime, fly at high speed and suddenly stop, hover, and launch like a rocket with clocked speeds of thirty thousand miles per hour."[13] In short, even if other evidence for the technology was ignored, the fallacy of *appeal to ignorance* (*argumentum ad ignorantiam*) is committed when the panel concludes that the cited effects are not caused by a technology that may be a product of scientific laws which are known by an alien intelligence.

The intelligence cannot be reasonably discounted. In consisting of scientists from prestigious universities, the panel of scientists is poignantly aware of this fact. Thus it may be supposed that the panel's report is partly a

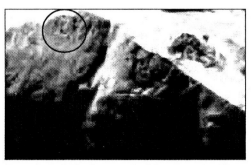

Has science fiction become fact? Is, or was, intelligent life that originated on Mars a source of some UFOs? "These images from the NASA/JPL Mars orbiters and rovers," notes US Air Force Maj. George Filer, has both "interested experts from around the country and helped point the way to more discoveries. It is my opinion, confirmed by several others, that there are alien symbols on the Martian rock nicknamed the 'E Rock'." Maj. Filer adds that "on the left upper corner is a symbol that appears to be similar to our capital letter E. The top of the E may be closed like a P. A second symbol similar to our letter G is also apparent."

stratagem, similar to past official UFO reports. Historically, they were designed simultaneously and paradoxically to both *salve* the public and condition it to a reality of alien UFOs that are officially denied. Paradoxically, there is clear physical evidence of UFOs behaving in ways that are a virtual mystery in terms of our laws. At the same time, our laws are not violated and exclude a technology of UFOs that may be extraterrestrial. The scientists admit of mysterious and inexplicable evidence. But again, at the same time, they are saying that the public should not worry because the evidence is not extraterrestrial.

Notwithstanding such ambiguous and inconsistent remarks, the public *will* undoubtedly become conditioned to being more aware of a credibility of UFO reports and not be suddenly traumatized when the day comes that UFOs are admitted to be products of a highly advanced alien technology. However, beyond the intention of gradually 'raising a consciousness' of the public, the report is evidently concerned with both scientists and government cover-ups in the same ambiguous double-message manner.

In holding that "the scientific community might learn something... if it can overcome the fear of ridicule associated with the topic [of UFOs] and get some funding for targeted research to... explain these occurrences,"[14] the report has a double message: Governments need not be alarmed because scientists are not going after their secrets and there are no secrets that will explode a conventional view of scientists. The panel, that is, has not 'gone off the deep end.' These words, of course, were used to ridicule Harvard psychiatrist John E. Mack by one of his colleagues.

A between-the-lines message of this news is obvious when it states that we should "carefully evaluate UFO reports to extract information about unusual phenomena currently unknown to science," with a reassuring caveat that "research could also improve understanding of or debunk, supposed UFO events."[15] Apparently, inexplicable UFOs, which paradoxically do not violate any known laws, are merely reminiscent of historical phenomena which were dismissed as either "folk tales" or "meteorites and certain types of lightning."[16]

A

In regard to *science*, why should the scientist be at odds with religious believers by a belief in such extraterrestrials? Many scientists may suppose that UFOs cannot be extraterrestrial, but the supposition involves philosophy as much as science.

Y

In the end, the report is hopeful because it both admits and laments that a serious study of UFOs has been ridiculed by the scientific community. And it states that this community should not relegate the phenomena to either mythology or folk lore about such things as fairies, leprechauns, and witches. At the same time the report is disappointing because the panel of scientists, in toning it down and making it ambiguous, do themselves clearly give in to a fear of ridicule. The ridicule, of course, brings to mind Col. Corso's reference to military projects that "all... had something to do with alien technology. But no one ever knew it." Hence, "when reporters were actually given truthful descriptions of alien encounters, they either fell on the floor laughing or sold the story to the tabloids... Again, everybody laughed."[17]

On the one hand, the report admits of a whole cascade of technological effects of UFOs that are both entirely inexplicable by conventional science and qualitatively different from the meteorites and lightening phenomena of an earlier time to which the panel disingenuously compares this new aerial phenomena. Though the previous phenomena were anomalous, they could be viably addressed by post-Copernican scientists who could explain them in terms of modern science without an extraterrestrial hypothesis. Today, a glib disregard of this hypothesis in view of a landslide of credible reports, from official documents and military pilots to civilian witnesses and eminent scientists, is not only scientifically untenable but intellectually dishonest.

On the other hand, the report encourages the very studies that, as the panel surely appreciates, could lead to a discovery of extraterrestrial UFOs and civilizations. The panel walks a fine line between the embracing arms of UFO researchers and alarmed governments that have secretly taken UFOs seriously for over sixty years. And who knows? Maybe the report has a blessing of the governments. This question about governments brings us to a point at which some implications for politics, God, and science can be summarized.

249

Politics, God, and Science

In terms of *politics*, there may be less to fear from a specter of ET than from political ideologies. Evolving most recently during the Cold War with an aftermath of ethnic cleansings and terrorism that pits one murderous group against another, the ideologies continue to foster divisive interpretations of the world. The worldviews are at odds with a biological fact that we are one human race. Dr. Irving Louis Horowitz, Hannah Arendt Distinguished Professor at Rutgers State University, notes that liberal values have been distorted by a divisive "counting of race, gender and class (running a distant third with the demise of Marxism as a universal faith)."[18]

Having seen how this faith in blind ideology was exalted

There is a long history of photographs and reports that prompted Stanford Physics Professor Peter Sturrock to head a panel of scientists to investigate UFOs. *Above*: UFO photo given to the US Air Force over the Capitol in Washington, D.C., on 4 February 1959 by A.S. Frutin. *Below*: Enlargement.

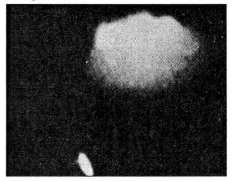

unscientifically while a scientific study of UFOs was derided, the prospect of aliens provides an unprecedented opportunity to avoid a politicization of our major religions by nurturing their more inclusive insights. Is it not possible that human beings as well as intelligent extraterrestrials, if we both have the capacity to love unconditionally, are made in the image of God?

In terms of *God*, it is consistent with Scripture to say that if ET is intelligent, the Creator could have created *it* with a capacity to discover truth in religion and science. And in regard to *science*, there is no reason why most scientists need be at odds with religious believers who believe in ET. Many scientists may suppose that UFOs cannot be alien. But this supposition involves philosophy as much as science — metaphysics as much as physics. The SETI Institute reveals that most scientists believe in alien civilizations even if they do not believe they visit us. However, even basic notions in the philosophy of science make clear that alien visits are possible.

Does this amount to "It's possible, therefore it is"? This would be as silly the scoffer's inference "It's impossible, so it isn't." While the latter appeals

NASA will search for extraterrestrials in 2014 by these Planet Finders. If these Finders can find ET, then ET can find us. Have we already been found? Is contact desired with a morally undesirable human species? In the film *Starman*, we are likened to cannibals who invite ET to dinner by our Voyager Probe!

naively to an infallibility of our science, the former is a *straw man* that distorts the *real man*: "It *is*, so it's possible." This inference rests on reliable witnesses as well as the history of science wherein what was impossible by former theories became possible by later ones. Aarthus University Physicist Helge Kragh reminds us that, in the nineteenth century, many "physicists believed that the basic principles underlying their subject were already known, and that physics in the future would only consist of filling in the details." He adds that they could not have been more wrong. The rise of quantum, relativistic, particle and solid-state physics "have fundamentally changed our understanding of space, time, and matter." They have, he adds, "transformed daily life, inspiring a technological revolution that has included the development of radio, television, lasers, nuclear power, and computers."[19]

Bringing to mind the country folk in the film *Oklahoma!*, do we now sing "Science has gone about as fer as she can go"! It is not only possible but probable that our future science will afford startling developments such as bypassing or exceeding the speed of light for feasible interstellar travel. And the travel proceeds *pari passu* with both visits here by a more advanced ET whose existence is accepted by the majority of scientists and a scientific assent to cogent cases made for the visits by expert witnesses. To reject them since there is no material evidence, that is evidently false in any case, is to reject testimony for drawing reasonable conclusions in the courtrooms of all open societies that seek the truth.

Notes

[1] Some paragraph indentions were added for clarity and some sentences were deleted for brevity as noted by '...'. Though referring to Rutger's UFO Archive at Rutgers – The State University of New Jersey, this data from the AF Cadet Manual is prefaced by: rkrouse@netcom.com. (Robert K. Rouse) Subject: Air Force Cadet Manual (part 1) Message-ID: <rkrouseCKJBLL.IuC @netcom.com.>Organization: NETCOM On-line Communication Services (408 241- 9760 quest) Date: Tue, 1 Feb 1994 07:32:06 GMT.

[2] See Dr. Frank Tipler, *The Physics of Immortality: Modern Cosmology, God & the*

Resurrection of the Dead (NY: Anchor Books, 1995).

[3] See Timothy Good, *Above Top Secret: The Worldwide UFO Cover-Up*, Foreword by the Former Chief of Defense Staff, Lord Hill-Norton, G.C.B. (NY: William Morrow, 1988), p. 297, emphasis deleted.

[4] Timothy Good, *Alien Base: The Evidence for Extraterrestrial Colonization of Earth* (London: Arrow Books Ltd., 1998), pp. 286, 325, and 431.

[5] For St. Thomas' proof in the *Summa Theologica* (I, 2, 3), see Peter Kreeft, Ed., *Summa of the Summa* (San Francisco: Ignatius Press, 1990).

[6] Professor Theodore Schick, who rejects rational belief in UFOs in *How to Think About Weird Things* (1995), also rejects the First Cause argument for erroneous reasons. His case that the argument is self-refuting because that Cause *itself* must be caused misrepresents the actual proof whose semantics specify that only causes other than the First Cause are causally dependent since they are not purely voluntary. At the same time, a voluntariness of second causes renders coherent the phenomena of human inquiries insofar as free will, supposed by purposive behavior, is a necessary condition for the intelligibility of 'truth' — an important point in light of the fact that humanists, with whom Professor Schick identifies, typically deny free will because it is allegedly inconsistent with a causal principle presupposed by scientific inquiry. If there were only natural causes, as implicit in the modern causal principle (given a so-called 'critical expression' by Hume and Kant), then an entirely determined reality would be the truth-condition for 'truth'. *Pari passu* truth-claims that are logically inconsistent would, incoherently, be equally true because equally caused. Exactly for this reason a naive *philosophical determinism* came to be distinguished from a *methodolocial one*. Whereas the former determinism holds dogmatically that human nature and Nature are exhaustively determined, with no caveat, the latter holds that voluntary realities such as free will are diregarded but not denied for the limited purposes of scientific inquiry. In the medieval philosophic theology going back to St. Augustine — studiously ignored by the humanist heirs of the Humean-Kantian tradition, the reality of free will was defended in terms of our being immediately, incontrovertibly and phenomenologically aware of our will. Indeed, Martin Heidegger's phenomenology echoed this point when he noted that unless we are *free from* a deterministic spatiotemporal realm, there can be no scientific truth — the very 'truth', ironically, that atheistic humanists champion exclusively! And a dilemma of this exclusive appeal to scientific truth is exacerbated by the denial of a First Cause since if there is not this Cause, the cosmos could have caused itself or come from nothing. In these cases the cosmos would be fundamentally unlike all the phenomena that compose it and not be subject to scientific inquiry. In short, the humanists are faced with a *constructive dilemma*: If there is no First Cause, then the humanists who exalt scientific inquiry must admit that this inquiry does not bear on the cosmos and if there is a First Cause, then they must admit of a Creator! One might almost feel sorry for their dilemma, except for the dumbing-down scoffer influence they have had on students and society in the name of 'science'. See Bernard Haisch's letter to his colleagues, in this regard, in Appendix I. See Schick's "The 'Big Bang' Argument for the Existence of God" in *Philo: J. Society of Humanist Philosophers* (1998) at www.infidels.org/library/modern/theodore_schick/bigbang.

[7] This paragraph and the next are indebted to my correspondence 10 May 1999 with Prof. John Nolt, though I am responsible for using $S\square \rightarrow F$ as a solution to an anomalous universe

u where ~*F* obtains. See Nolt's chapter "Kripkean Modal Logic," in *Logics* (London: Int. Thomson Publishing Inc., 1997), pp. 351-356. See also my articles "Thomas' 2nd Way: A Defense by Modal Scientific Reasoning" in *Logique et Analyse: Belgium National Center for Research in Logic* 146 (1994) 145-168 and "A First Cause and the Causal Principle: How the Principle Binds Theology to Science" in *Philosophy in Science* X, edited by William Stoeger, S.J., Ph.D. Astrophysics, University of Cambridge (Kraków Papal Academy of Theology, Vatican Astronomical Observatory & University of Arizona), December 2003, 7-35. In these articles I note that some of the most seemingly serious criticisms of the First Cause argument have confused modalities, that have truth-value, with a truth-valueless metaphysics — bringing to mind Yale University Professor Ruth Barcan Marcus' sarcastic quip "No metaphysical mysteries" in *Modalities* (Oxford University Press, 1993), p. 69.

[8] Professor Nolt, correspondence, May 10, 1999.

[9] See Theodore Schick and Lewis Vaughn, *How to Think About Weird Things: Critical Thinking for a New Age* (CA: Mayfield Publishing Co., 1995). A euphemism of "critical thinking" stems from the empiricism of David Hume. An upshot of his influence is a rejection of any "necessary" causal connection between events. By ignoring a modal *necessity*, there was a rejection not only of reasoning to a First Cause, which relates theology to science, but of having scientific knowledge since there would be no knowledge *a priori* of a causal principle presupposed by scientific inquiry. For a discussion of this dilemma, see my work *From Physics to Politics* (Transaction Press – Rutgers, The State University, 1999).

[10] These considerations are entangled with others. One might object that it is anthropocentric to expect that aliens are like us—much more to have a rational nature or similar religio-scientific belief system. As noted in Ch. 2, however, a principle of homogeneity in general science supposes a uniformity in the structure and composition of the universe. Thus with due respect to many scientists and science fiction writers themselves, it is quite rational to believe that many species of intelligent aliens have a similar psychobiological nature even if it is more or less developed.

[11] Good, *Alien Base: The Evidence for Extraterrestrial Colonization of Earth*, p. 286. The point about the ethics of one given advanced extraterrestrial species would hold *ceteris paribus* for others. In point of fact, most serious researchers believe that UFO visits suggest several species.

[12] See "Report Urges Study of UFO Sightings," *Washington Post*. From *The Cincinnati Enquirer*, June 29, 1998, p. A2. The following quotes are from this article.

[13] US Air Force Maj. George Filer (Ret.), "The Filer Report," from *UFOs are Real: Here's the Proof*, by E. Walters and Dr. B. Maccabee (NY: Avon Books, 1997), p. 7.

[14] "Report Urges Study of UFO Sightings," p. A2.

[15] *Ibid.*, p. A2.

[16] *Ibid.*, p. A2.

[17] Corso, *The Day After Roswell*, p. 78.

[18] Dr. Irving Louis Horowitz, "Sociology & the Common Culture," Sent to author by Dr. Horowitz who presented the paper at his retirement dinner, the Eagleton Institute of Politics, 14 May 1998.

[19] Dr. Helge Kragh, *Quantum Generations: A History of Physics in the 20th Century* (NJ: Princeton University Press, 1999), author's own PUP flier.

Appendix 1

Bernard Haisch's Open Letter to His Colleagues

POSITIONS: *Chief Science Officer* – ManyOne Networks,Inc.(2002-) *Director* – California Institute for Physics and Astrophysics, Palo Alto (1999-02); *Staff Scientist* – Lockheed Martin, Solar and Astrophysics Laboratory, Palo Alto (1979-99); *Scientific Editor* – The Astrophysical Journal (1993-02); *Deputy Director* – Center for EUV Astrophysics, Univ. Calif., Berkeley (1992-94). *Visiting Fellow* – Max-Planck-Institut fuer Extraterrestrial. Physik, Garching, Germany (1991-94); *Editor-in-Chief* Journal of Scientific Exploration (1988-99); *Visiting Scientist* – The Astronomical Institute, Rijksuniversiteit Utrecht, the Netherlands (1977-78); and *Research Associate* – Joint Institute Lab. Astrophysics, University Colorado, Boulder (1975-77 and 1978-79). **EDUCATION:** *Ph.D.* of Wisconsin, Madison, Astronomy (1975); *B.S. with High Distinction* Indiana University at Bloomington, Astrophysics (1971). **PRINCIPAL INVESTIGATOR EXPERIENCE:** *NASA Guest Investigator Programs* (IUE, Einstein, Exosat, ROSAT, EUVE, ASCA, XTE), *NASA Research Program* (Inertia and Gravitation in the Zero-Point Field Model), and *Hardware Program* (AFGL/AURA Program: Solar X-ray Multilayer Telescope). **MEMBERSHIPS:** International Astronomical Union, American Astronomical Society, Fellow of the Royal Astronomical Society, European Astronomical Society, Associate Fellow of American Institute of Aeronautics and Astronautics, Patron of American Association for the Advancement of Science, Astronomical Society of the Pacific, American Association of Physics Teachers, Society for Scientific Exploration, Phi Beta Kappa, Sigma Xi, and Phi Kappa Phi. **BIOGRAPHICAL REFERENCES:** Who's Who in America (52nd edition onward, 1998), American Men and Women of Science, and Who's Who in Science and Engineering. **OVER 130 PAPERS PUBLISHED IN:** Astrophys. Journal, Physical Rev., Physics Letters, Nature, Science, Astron. & Astrophysics, Astronomical J., American J. of Physics, J. Geophysical Research, Pub. of the Astronomical Society/Pacific, Annual Rev. Astron. Astrophys., Found. of Physics, Irish Astron. J., Solar Physics, J. of the Astronautical Sciences, Sky & Telescope, Monthly Not. Royal Astron. Society, Annalen der Physik, Advances in Space Research, Journal de Physique, Comments on Astrophysics, J. British Interplanetary Society, and Conference Proceedings. **BOOKS:** *Solar and Stellar Flares.* B. Haisch and M. Rodono (eds.), Kluwer Acad. Press, (1989); *The Many Faces of the Sun: A Summary of the Results from NASA's Solar Maximum Mission.* K. Strong, J. Saba, B. Haisch and J. Schmeltz (eds.), Springer Verlag, (1999).

Dear Colleagues,

I have been an active professional astronomer since earning my doctorate in 1975. I have published a respectable number of scientific papers in most of the right journals (including our favorites, *Science* and *Nature*), have been Principal Investigator on several NASA studies, have served as referee and proposal reviewer for NASA and NSF, belong to half a dozen professional societies, have chaired international conferences, i.e. I've engaged by and large successfully in all the usual activities of a busy professional scientist.

During my career I have had the responsibility and privilege as an editor of accepting or rejecting somewhere in the neighborhood of a thousand articles in a prestigious astrophysics journal. This does not conclusively prove, but certainly indicates, that I recognize good science when I see it. I have also had the responsibility of accepting or rejecting papers on the UFO phenomenon in a quite different refereed journal, the *Journal of Scientific Exploration* (JSE). For 12 years I served as editor of JSE (as an unpaid public service) because I believe that examining evidence that may challenge prevailing scientific dogma is good for science and a necessary part of searching for the truth. The road of discovery may have 99 dead ends in the thicket for every new path winding its way up the peak, but that is just how it is. Curiosity and tenacity are equal prerequisites for a scientist... as is an open mind.

I have learned quite a bit about the UFO phenomenon over the years (certainly more than I had bargained for) and have met many of the leading figures, some credible, some deluded. When Prof. Peter Sturrock, a prominent Stanford University plasma physicist, conducted a survey of the membership of the American Astronomical Society in the 1970s, he made an interesting finding: astronomers who spent time reading up on the UFO phenomenon developed more interest in it. If there were nothing to it, you would expect the opposite: lack of credible evidence would cause interest to wane. And the fact of the matter is, there does exist a vast amount of high quality, albeit enigmatic, data. UFO sightings are not limited to farmers in backward rural areas. There are astronomers and pilots and NASA engineers — and others who have been around the block a few times when it comes to observing natural phenomena — who have witnessed events for which there is no plausible conventional explanation.

Recently, astrophysicist Ken Olum at Tufts University argued (*gr-qc/0303070*) that anthropic reasoning applied to inflation theory predicts that we should find ourselves part of a large, galaxy-sized civilization, implying that the "We-are-alone" solution to Fermi's paradox is inconsistent with our best current theory of cosmology. Beatriz Gato-Rivera, a physicist at the

Instituto de Matematicas y Fisica in Madrid, followed up on this (*physics/ 0308078*) with the hypothesis that Olum is correct, but that by design we would be kept unaware of a greatly advanced surrounding civilization. She also argues that modern superstring and M-brane theory further aggravate Fermi's "missing alien" problem.†

There is another aspect to the UFO phenomenon that involves politics and secrecy rather than observational evidence. I do not currently have a ticket to any SCI program, but over the years I have gotten to know individuals who for one reason or another would be aware of the existence of relevant black programs. From such sources, certain possibilities have made it through my credibility filter and now reside — like Schrödinger's cat — in kind of an unresolved mental superposition of quantum states having both the eigenvalues "true" and "false" and no operator around to collapse the wave function. My credibility filter is a function of several parameters such as my own knowledge of physical laws, state of technology and history of its origin, some personal experience with government agencies and security classification systems, but mostly the filter is tuned to the questions: Which people have I learned over the years to be trustworthy, sensible and knowledgeable? How would they be in a position to know the things they do? Why and to what extent would they tell me anything, even based on longtime friendship? Do they have anything to gain by telling stories or making claims? What consistency and convergence is there among various people's claimed information?

I see myself a bit like the kid standing next to the kid looking through the hole in the big tall fence at the baseball game. This means that the closest I am getting to inside information will be a recounting of what is going on in there. I myself am definitely not an insider, but contacts I have acquired and/ or befriended over a long period of time seem to be on the periphery of some kind of inside which appears to contain at least remarkable information, and apparently more than that.

Let me be (somewhat) more specific. I now have three completely independent examples of individuals whom I trust reporting to me that individuals they trust have admitted to handling alien materials in "our" possession in the course of secret official duties [emphasis]. (The special access level in the one case for which I know it is R, a not widely known SCI level whose existence was finally verified for me by someone who himself had a very high access level, though short of that one, as being "reserved for someone at the very top." I do not know, however, whether it is specifically reserved or designated for this topic.) *And in yet two mores cases, I am similarly one (trustworthy) step removed from a former head of a federal government agency who was involved with a special access program reporting decades-*

long extraterrestrial reverse engineering efforts and a head of state of a G8 country who also said he had been officially briefed on that program [emphasis]. Now the Air Force Project Blue Book of the 1950s and 1960s did have both a public and a classified side. I suspect that after the public half of Blue Book closed up shop following the Condon Report, its classified half continued, existing today as a black special access program.

Could such things possibly be true? While I am intrigued by what I have learned over the years, I can't be absolutely certain. It is interesting that from the clandestine intelligence-world perspective, the scientific community — for all of its technical and theoretical sophistication — is viewed as remarkably naive in certain respects. *We scientists tend to think that we know better than anyone else what is possible and what is impossible, and that we of all people could surely not be kept in the dark for very long. Over the course of time I have learned how it would indeed be possible to maintain decades-long secrecy on this topic and why this might be justified, concepts I myself once dismissed* [emphasis]. (See "Black Special Access Programs," also "Some Thoughts on Keeping It Secret." And for some insight on the origin of this situation see the book UFOs AND THE NATIONAL SECURITY STATE: AN UNCLASSIFIED HISTORY. VOL. 1: 1947-1973 by Richard Dolan; also THE MISSING TIMES by Terry Hansen that documents the history of ties between the national media and the intelligence community. I am aware that these two books have been criticized for overreliance on secondary sources. More scholarly work is available, such as that of Jan Aldrich, but I think that Dolan and Hansen present a useful and eye-opening introduction to the situation in general, especially for someone first approaching this topic.)

The above is, of course, short of any kind of proof, but all in all I have now gotten to the point in my exposure to the subject at which I think it somewhat more likely than not that something not merely delusional, but real and important may be going on in regard to the UFO phenomenon. If so, I would like to discover what it is, or what the ensemble of phenomena are if it is a multiplicity of things. My estimation of the probable reality of the subject puts me somewhere between the majority rejectionist view of the mainstream scientific community and the majority accepting view of the general public (depending on how the issue is presented in opinion polls).

I propose that true skepticism is called for today: neither the gullible acceptance of true belief nor the closed-minded rejection of the scoffer masquerading as skeptic. One should be skeptical of both the [naive] believers and scoffers. The negative claims of pseudo-skeptics who offer facile explantations must themselves be subject to criticism. If a competent witness reports having seen something tens of degrees of arc in size (as happens) and the scoffer — who of course was not there — offers Venus or a high altitude weather balloon as an explanation, the requirement of extraordinary proof for an extraordinary claim falls on the proffered negative claim as well. That kind of approach is also pseudo-science. Moreover, just being a scientist confers neither necessary expertise nor sufficient knowledge. (I wish it did, *sigh.*) Any scientist who has not read a few serious books and articles presenting actual UFO evidence should, out of intellectual honesty, refrain from making scientific pronouncements. To look at the evidence and go away unconvinced is one thing. To not look at the evidence and be convinced against it nonetheless is another. That is not science. Do your homework!

This [letter] is a work in progress. It is certainly no statement of any 'truth' but, in that regard, it is worth keeping in mind something Winston Churchill once said on that topic: "Men occasionally stumble over the truth, but most pick themselves up and hurry off as if nothing had happened."

Bernard Haisch*
March 10, 2004
Palo Alto, California
(www.ufoskeptic.org)

† *Editorial note:* Astronomer Ian Crawford's diffusion model for ET led to its galactic colonizations in 5 to 50 million years (*Sci. Am.*, Nov. 2000), a fraction of our Galaxy's age. Even if only a few alien civilizations arose in the last 10 billion years, most of our Galaxy would be colonized by now. The famous physicist Enrico Fermi referred to this when he asked "Where are they?" *Cf.* astrophysicist Eric Davies, www.ufoskeptic.org/davis.

* *Editorial note:* I take the liberty of noting that, in reply to my request to use this letter, Professor Haisch stated: "Yes, go ahead.... Doing this is certainly not going to help me professionally, but is the right thing to do." His reply, reminiscent of Harvard Psychiatrist John E. Mack's harassment because of his inquiries into UFO abductions, is a pitiful commentary on the academic community. Indeed, the famous Russian and Middle East scholar Dr. Daniel Pipes, who taught at Harvard, the University of Chicago and U.S. Naval War College, has noted repeatedly the *cowardice* and *vanity* of a sizeable percentage of American professors. One can scarcely imagine a more perverse set of vices for corrupting the essential mission of higher education that alone justifies tenure: the pursuit of truth. But a studious neglect of truth makes sense if it threatens to unravel the status quo by unprecedented mind-boggling possibilities, worsened by a fearful pandering to academic Don Quixotes for whom 'truth' is relative to different races, genders and cultures in terms of a multicultural political correctness. Here, what is 'correct' can be empirically false. Its being 'false' *a priori* — prior to any investigation — that a superior extraterrestrial civilization is visiting Earth would both exclude any evidence at odds with anthropocentric political agenda and boost a professorial 'self-esteem': a deluded antidote to a humbling cultural trauma of Copernican, if not cosmic, proportions. In this regard, see the scathing criticism of Harvard Professor Harvey Mansfield at the outset of this book in "Emerging Awareness of the Problem."

Appendix 2

Press Release – "Spread of Life Through the Galaxy"

Contact: John McFarland
Armagh Observatory
Tel : 028-3752-2928
Fax: 028-3752-7174

College Hill
Armagh
Northern Ireland
BT61 9DG

Armagh Observatory

Press Release

SPREAD OF LIFE THROUGH THE GALAXY

Armagh Observatory, 31 October 2003: Astronomers at Armagh Observatory and Cardiff University have independently discovered closely similar mechanisms by which micro-organisms may have spread throughout the Galaxy. Scientific papers on the topic by Professor Bill Napier at Armagh, and by Dr Max Wallis and Professor Chandra Wickramasinghe at Cardiff, will appear simultaneously in the Monthly Notices of the Royal Astronomical Society. The discovery of these new interstellar routes for transmission of micro-organisms strengthens the view that life did not originate on Earth but arrived from elsewhere.

It is known that boulders and other debris may be thrown from the Earth into interplanetary space as the result of collisions with asteroids or comets, and that micro-organisms within the boulders could survive the enormous accelerations involved. Life could easily have crossed the few astronomical units separating Mars and Earth in this way. To colonise the Galaxy, however, thousands of light years must be traversed. These enormous distances have always seemed an insurmountable barrier because of the lethal effects of cosmic radiation and the low probability that an ejected boulder would ever land on a planet in another star system.

However, Napier finds that collisions with interplanetary dust will quickly erode the ejected boulders to much smaller fragments and that these tiny, life-bearing fragments may be thrown out of the solar system by the pressure of sunlight in a few years. The solar system is therefore surrounded by an expanding 'biosphere', fifteen or more light years across, of dormant microbes preserved inside rock fragments. In the course of Earth history there have been a few dozen close encounters with star-forming nebulae, during which microbes will be injected directly into young planetary systems. A single microbe falling onto a receptive planetary surface could populate the planet within a year. If planets capable of sustaining life are 'sufficiently common in the Galaxy, Napier concludes that this mechanism could have populated over 10,000 million of them during the lifetime of our Galaxy.

Wallis and Wickramasinghe have identified another delivery route. They point out that fertile ejecta would, on impact, bury themselves in the radiation-shielded surface layers of frozen comets. A belt of such comets, the Edgeworth-Kuiper belt, lies beyond the planetary system. This belt gradually leaks comets into interstellar space, some of which will eventually reach proto-planetary discs and star-forming nebulae. There they are destroyed by sputtering, releasing the trapped micro-organisms and seeding the formative planetary systems.

ECHOING AN ARTICLE by the famous astronomers Sir Fred Hoyle and Chandra Wickramasinghe, "The Unity of Cosmic Life and Inevitability of Evolved Life Forms" in the *SPIE Proceedings of the 2nd International Conference on Optical SETI* (Vol. 4273-01 in 2001), this Public Press Release from the UK's Armagh Observatory on 31 October 2003 states that "astronomers at Armagh Observatory and Cardiff University have independently discovered closely similar mechanisms by which micro-organisms may have spread throughout the Galaxy. Scientific papers on the topic by Professor Bill Napier at Armagh, and by Dr. Max Wallis and Professor Chandra Wickramasinghe at Cardiff, will appear simultaneously in the *Monthly Notices of the Royal Astronomical Society*." In challenging the predominate theory of evolution that human life started on Earth, they add that "life could easily have crossed the few astronomical units separating Mars and Earth in this way" and that this "mechanism could have populated over 10,000 million [planets] during the lifetime of our Galaxy." This publication supports the hypothesis of physicist Beatriz Gato-Rivera at the Instituto de Mathematicas y Fisica in Madrid, in response to physicist Enrico Fermi's "missing-alien" problem, that by design human beings might be kept unaware of the surrounding civilization of a greatly advanced extraterrestrial intelligence. Given that an intelligence of humans has been used homicidally against humans throughout their history, culminating in a twentieth century that has been called the most murderous of all centuries, there can be little wonder why contact would be avoided. A cooperative contact with humans in which they acquired advanced technologies would be analogous to giving a loaded handgun to a child — an immanent danger not only to the child and other children but to the adults (the extraterrestrials in this analogy). See "33.4 Human Fear and Hostility," by the U.S. Air Force Academy, and the author's commentary on page 235.

Appendix 3

A Space-Craft Operator's Stargate Chronicles

The following is one of several incidents from "The Stargate Chronicles" — a small part of a far greater story, but with them, the Big Picture begins to unfold.

Dr. Von Braun at Roswell

By Clark C. McClelland, former ScO, Space Shuttle Fleet, Kennedy Space Center (1958-1992): The Real Fox Moulder?

IN 1947 A controversial event took place in New Mexico near the town of Roswell. The "Roswell Incident," as it is now called, remains the paramount case in UFO crash / retrieval history. In addition to the claims of a downed alien ship, alien bodies were said to have been recovered from the debris. The USAF and Federal Government now maintain steadfastly that the object in question was a high altitude balloon project, code named: Mogul. The project was designed to detect nuclear blasts in the USSR. The bodies that were recovered, according to the USAF, were parachute test dummies that had been released high above the desert, and had eventually drifted into the "balloon" crash area. The USAF finally settled on this fabricated version of

ScO Clark McClelland in the cockpit of the space shuttle.

events and passed it off to the American public as truth.

During my long years of service in our national space program, I was very fortunate to come to know and exchange some very exciting data with former German scientists, who had been brought to the USA under Operation Paper Clip following WW II. These men were the elite of the German rocket programs controlled by Adolph Hitler. On many occasions I had the distinct privilege of speaking with Dr. Wernher von Braun, the leader of the elite group, and several other scientists who were assigned to the ABMA (Army Ballistics Missile Agency) launch crews at the Cape Canaveral launch sites. Eventually, these same men were incorporated into the new National Aeronautics and Space Administration (NASA) organization. During the periodic MFA (Manned Flight Awareness) meetings that were held at Cocoa Beach, I was able to talk freely and briefly with such scientists, particularly Dr. von Braun.

On one such occasion, he and I had taken a break and stepped out of the Cocoa Beach Ramada Inn into the back patio. I admitted that I was aware that he and his German Scientific team were located not too far from the crash site at that time. They were launching captured V-2 rockets from the White Sands Testing Range. On this night, I asked him a question concerning the Roswell Incident that caused his eyebrows to raise.

Did the Roswell Incident in fact happen, was an alien craft recovered along with alien bodies? Did you have a chance to go to the crash site?"

Dr. von Braun was a cigarette smoker and he lit one up. He thought for a second, then proceeded to talk freely about his inspection of the crashed craft.

He trusted me to hear such astonishing events because I vowed to not report it to newspapers, magazines, television, etc. I never broke that vow. Since he is deceased, and the incident happened over fifty years ago, I am now disclosing what I heard. I have a right to speak about anything — even things that, according to certain agencies, "do not exist."

Dr. von Braun explained how he and his (unnamed, for now) associates had been taken to the crash site after most of the military were pulled back. They did a quick analysis of what they found. He told me the craft did not appear to be made of metal as we know metal on earth. He said it seemed to be created from something biological, like skin. I was lost as to what he indicated, other than thinking perhaps the craft was "alive." The recovered bodies were temporarily being kept in a nearby medical tent. They were small, very frail and had large heads. Their eyes were large. Their skin was grayish and reptilian in texture. He said it looked similar to the skin texture of rattle snakes he'd seen several times at White Sands. His inspection of the debris had even him puzzled: very thin, aluminum colored, like silvery chewing gum wrappers. Very light and extremely strong. The interior of the craft was nearly bare of equipment, as if the creatures and craft were part of a single unit.

That's when I became lost in the moment. We returned to the awards ceremony, in which he participated, later bidding farewell. I went home with my head spinning from all I had heard. Keeping this quiet for many years was very difficult, especially with the temptations of having many friends and associates who believe in UFOs, ETs, etc. I never released this amazing data to Major Keyhoe and NICAP, or the public, until now. I considered my honor sacred when a vow was made.

Appendix 4

Abductees' Eerie E-Mails

IN RESPONSE TO an interview of the author in a newspaper, a host of e-mails were received from eminent scientists who encouraged my research, such as former ScO Clark McClelland, affiliated with the Space Shuttle, and Prof. Emeritus Dr. James Deardorff. Also, e-mail was received from a social worker in Canada who claimed to have been abducted. This person asked to remain anonymous for fear of losing his/her job. Not having met this person, and not possessing the proper credentials, such as those of Harvard psychiatrist Dr. John E. Mack, who has defended the veracity of abductee reports, the plausibility of this report cannot be assessed. Still, the person struck me as sincere, is not noted in any UFO report, and checks out in regard to profession and marital status. Thus, with a caveat that two e-mails were edited for clarity, they are presented below:

----Original Message----
From: ▮▮▮▮▮▮▮▮▮▮▮▮▮▮▮▮
Sent: Sunday, July 18, 2004 8:26 PM
Subject: Re: alien life forms among us...

Sir... You're right in your conclusion [that ET is here]. I believe that there are three questions that will confirm the presence of Alien visitation and abduction among abductees.

Question 1: As they are led into the proverbial examining room by the aliens, what is the shape of the examining table that they are being taken toward? [Questions 2 and 3:] What are the nature and properties of the light usually focused onto the table. And the smell of the air that surrounds them?

In answer to One: The light that encompasses the examining table comes from the ceiling and stops abruptly on top of the table. It does not beam to the floor. As one is led toward this light one cannot see *through* it. But when one is on the examining table, one can see *out* of the light.

I will attempt to draw the shape of the examining table...

I__L__I

) (

The L is the light focused on the table, the marks ") (" represent the base, and "I_____I" is the light.

The air is pure oxygen, heavy with ozone. I am not a man of science regarding the air formulation but that is the impression that I get.

Perhaps during your research if any of your clients have the same answers, there is only one conclusion — they have been abducted. How else would they know what the interior of the craft looks like especially the properties of the light (unable to see through it, but once you are in it you can see out).

I have only read one book on abduction that was [Harvard psychiatrist] John Mack's first book because I do not want to corrupt my memories. Thank you for your time. This information is for you. I hope, perhaps, that it helps in your endeavors.

Regards.

PS: You can't make anyone believe in UFOs, Aliens or Abductions. If a real disc was placed on display and people were officially told it was real, they would still deny the reality. I have said enough. Thank you for taking the time to read this. You are the only person other than my spouse that I have told this to.

----Original Message----
From: ███████████████████████
Sent: Tuesday, 20 Jul 2004 01:11:28-0400
Subject: Re: alien life...

Details —

As I walked toward the table, it was strange how the light sort of stopped at the table. I made note of its shape, but it was as if I could not turn my head to the left or right... sort of like I was being drawn to the table — the closer I got the more perplexed I was about not being able to see through the light. Arriving at the table, I turned 180 degrees with my back to it. It was about 8 inches higher than my tail bone. I know this since my back was against it, and at this time I had severe back problems.

When I turned with my back to the table, I saw straight ahead through the entrance where I had come into the room. To my right there was a Grey with his back to me working on a control panel that was a little higher than his waist. After about 30 seconds this Grey was joined by a second. They turned to each other as if they were conversing. At that instant I placed the heels of my hands on the table's edge to hoist myself up.

The first thing I noticed was that the table's top was like a *fluid*. Around the edge was a rounded lip. As I went through the light that was coming from the ceiling around the edge, I could actually feel the light. It wasn't warm, it just felt *heavy*. Sitting on the edge, my hands and butt sank about 3 inches into the table's top. The top *shimmered*. I am not sure why, whether it was the light on it or the makeup of the top itself.

Then, the second Grey walked over to me. I was not scared at all. In my mind this Grey told me to lie down on the table. Funny thing, *her* voice in my head was distinctly that of a *female*, sort of a *relaxing, calm friendly voice*.[1] Before lying, I asked her why I was here. She said I was special and that they wanted to examine my knee. I replied that I had a severe back problem and they should look at that instead.

I was flat on my back, by this time, on the table. Its top seemed like a pearlescent liquid that I floated in. I remember looking up at the light overhead... I could not see where it was coming from, only that it *stopped* around the edge of the table — as if I was sandwiched between four walls of light. As I lay there, I looked out of the light to my right and saw another Grey working on another control panel.

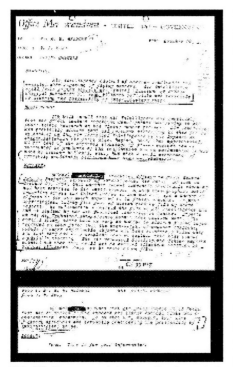

This US Government document, released *via* the FOIA, reports that the Executive Officer to the Intelligence Director stated: "another recent extremely creditable sighting had been reported to Air Intelligence." A Navy photographer took 35 ft of motion-picture film of preternatural flying machines. "Experts at the Air Technical Intelligence Center have advised that, after careful study, there were as many as twelve to sixteen flying objects... that the possibility of weather balloons, clouds or other explainable objects has been completely ruled out." Given such reports of UFOs by intelligence experts themselves, it seems reasonable to suppose that UFOs are not craft from this planet, that they are piloted by alien beings and that they might sometimes abduct humans.

This made a total of 3 Greys in this area. As I look to my left, the second Grey was standing by my waist. I looked at its face, and then immediately looked to my right where the wall slid sideways. A Tall Grey dressed in a cassock *floated* from the door over to the table.[2] I sensed that this Grey meant business. As he came to the edge of the table, the table parted between my legs from the base of my heel to my crotch. My right leg was lowered and my left leg was level, allowing him access to the right side of my left knee.

I was amazed by how the table parted and, looking down toward my knee, noticed to my right that the Tall Grey was opening what seemed to be a drawer at the base of the table. He removed an instrument with a glowing tip. The tip's light was yellow and from that light there was a central shaft about the thickness of a pencil, although longer, that terminated in a handle.

Fear overtook me as I tried to move away from the Tall Being, seeking to slide across the table to get as far away as possible from that instrument. But I could not budge — it was like I was glued to the table. Then, the smaller Grey, who had been at my waist, walked up to my head and placed her hand on my forehead. A calm came over me as I distinctly heard a voice in my head telling me that I would not feel any pain.

Looking down, I could see the Tall Grey insert the instrument into my knee — I could see the lighted end disappear into it, but could not feel any pain. Not at any time did the Tall Grey look at me. He just focused on the job at hand. When he finished, he put the tool into the draw and the table where my right leg was lowered came up parallel to the other. Then, this Tall Grey did a 180 and *floated across the room*, out the door he had come in through. The Grey who was working on the control panel against the wall did not even look at that alien as he left. He just focused on his work on the panel.

At this time the other grey came back to the table and told me in my mind to get off the table. I replaced my pajama bottoms and headed toward the opening that I came in through initially. At the opening, the two Small Greys that had greeted me when I came into the craft took me by the hand, one on each side, and led me down a circular corridor.

The hands of these Greys felt like cool neoprene. The corridor was a dark grey color and the floor we walked on seemed like it was backlit from below by a green phosphorescence that flowed from beneath the floor up the side of the walls. When we reached a specific spot, the Greys let go of my hands and I was surrounded by a white bluish light. It was so bright that I could not see out of it, and I found myself in my bedroom at home!

I tried to wake my spouse, but to no avail. As I was driving my spouse to work in the morning, I mentioned what had happened. Remembering the operation, I put my foot on the dash and pulled up my pant leg. Sure enough, there was a 3/4-inch crescent shaped scar. It was red like new skin that is healing. Now it is white, but still hurts when I press it.

This incident occurred 15 years ago. The inside of the alien room was definitely round, about 50 feet in diameter. The table was in the centre of the room. The room was illuminated by a greenish gray phosphorescence, broken up by the different colors of the control panels that were around the walls. The controls glowed predominantly red and yellow. This is all I remember. One final thing: My distinct impression was that *that* would be the last time I would be abducted. Perhaps, though, that was wishful thinking.

Regards

Notes

[1] Should these female traits be dismissed *a priori* as anthropocentric or sexist? Given a principle of homogeneity in modern physics in which the cosmos is generally uniform in composition and structure, and that a structured order has been underestimated (as one government document states in regard to UFOs), are there generally uniform evolutions in the cosmos that render relatively uniform a biological difference of males and females? In this case, females would have different biological traits that make a difference psychologically. And this psychological difference would explain the feminine behavior described by many

abductees. Is all behavior largely a matter of *nurture* in terms of its being a mere social construct, as alleged by many gender feminists in universities? Universities have often allowed science to be trumped by politics. Should political ideology or science be used to consider abductee claims? Might *nature*, not *nurture,* be the central factor for explaining the behavior of ET as well as that of humans?

[2] To paraphrase another e-mail in regard to this alien: The Small Greys do not harm us, they are the 'menials'. The harm is done by the Six-Foot Greys. They have pronounced facial features and wear silvery champagne colored cassocks. These extend from the neck to the floor in a somewhat A shape, trimmed at the collar with gold around the edges of the sleeves. They float across the floor rather than walk. "THESE are the ones that have you here for a purpose. And [they] go about their work with extreme efficiency... unemotional and focused."

Appendix 5

Extraterrestrial Biological Entity: Autopsy Report

IN A MESSAGE to the author, Harry Jordan wrote:

Professor Trundle,

"You have my permission to use the drawings and autopsy reports in your book *(documents not authored by me)*. Please credit the originals as being sent/released to me by Mr. John Leer, son of the Leer Corporation founder to use as I wish, so under that qualification I release them to you."

Mr. Jordan notes about the autopsy data (and all other original materials), that they "are kept in a safe location for future reference and to serve as a public information source whenever physical proof is required to reinforce conclusions. No original materials are ever released to the media nor to provide intelligence to any government agency for purposes of financial gain or other compensations."

Reports and photos posted [on www.creightonprep. creighton.edu/ depts/finearts/Architecture/harryjordan/mufon/ufojordanfiles.htm] have 'real names' removed or replaced with John Doe's for purposes of confidentiality. Whenever possible real names are used when permission is given. All information collected was witnessed by me while working with Dr. Jack Kasher or other field investigators some who wish to remain anonymous. Dr. Kasher is a physicist who worked for NASA taught physics at the University of Nebraska at Omaha for many years and is now retired. Dr. Kasher still serves as Nebraska State Director for MUFON.

We collected physical evidence and collated data in many forms from South Dakota, North Dakota, Nebraska, Iowa, Kansas and Missouri.

One incident involved us with a crash report in Pretoria, South Africa. Dr., Jack Kasher flew to Africa at his own expense to cover the incident while I had to travel to a town in southern Iowa. NASA had an STS mission ongoing in orbit during the same period of time these events were happening on the ground.

We conducted interviews in the field over a period of 17 years with persons from age 5 up to 85 years old, collected samples on site, and gave conferences on the subject of UFO's with this data. We were directly involved with NASA's STS-48 UFO flap which concerned video tapes generated by five cameras mounted in the cargo bay of the shuttle. We heard astronauts taking about 'alien spacecraft' over shortwave radio DX-440 at 28.650 hrtz 10 meter band width. NASA astronauts forgot to turn on their scrambler vox-mike. We were subsequently interviewed by the producer Tim White of the program Sightings'. The SIGHTINGS program on which we appeared was aired on television the year of the incident. Military intelligence officers from the pentagon visited us in Omaha to inspect our Ham radio equipment and to see tapes remotely recorded using satellite downlinks from NASA's own transmissions. All of our activities were/are legal and continue to be with monitoring through other Ham friends from all over the mid-west.

All the photographs posted here are authentic and not retouched by any computer software...

I believe in an all… knowing, loving God, the Holy Spirit, the Communion of Saints and life everlasting. I do not believe in "UFOs" as though they were a religion, but I have no doubt of their existence .
NEBRASKA mufon FILES WEBMASTER
~ Harry A. Jordan *mfa/bfa*

THIS AUTOPSY DATA IS PURPORTEDLY FROM
Walter Read Army Hospital in Washington, DC.

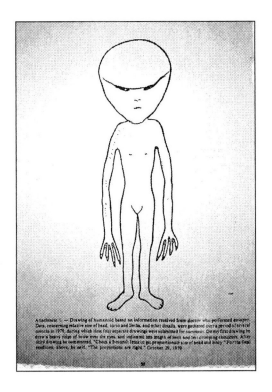

Attachment 1. — Drawing of humanoid based on information received from doctor who performed autopsy. Data, concerning relative size of head, torso and limbs, and other details, were gathered over a period of several months in 1979, during which time four separate drawings were submitted for comment. On my first drawing he drew a heavy ridge of brow over the eyes, and indicated less length of neck and less drooping shoulders. After third drawing he commented, "Check a 3-month fetus to get proportional size of head and body." For the final rendition, above, he said, "The proportions are right." October 29, 1979.

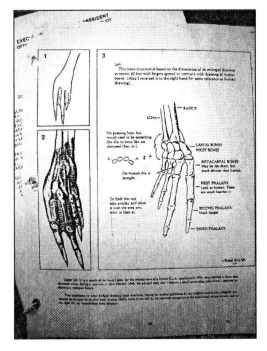

EBE Autopsy

- The approximate height of the alien humanoid is 3½ to 4½ feet tall. One source approximated 5 feet. The weight is approximately 40 lbs.

- Two round eyes without pupils. Under heavy brow ridge, eyes described variously as large, almond-shaped, elongated, sunken or deep set, far apart, slightly slanted, appearing "Oriental" or "Mongoloid."

- The head, by human standards, is large when compared with the size of the torso and limbs. "Take a look at a 5-month human fetus," I was told.

- No ear lobes or protrusive flesh extending beyond apertures on each side of head.

- Nose is vague. Two nares are indicated with only slight protuberance.

- Mouth is indicated as a small "slit" without lips, opening into a small cavity. Mouth appears not to function as a means of communications or as an orifice for food ingestion.

- Neck described as being thin; and in some instances, not being visible because of garment on that section of body.

- Most observers describe the head of the humanoids as hairless. One said that the pate showed a slight fuzz. Bodies are described as hairless.

- Small and thin fits the general description of the torso. In most instances, the body was observed wearing a metallic but flexible garment.

- Arms are described, long and thin and reaching down to the knee section.

- One type of hands has four fingers, no thumb. Two fingers appear longer than others. Some observers had seen fingernails; others without. A slight webbing effect between fingers was noted by three authoritative observers. (See Attachment 3.) Other reports indicate types with less or more than four fingers.

- Legs short and thin. Feet of one type described as having no toes. Most observers describe feet as covered. One source said foot looked like an orangutan's.

- Skin description is NOT green. Some claim beige, tan, brown, or tannish or pinkish gray and one said it looked almost "bluish gray" under deep freeze lights. In two instances, the bodies were charred to a dark brown. The texture is described as scaly or reptilian, and as stretchable, elastic or mobile over smooth muscle or skeletal tissue. No striated muscle. No perspiration, no body odor.*

- No teeth.

- No apparent reproductive organs. Perhaps atrophied by evolutionary degeneration. No genitalia. In my non-professional judgement, the absence of sexual organs suggests that some of the aliens, and perhaps all, do not reproduce as do the Homo sapiens, or that some of the bodies studied are produced perhaps by a system of cloning or other unknown means.

- To most observers the humanoids appear to be "formed out of a mold," or sharing identical facial characteristics.

- Brain and its capacity, unknown.

- Colorless liquid prevalent in body, without red cells. No lymphocytes. Not a carrier of oxygen. No food or water intake is known. No food found aboard craft in one known retrieval. No digestive system or GI tract. No intestinal or alimentary canal or rectal area described.

- More than one humanoid type. Life span unknown. Descriptive variations of anatomy may be no more diverse than those known among Earth's Homo sapiens. Other recovered alien types of human or other grotesque configurations are unknown to me. Origin unknown.

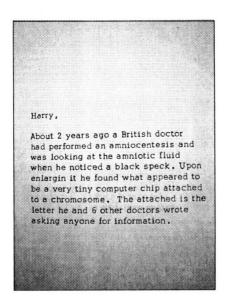

Harry,

About 2 years ago a British doctor had performed an amniocentesis and was looking at the amniotic fluid when he noticed a black speck. Upon enlargin it he found what appeared to be a very tiny computer chip attached to a chromosome. The attached is the letter he and 6 other doctors wrote asking anyone for information.

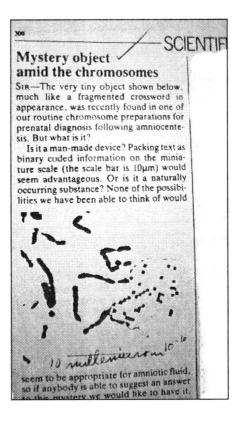

SCIENTIF

Mystery object amid the chromosomes

SIR—The very tiny object shown below, much like a fragmented crossword in appearance, was recently found in one of our routine chromosome preparations for prenatal diagnosis following amniocentesis. But what is it?

Is it a man-made device? Packing text as binary coded information on the miniature scale (the scale bar is 10µm) would seem advantageous. Or is it a naturally occurring substance? None of the possibilities we have been able to think of would

seem to be appropriate for amniotic fluid, so if anybody is able to suggest an answer to this mystery we would like to have it,

Acknowledgments

I AM INDEBTED TO former USA Air Forces Lt. Walter Haut who, on orders from his CO, was the first person in history to officially report a recovered UFO. A forced retraction of that order made him a fascinating interviewee since, despite his candor, he personifies an enigma of coverups for over a half century. And I remain grateful to the late research engineer Bob Achzehner for interviews on an evolution of thought since the golden days of UFOs in the 1950s; to former intelligence officer Maj. George Filer, USAF (Ret.); and to retired NASA scientist Dr. Richard Haines. He led the way in pioneering UFO studies as a bona fide field of scientific inquiry.

Others to whom I remain beholden include Richard H. Hall and John E. Mack, M.D. As Professor of Psychiatry at the Harvard Medical School, Pulitzer Prize winning author, and world renown investigator of alien abductions, Dr. Mack fortified my conviction that our sciences cannot adequately explain UFOs. My scholarship is not comparable to his, but he and I are kindred spirits of a sort in virtue of our grave misgivings about the current state of the academic community. And Mr. Hall's review was inspiring because of his exclamation marks next to my key points. For example, witnessed UFOs are not averted by their 'impossibility' in terms of our science. Also, Mr. Hall's support is significant in view of his being past Director of the National Investigations Committee on Aerial Phenomena (NICAP) and famed author of *The UFO Evidence*, an official report for the NICAP whose members included former CIA Director and US Navy Vice Admiral Roscoe Hillencoetter; Rear Admiral H.B. Knowles; Wilbert Smith, Chief Engineer of Canada's Project Magnet; Jack Brotzman, Electronics Physicist for the US Naval Research Laboratory; and Dr. Leslie Kaeburn, Professor of Biophysics at the University of Southern California.

Special thanks is owed to Dr. Roberto Di Quirico at the University of Pisa, Italy, for counsel in promoting my new work; Dr. Terry Pence at NK University for scholarly and editorial assistance; and Professor Emeritus Jim Hopgood. An anthropologist noted in *Marquis Who's Who in America*, Dr. Hopgood's esoteric investigations range from a cultural import of UFOs to a cult phenomenon of the late film star James Dean that won him recognition in *The Wall Street Journal*. And besides faculty and students who put paranormal news in my mail, I am beholden to my wife Janeanne for discerning better ways to both express my ideas and use illustrations.

In regard to quotations, gratitude is owed to Simon & Schuster for LTC Philip Corso's *The Day After Roswell*; Marlowe & Co. for Stanton Friedman's *Top Secret/ MAJIC*; William Morrow for Timothy Good's *Above Top Secret: The Worldwide UFO Cover-Up*, with a Foreword by former Chief of Defense Staff, Lord Hill-Norton GCB; and William Morrow for Good's *Alien Contact: Top-Secret UFO Files Revealed*, with Commentary by former Admiral of the Fleet, The Lord Hill-Norton GCB, Chief of Defense Staff, 1971-73.

And in regard to illustrations, I am above all grateful to artist and www.rense.com webmaster James Neff for his extraordinary generosity in providing his art to me gratis. His art captures a UFO enigma with the imagination of a Salvadore Dali and witness-based realism of a documentary film. Other images provided gratis are beholden to Filer's Research Institute directed by Maj. George Filer, US Air Force Intelligence (Ret.), UFO Casebook Production President B.J. Booth, LUNA's Black Project Analyst Bill Lohmeier, artist Larry Elmore at www.larryelmore.com, the esteemed artist Monarca L. Merrifield at UFO-Bible.homestead.com, and the US Government — from the Air Force to NASA. And although some images in the public domain are indebted to Project RRO Information System at www.rr0.org, open to the public and operated by Jérôme Beau, my apology extends to anyone who is not properly acknowledged. Below, credits for images are acknowledged chronologically from the cover through the chapters:

Cover:

Abduction: Unwanted Interlopers © 2004 James Neff, www.rense.com

Contents:

Photo Peter Gersten, J.D., courtesy Dr. Gersten

Photo Peter Redpath, Ph.D., courtesy Dr. Redpath

1 ET and Terrestrial Politics:

Image of Stalin, courtesy (© 2004) Funet Russian Archives, www.funet.fi/pub/culture/russian/html

Photo Dr. Bernard Haisch, courtesy Dr. Haisch. From "Weird Science," *Metro* (Nov. 4-10 Nov
 1999), photo by George Sakkestad, www.metroactive.com

Photo Dr. Neal Grossman, courtesy Dr. Grossman

Image Negative Energy Induction Ring, US Gov. NASA, Les Bossinas, www.grc.nasa.gov, 22 Jan 02

2 Science on Angels and Aliens:

Author Robert Trundle's computer enhancement of Turin Shroud, courtesy Akvilon Trident, D.
 Yurchey's "Exploring the Shroud of Turin" 12 Dec 02, www.paranormal.com

Image Lt. Corso's DA Form 66, courtesy US Gov, Dept. of the Army, Ange Nash, FOIA Officer, via
 Computer UFO Network, www.cufon.org

Image "ET" from Child Tuska, courtesy US Gov, kids.msfc.nasa.gov/gallery/Pic

Photo John Mack, M.D., courtesy Dr. Mack, displayed at www.centerchange.org

Image of book CE-5 by NASA scientist Dr. Richard Haines (Ret), courtesy Dr. Haines

Art – Elijah, courtesy (© 2004) Monarca Lynn Merrifield, ufobible.homestead.com

Photo of the Church of Ireland's W. Lendrum, courtesy (© 2004) Akvilon Trident, Beta Blk, "The Exorcist," lolik4, 19 Feb 03, www.paranormal.com

Photo of extraordinary UFO that, while taken near Fostoria, Ohio, is not atypical of bizarre phenomena seen worldwide. Courtesy of George Ritter, 2003, see www.abduct.com

Image Aliens with Child – Logo paranormal.com, courtesy (© 2004) Akvilon Trident

Photo Professor William Lane Craig, courtesy Dr. Craig

Image of book *Many Worlds* by Steven Dick *et al* courtesy Templeton Foundation Press

Photo Dr. James E. McDonald (deceased), public domain, *cf.* www.cohenufo.org

Photo Dr. Allen Hynek (deceased), public domain, *cf.* RRO Information Systems at www.rr0.org

3 Alien Anatomy and Technology

Photo George Adamski (deceased), public domain, *cf.* RRO Information System, www.rr0.org

Infrared UFO Photo released to the public by the Mexican Department of Defense

Author's computer enhanced image of Saucer, public domain and origin unknown, *cf.* Akvilon Trident, UFO Theatre, April 2003

Image on Big Bang, courtesy US Gov., NASA, liftoff.msfc.nasa.gov/academy/

Photo "Collaborative Virtual Environment," courtesy US Gov., www.sandia.gov/media/ NewsRel/ NR2000/virtual.htm

Photo Dr. Robert Sarbacher (deceased), public domain, *cf.* RRO Information System, www.rr0.org

Image *Operation Animal Mutilation* (June 1980), Funded by the Department of Justice, courtesy US Gov., *cf.* www.noveltynet.org.

Photo US Air Force Museum, courtesy US Gov., www.wpafb.af.mil/museum/info.htm

Photo Dr. Feliks Zigel (deceased), public domain, *cf.* RRO Information System, www.rr0.org

Computer Enhanced Image Alien Woman, public domain, *cf.* RRO Information System, www.rr0.org

Photo Robert Lazar, public domain, *cf.* RRO Information System, www.rr0.org

Image Saucer Section, sketch by Robert Trundle

Image Antimatter Reactor, sketch by Robert Trundle

Image from Patent #3,626,605, Method & Apparatus For Generating a Secondary Gravitational Force Field by Henry W. Wallace, courtesy US Gov., Patent Office, 14 Dec 1971, *cf.* www.aw-verlag.ch/Leckerbissen.htm#LB_Wallace_Pat3626605.

Photo President Harry Truman, courtesy US Gov., Defense Visual Information Center, http:// nfo.dodmedia.osd.mil/CGI-BIN/

Photo Radio Telescopes, courtesy US Gov., deepspace.jpl.nasa.gov

Image Pulse Propagation Using Gas, drawn by Robert Trundle, based on Diagram by NEC physicists L.J. Wang, A. Kuzmich, and A. Dogariu, www.neci.nj.nec.com

Image Worm Hole Tunnel, drawn by Robert Trundle, based on Morse & Thorne 1988, courtesy US Gov., www.grc.nasa.gov

4 Singular People and Strange Places

Photo Robert Achzehner (late 1990s), courtesy Beverly Harding of Harding Graphics and Petlovers Publications

Photo John Lear, public domain, *cf.* RRO Information System, www.rr0.org

Photo William (Mac) Brazel, public domain, *cf.* RRO Information System, www.rr0.org

Photo Men-in-Black (Pretend), courtesy US Gov., www.nasa-academy.org

Photo Congressman Hugh Addonizio, US Gov., Cong. Port. File, Library of Congress.

Photo Capt. Robert Adickes, public domain, *cf.* RRO Information System, www.rr0.org

Computer enhanced Image Adickes' UFO, *cf.* RRO Information System, www.rr0.org

Photo USAF Officers, Project Blue Book, US Gov., *cf.* www.polkawitch.com/main/ little/images/ nasa/projectbluebook.asp

Author Robert Trundle's Sketch of UFO seen by JB Brooks

Author Robert Trundle's Sketch of UFO described by Daniel Fry

Image Mechanical Device, *circa* 200 BC, From *The Growth of the Steam Engine* (NY: Appleton & Co., 1878)

Image KGB Logo, courtesy US Gov., www.nasa.gov

Photo Lord Clancarty, British Parliament, public domain, *cf.* RRO Information System, www.rr0.org

Image Dewilde Alien, courtesy Akvilon Trident, lolik4, 5 Mar 03, Alpha Block: "The meeting of Marius Dewilde - France 1954"

Photo Crashed UFO in USSR (*circa* 1940s), public domain, *cf.* www.burlingtonnews.net

Image Dept. of the Army Intelligence Report No. 2130720, on Afghanistan UFO, Declassified via FOIA, courtesy US Gov. (DoA), cf. www.bvalphaserver.com

Author Robert Trundle's Image of Fred Ayer's "Instrumentation for UFO Searches," *Scientific Study of Unidentified Flying Objects* by University of Colorado (1968), contract 44620-67-C-0035 with the US Air Force, courtesy US Gov.

Author Robert Trundle's Sketch of UFO seen by pilots John Witted and Clarence Chiles

Photocopy FBI Dallas Memo 7-8-47, declassified per FOIA, courtesy US Gov.

Photo UFO Spokane, Washington, 17 Oct 02, courtesy Maj. George Filer, USAF (Ret)

Photo Pine Gap, US Military Installation, US Gov. via Akvilon Trident, Gamma Block: "The mystery of Pine Gap secret facility," lolik4, 8 Oct 03

Photo Laguna Cartagena Refuge, US Gov., Fish & Wildlife Service, southeast.fws.gov

Image Cigar-Shaped UFO, US Gov.www.fas.org/man/dod-101/sys/ship/nssn

Image Monarca Merrifield's *Red Sea Miracle*, ©2003 www.ufosbible.homestead.com

5 Official Reports and Odd Encounters

Photo Gen. Nathan Twining and SoAF Harold Talbott, courtesy US Gov. *via* www. majesticdocuments.com

Photo US Air Force Test Dummies, courtesy US Gov., www.af.mil/lib/roswell/

Images of Twining's White Hot Intelligence Estimate, claimed US Gov. *via* Salina Cantwheel – daughter of late Lt. Col. Cantwheel. *Cf.*, and courtesy, Dr. Robert Wood's Majestic Documents, www.majesticdocuments.com

Image FBI Letterhead and Winston Churchill, US Gov., FBI Files, foia.fbi.gov/churchill

Image Plaque, USAF Office of Special Investigations, courtesy US Gov., www.wpafb. af.mil/museum

Image Logo – US Air Force Security Service, courtesy US Gov., www.raymack.com/ HomePage/ usafss.html

Photo Future Warrior, courtesy US Gov., www.sbccom.army.mil/aia.lackland.af.mil/ aia/index.cfm, and www. deploy.eustis.army.mil/Photos/AAN/AAN

Photo Saucer by Rex Heflin, courtesy US Gov., Air Force Security Service, cf. www.chez.com/lesovnis

Image "The Real Men in Black," courtesy of Contact Info Area-X-Conspiracy World

Photo Saucer UK by Sharon Rowlands, UFO, August 29, 2001, "Another Sighting!," *Matlock Mercury*, www.matlockmercury.co.uk

Photo UFO, courtesy US Gov., NASA STS-75 Mission, *cf.* B.J. Booth's UFO Casebook

Artwork Alien Survivor, courtesy © 2004 Larry Elmore, www.larryelmore.com

Photo Graduates of AF Institute of Technology, US Gov., www.afit.edu/gallery/Morgue

Image 2 Alien Kidnappers, public domain, Centro Italiano Studi Ufologici, www.ufo.it, thanks to Maurizio Verga

Image 1 Rosa Lotti and Aliens, public domain, Centro Italiano Studi Ufologici, www. ufo.it, thanks to Maurizio Verga

Photo Professor Michael Zimmerman, courtesy Dr. Zimmerman

Illustration of Rocket-Powered Experimental Plane, *circa* 1940s, by Alex S. Tremulis, courtesy US Gov., www.ufx.org/tremulis/tremulis.htm

Illustration of Aliens being Interrogated, *circa* early 1950s, by Alex Tremulis, courtesy US Gov., www.ufx.org/tremulis/tremulis.htm

Image Alien Dwgs by Suttons, public domain, *cf* B.J. Booth, www.ufocasebook.com

Image 1954-Nouatre Saucer, courtesy © 2004 B.J. Booth, ufocasebook.conforums.com

Artwork Scary Alien, courtesy © 2004 Larry Elmore, www.larryelmore.com

6 Outlandish Organizations and Intriguing Issues

Photo Coral and Jim Lorenzen, public domain via RRO Info. System, rr0.free.fr

Headlines *Los Angeles Times* (Nov. 6, 1957), courtesy *LA Times via* Frank Altomonte's "Six Saucer-Shaped' UFOs..." (18 March 2001) www.rense.com

Photo Gen. Spaatz, AF Sec. Symmington, and Gen. Vandenberg, courtesy US Gov, US Air Force photographer James Evans *per* www.majesticdocuments.com

Image Alien sun IRC+10216,, courtesy US Gov, nai.arc.nasa.gov

Photocopy Official Gov. Letter from Rep. Steven Schiff to Robert Barrow, US Gov. *via* www.presidentialufo.com

Photocopy *Government Records: Results of a Search for Records Concerning the 1947 Crash Near Roswell, New Mexico*, courtesy US Gov. *via* www.project1947.com/roswell

Image of AFOSR Logo, courtesy US Gov., www.afosr.af.mil

Photo Star Team at AFOSR, courtesy US Gov., www.pr.afrl.af.mil/97-98/ar98-1

Photo Col. Harold Watson & Team (ATIC), courtesy US Gov., www.wpafb.af.mil/naic/

Photocopy of Official Gov. Letter to John Greenewald Jr., US Gov. document *via* Mr. Greenewald, , www.blackvault.com

Photocopy Top Secret Majestic 12 Documents, US Gov. via Dr. Robert Wood at www. majesticdocuments.com

Photocopy Truman Letter 24 Sep 47, US Gov. Document (alleged), *cf.* www.realufos.com/artbellhoax

Photo Dr. Lloyd Berkner, courtesy US Gov. *via* Nat. Acad. Sci., www.nap.edu

Photo Dr. Detlev W. Bronk et al, courtesy US Gov. *via* National Library of Medicine, profiles.nlm.nih.gov

Photo Dr, Vannevar Bush, courtesy Jenny O'Neil, Library Admin., MIT Museum, www.eecs.mit.edu/AY95-96

Photo SoD James Forrestal & Pres. Harry Truman, courtesy US Gov, public ceremony *via* RRO Information System, www.rr0.org

Photo Gordon Gray, courtesy US Gov. *via* Truman Lib, Nat Arch. Admin, www. trumanlibrary.org

Photo Adm. Roscoe HillenKoetter, courtesy US Gov. *via* www.cia.gov/cia/publications

Photo Dr. Jerome Hunsaker, courtesy US Gov. *via* Nat Acad Sci, http://stills.nap.edu

Photo Dr. Donald Menzel et al, courtesy US Gov. *via* UCAR/NCAR, www.ucar.edu/
communications/staffnotes

Photo Gen. Robert M. Montague, courtesy US Gov. *via* www.majesticdocuments.com

Photo Adm. Sidney Souers, US Gov. *via* National Archives, www. archives.gov

Photocopy Memo of Adm. Souers, US Gov. *via* www.cia.gov/csi/books/venona

Photo Gen. Nathan Twining, courtesy US Gov. and *Kansas City Star* 1 Nov 44 & 450th BG Assoc,
www.450thbg.com

Photo Gen. Hoyt S. Vandenberg, US Gov. *via* National Archives, www.archives.gov

Photo UFO-like Probe, courtesy US Gov., Roswell Report, www.af.mil/lib/roswell

Image ET in Container, courtesy Akvilon Trident: More Evidence of Roswell Coverup? lolik4, 2 July 03

Photo Nano-technological Microchain, courtesy US Gov, Sandia National Laboratoies, http://
www.sandia.gov/media/NewsRel/NR2002/chain.htm

Photocopy FBI Memo 22 March 1950, courtesy US Gov., released by the Freedom of Information
Act (FOIA) foia.fbi.gov/ufo/ufo8.pdf, p.34

Image Flying Saucer, courtesy US Gov., aether.lbl.gov/www/projects/neutrino/UFO.html

Photo Pope John Paul II and Pres. George Bush, courtesy US Gov., www.whitehouse. gov/
president/europe

Photo Crescent-Shaped UFO 7 July 1947, courtesy US Gov, US Air Force Project Blue Book, via
www.think-aboutit.com

Sketch Triangular UFOs, courtesy Peter Gersten, J.D., Citizens Against UFO Secrecy (CAUS),
Attachment Five' for a Freedom of Information Act (FOIA) request, 5 USC § 552., submitted to
the US Department of Defense, www.caus.org/img/dod02-18.GIF

Photo Multi-Colored UFO, courtesy Maj. George Filer, Filer's Research Institute 16 July 2003,
www.nationalufocenter.com

UFO Outside NASA Space Shuttle, courtesy US Gov., *cf.* images.jsc.nasa.gov/... earth/ STS001/
lowres

Sketch by H.G. Wells, courtesy US Gov., vesuvius.jsc.nasa.gov/er/seh/daysSep3

Image NASA Plaque, courtesy US Gov., grin.hq.nasa

Photo Black Hole at Core of Galaxy M87, courtesy US Gov., maxim.gsfc.nasa.gov/docs

Diagram of Alpha Centauri by author Robert Trundle, *via* K. Croswell's "Does Alpha Centauri have
intelligent life?," *Astronomy* 19 (1991), No. 4, 28–37, and homepage. sunrise.ch/homepage/
schatzer/Alpha-Centauri

Image Warp-Speed Cockpit, US gov, www.grc.nasa.gov/www /pao/html/warp/warpstat

Image Alcubierre Warp Drive Bubble, courtesy Dr. John Cramer, Dept. of Physics, University of
Washington, www.npl.washington.edu/AV/altvw81.html

Image Rear View Mirror, courtesy US Gov. (Dept. Transportation), safety.fhwa.dot. gov/
fourthlevel/10/rroberts.htm

Image Face-Optical Illusion, courtesy US Gov., www.niehs.nih.gov/kids/illusions

Image Cylindrical UFO, courtesy RUFORS (Nikolay Subbotin) *via* www.rense.com/ ufo/
tianshan.htm and www.psu.ru

7 Military Projects and Moon Missions

Sketch for Proj. Grudge of Torpedo-like UFO by John Whitted, courtesy US Gov. (DoD) via
www.ufx.org/wcs/chileswhitted.htm

Photocopy News Story on Pilots John Whitted and C.S. Chiles, courtesy Atlanta Journal (July 25, 1948) via UFX: Technology, Culture, Cold War, UFOs at www.ufx.org/wcs

Photo Green Fireball, courtesy US Gov (Air Force – Proj. Grudge/Blue Book) via www.ufx.org/gfb/chron.htm

Photocopy Project Grudge Memo 10-11 Sep 1951, courtesy US Gov. (Air Force) via NICAP, www.nicap.dabsol.co.uk

Photo Hotel of Bilderberg, RRO Information Systems, www. rr0.org

Photocopy CIA Recruiting Poster, courtesy US Gov. (CIA)

Photo UFO Bluebook Case 9666, courtesy US Gov., Natl. Admin. for Recs. & Archives (NARA) *via* www.chez.com

Photo UFO Bluebook Case 9666 (Blowup), courtesy US Gov., Natl. Admin. for Recs. & Archives (NARA) *via* www.chez.com

Photocopy US Air Force Project Blue Book Special Report 14, courtesy US Gov (AF)

Photo Truman & Collier Robert, US Gov., history.nasa.gov/SP-4219/ Chapter2.html

Photo Mars Rock, courtesy US Gov.(NASA/JPL/MSSS) *via* Filer's Research Institute

Image Thomas Paine, courtesy US Gov., Natl Arch. Recs Admin., www. archives.gov

Image 18th-C. UFO Sighting, courtesy Maj. George Filer, Filer's Research Institute, www.nationalufocenter.com

Sketch of UFO sighted by Betty Cash, by author Robert Trundle

Photo Nuclear-Powered Rocket Engine (XECF, courtesy US Gov., grin.hq.nasa.gov

Photocopy News Headline on Mutilations and UFOs, courtesy US Gov., foia.fbi.gov/

Sketch by Betty Cash of UFO, courtesy US Gov., USAF (FOIA) *via* www.cufon.org

USAF Form 95 for Betty Cash, courtesy US Gov., (FOIA) *via* www.cufon.org

Photo Tank for Ion-Engine Test, courtesy US Gov., grin.hq.nasa

Photo President John F. Kennedy, courtesy US Gov., grin.hq.nasa

Image NASA Space Station, courtesy US Gov., grin.hq.nasa.gov

Image NASA Aerover Blimp at Titan, courtesy US Gov., www.jpl.nasa.gov/technology

Photo Blimp at Edwards AFB, courtesy © 2004 Todd Fluhr, www.win.net/~brother/fat/edwards.html

DARPA Document, courtesy US Gov (FOIA) *via* Todd Fluhr, www.win. net/~brother/

Photo Moon Phobos, courtesy US Gov., observe.arc.nasa.

Photo UFO at Nellis Test Range, courtesy US Gov. *via* he AnomaliesNetwork, www. anomalies.net/ufo/gov/s30/

Photo Saucer, US gov. (USM Air Group) *via* www.rense.com/general39/ tm_aug.htm

8 Since the Missions: Controversy over Religion and Science

Logo Physics Dept., US Air Force Academy, courtesy US Gov., www.usafa.af.mil

Image Dark Earth Data, courtesy US Gov., heasarc.gsfc.nasa.gov/docs/asca/rbm_cont

Image The Book of Dzyan, Miskatonic University Archives, A Chaosium Book, Edited and Introduced by Tim Maroney, www.chaosium.com

Photocopy News Headline: 'UFO' on Nasa camera, US Gov. soho.nascom.nasa.gov/hotshots2003_01_17

Image Alien Face (Typ.), courtesy US Gov., stardust.jpl.nasa.gov/classroom/kids

Photo Dummy Body Bag, courtesy US Gov. (Air Force), *via* www.cnn.com/TECH/ 9706/24/ufo.presser/gallery1.html

Painting of Zamora Incident, courtesy B.J. Booth, UFOCasebook.com

Photo UFOs at Salem MA by US Navy Seaman Shell Alpert, courtesy US Gov. (Air Force Blue
Book *via* www.chez.com

Photo Tunguska Effect, courtesy US Gov., observe.arc.nasa.gov/nasa/exhibits/craters/impact_tunguska

Photo Nazi Victim by Sgt. E.R. Allen, courtesy US Gov., www.archives.gov/research

Photo USSR Guard Tower, courtesy US Gov. Archives *via* cidc.library.cornell.edu/dof/

Poster on Genocide, courtesy Poster – US Gov. (Department of State) www.state.gov

Photo UFO, courtesy © 2003 George Ritter at Fostoria, Ohio, via Maj. George Filer's Filer's
Research Institute

Photocopy Letter in FBI File, titled "The New Project Blue Book," foia.fbi.gov

Photo Voyager Record, courtesy US Gov., voyager.jpl.nasa.gov/mission/mission

Diagram Fuzzy Logic, courtesy US Gov., sbir.gsfc.nasa.gov/SBIR

Image Big Bang, courtesy US Gov., universe.gsfc.nasa.gov

Photocopy Declaration of Independence, courtesy US Gov., Library of Congress via www.loc.gov

Photo UFO, Scarborough, NY, 10 Oct 1966, Blue Book Case 11073, courtesy US. Gov. (Air Force)
via www.temporaldoorway.com

Photo E Rock on Mars, courtesy US Gov. NASA/JPL via www.nationalufocenter.com

Photo and Blowup of Washington D.C. UFO, courtesy US Gov. (Air Force Blue Book) *via*
users.ev1.net/~seektress/censor

NASA Planet Finder, courtesy US Gov., www.jpl.nasa.gov/missions/proposed

Appendix I: Bernard Haisch's Open Letter to His Colleagues

Photo of Astronomer Bernard Haisch, courtesy of Dr. Haisch, www.ufoskeptic.org

Appendix II: Press Release – "Spread of Life Through Galaxy"

Public Archive Press Release, Armagh Observatory: "Spread of Life Through Galaxy" by John
McFarland on 31 Oct 2003, http://seti.sentry.net/archive/public/2003 and http:// panspermia.org/
napierblurb.pdf

Appendix III: A Space-Craft Operator's Stargate Chronicles

Photo Dr. Von Braun US Gov., images.ksc.nasa.gov/photos

Photo Clark McClelland, www.stargate-chronicles/background.htm

Appendix IV: Abductees' Eerie E-Mails

Image "Beam" © 2005 and courtesy James Neff, www. rense.com

Photocopy Office Memorandum, courtesy US Gov., *via* FOAI

Appendix V:

Images (all) www.creightonprep. creighton.edu/depts/finearts/Architecture/harryjordan/mufon/
ufojordanfiles.htm

Bibliography

Alcubierre, Miguel (Dept. Physics & Astronomy, Univ. College of Cardiff, UK). "The Warp Drive: Hyperfast Travel Within General Relativity," *Class. Quant. Grav.* 11 (1994) and *gr-qc/0009013* (2000).

Ambrose, Stephen E. *Eisenhower: The President*. Vol. Two. NY: Simon & Schuster, 1984.

Ansell-Pearson, Keith. "Toward the *Übermensch.*" *Nietzsche-Studien*, Vol. 23 (1994), 123-145.

Aquinas, St. Thomas. *Summa Theologica*. In *A Summa of the Summa*. Ed. by Peter Kreeft. San Francisco: SIgnatius Press, 1990.

Aristotle. *Generation of Animals*. From Richard McKeon, Ed., *The Basic Works of Aristotle*. NY: Random House, by arrangement with The Oxford University Press, 1941.

Armagh Observatory and Cardiff University Astronomers. "Spread of Life throughout the Galaxy," *Press Release* 31 October 2003.

Augustine, St. *On Free Choice of the Will*. Tr. by A. S. Benjamin and L. H. Hackstaff. NY: Macmillan Publishing Co., 1985.

Benokraitis, Nijole and John Macionis, Eds. *Seeing Ourselves: Classic, Contemporary, and Cross-Cultural Readings in Sociology*. NJ: Prentice-Hall, 2001.

Bento, C. and Bertolami, O. *Classical and Quantum Gravity*. NY: World Scientific Publishing, 1993.

Bohm, David. *Wholeness & the Implicate Order*. NY: Routledge, 1996.

Bradbury, Robert. "Life at the Limits of Physical Laws," *The Search for Extraterrestrial Intelligence in Optical Spectrum III – SPIE Proceedings of the 2nd Int. Conf. on Optical SETI*, Vol. 4273-03, 2001.

Bragg, Desmond and Joslin, Paul. *Science Meets the UFO Enigma*. NY: Nova

Kroshka Books, 2000.

Brody, J. "Notions of Beauty Transcend Culture." Interview of neuropsychologist Dr. N. Etcoff, Dr. D. Perrett, and Dr. J. Langlois, *Themes of the Times*. NY: Prentice Hall, The New York Times, 1994.

Burk, A. E., Ed. *Political Handbook of the World*. NY: CSA Publication, State University of New York, 1993.

Carlotto, M.J. "Imagery Analysis of Unusual Martian Surface Features," *Applied Optics* 27 (1998).

Carpenter, D. Maj. and Lt. Col. E. Therkelson. *Introductory Space Science, Vol. 11*. Col Springs, CO: US Air Force Academy, 1968.

Chalmers, Alan. *Science and Its Fabrication*. Minneapolis: University of Minnesota Press, 1990.

Cohen, Jack and Stewart, Ian. *What Does a Martian Look Like? The Science of Extraterrestrial Life*. NY: John Wiley & Sons, 2002.

Cooper, William *Behold a Pale Horse*. Sedona, AZ: Light Technology Publications, 1991.

Corso, LTC Philip J. (Former Chief US Army Foreign Technology). *The Day After Roswell*. NY: Pocket Books, Simon & Schuster, 1997.

Courtois, Stephane and A. Paczkowski et al. *Black Book of Communism: Crimes, Terror, Repression*. Tr. M. Krammer and J. Murphy. Cambridge: Harvard University Press, 2000.

Craig, William Lane. "Must the Beginning of the Universe Have a Personal Cause?: A Rejoinder," *Faith and Philosophy*, Vol. 19, 2002.

Craig, William Lane. "The Caused Beginning of the Universe: a Response to Quentin Smith." *British Journal for the Philosophy of Science*, Vol. 44 (1993): 623-639.

Crews, F. "The Mindsnatchers." *The New York Review of Books*. Vol. XLV, No. 11, June 25, 1998, pp. 18, 19.

D'arc, Joan. *Space Travelers & Genesis of the Human Form: Evidence for Intelligent Contact in the Solar System*. NY: Book Tree, 2000.

Darling, David. *Life Everywhere: The Maverick Science of Astrobiology*. NY: Basic Books, 2002.

Darnay, A. J., Ed. *Manufacturing USA: Industry, Analyses, Statistics, and Leading Companies*. 4th Ed. Vol. 2. London: Gale Res., 1994.

Dean, Jodi. *Aliens in America*. NY: Cornell University Press, 1998.

Delgado, P. and Andrews, C. *Circular Evidence*. Grand Rapids, MI: Phanes Press, 1989.

Deutsch, David and Michael Lockwood. "The Quantum Mechanics of Time Travel."

Scientific American. Vol. 270, March 1994, pp. 68-74.

Devereux, Paul and Peter Brookesmith. *UFOs and Ufology: The First 50 Years.* NY: Facts on File, 1997.

Dick, Steven J., ed. (Astronomer, US Naval Observatory). *Many Worlds: The New Universe, Extraterrestrial Life, and the Theological Implications.* Templeton Foundation Press, 2000.

Dolan, Richard M and Vallée, Jacques F. *UFOs and the National Security State: Chronology of a Coverup, 1941-1973.* NY: Hampton Roads Pub. Co., 2002.

Erdmann, Erika and David Stover. *A Mind for Tomorrow: Facts, Values, and the Future.* Westport, CT: Praeger Publishers, 2000.

Evans, Hilary. "Where Do We File 'Flying Saucers'? The Archivist and the Uncertainty Principle," *Journal of Scientific Discovery,* Vol. 15, No. 2, 2001, article 8.

Feinberg, G. and Shapiro, R. *Life Beyond Earth.* NY: Morrow, 1980.

Filer, Maj. George (US Air Force Intelligence, Ret.). "The Filer Report." From *UFOs are Real: Here's the Proof,* by E. Walters and B. Maccabee, Ph.D. NY: Avon Books, 1997.

Flandern, Thomas Van (former Chief Celestial Mechanics, US Naval Observatory). "The Speed of Gravity – Repeal of the Speed Limit," *Physics Meta Research www.metaresearch.org,* 2004

Flandern, Thomas Van (former Chief Celestial Mechanics, US Naval Observatory). "The Top 30 Problems with the Big bang," *Apeiron,* Vol. 9, No. 2, 2002, 72-90.

Flandern, Thomas Van (former Chief Celestial Mechanics, US Naval Observatory). "New Evidence of Artificiality at Cydonia on Mars," *Physics MetaRes Bull.* 6 (1991).

Fox, Matthew and Sheldrake, Rupert. *The Physics of Angels: Exploring the Realm Where Science & Spirit Meet.* SanFrancisco: Harper, 1996.

Friedman, Stan (nuclear physicist, former gov. clearance). *Top Secret/ MAJIC.* Foreworded by Whitley Strieber. NY: Marlow & Co., 1996.

Fry, Daniel. *Steps to the Stars.* Lakemont, Georgia: CSA, 1965

Gato-Rivera, Beatriz (Instituto de Matemáticas y Física Fundamental, CSIC, Serrano, Madrid, Spain), "Brane Worlds, the Subanthropic Principle & Undetectability Conjecture [*per* Enrico Fermie's 'Missing-Alien' Thesis]," *arXiv:Physics/0308078,* v3, 17 Jan 2004, pp. 1-11.

Gillmor, D.S., Ed. *Final Report of the Scientific Study of Unidentified Flying Objects*: Univ. of Colorado and the US Air Force. NY: Bantam Books, 1969.

Goldberg, Jonah. "X Marked the Spot." *National Review.* Vol. LII, No. 12, 2000, pp. 27-28.

Good, Timothy. *Unearthly Disclosure*, Forewarded by Admiral of the Fleet The Lord Hill-Norton, Former Chief of the Defence Staff and Chair of the NATO Military Committee. NY: Random House, 2000.

Good, Timothy. *Alien Base: The Evidence for Extraterrestrial Colonization of Earth.* London: Arrow Books Ltd., 1999.

Good, Timothy. *Alien Contact: Top-Secret UFO Files Revealed.* With Commentary by Admiral Lord Hill-Norton G.C.B., Chief of Defense Staff. NY: William Morrow, 1993.

Good, Timothy. *Above Top Secret: The Worldwide UFO Cover-Up.* With a Foreword by the Former Chief of Defense Staff, Lord Hill-Norton, G.C.B. NY: William Morrow, 1988.

Gotz, I. L. *The Culture of Sexism.* Westport, CT: Praeger Pub., 1999.

Grinspoon, David. *Lonely Planets: The Natural Philosophy of Alien Life.* NY: HarperCollins, 2003.

Greer, Steven, M.D., *Disclosure: Military and Government Witnesses Reveal the Greatest Secrets in Modern History.* NY: Crossing Points Publications, 2001.

Grossman, Neal (Dept. Philosophy, University of Illinois). "On Materialism as Science Dogma," *ufoskeptic.org*, 2004.

Guerin, P. "Are the Reasons for the Cover-Up Solely Scientific?" *FSR.* Vol. 28, 1982, pp. 2-6.

Haines, Richard. *Advanced Aerial Devices Reported During the Korean War.* CA: LDA Press, 1990.

Haines, Richard. *CE-5: Close Encounters of the Fifth Kind – 242 Case Files Exposing Alien Contact.* Naperville, IL: Sourcebooks, 1999.

Haisch, Bernard (Director, California Institute for Physics & Astrophysics). "UFOs and Mainstream Science," *The MUFON UFO Journal* No. 335, 1996, pp. 14-16.

Haisch, Bernard, A. Rueda, L.J. Nickisch, & J. Mollere. "Update on an Electromagnetic Basis for Inertia, Gravitation, The Principle of Equivalence, Spin & Particle Mass Ratios," *Amer. Inst. Physics Conf. Proc., Space Technology & Applications International Forum* (STAIF-2003), Ed. Mohamed S. El-Genk, pp. 922-931, *gr-gc/0209016* (2003).

Haisch, Bernard and Rueda, A. "Toward an Interstellar Mission: Zeroing in on the Zero-Point-Field and the Einstein-De Broglie Formula," *Physics Letters A*, 268, 224 (2000).

Hamilton III, William F. *Cosmic Top Secret: America's Secret UFO Program – New Evidence.* NY: Inner Light Publications, 2002.

Hall, Richard H., *UFO Evidence: Vol. I.* Barnes & Noble Books, 1997.

Haney, D. "Mind Found Fertile Soil for False Memories." Interview of psychologist Dr. E. Loftus. *Associated Press*. Feb.16, 1997.

Hanley, C. "Spotlight: Pacific Bombing Fallout." *Associated Press*. July 14, 1996. From The Cincinnati Enquirer, p. A.

Hanley, Richard. *The Metaphysics of Star Trek*. NY: BasicBooks, 1997.

Hauke, Manfred. *Women in the Priesthood? A Systematic Analysis in the Light of the Order of Creation and Redemption*. Tr. by David Kipp. San Francisco: Ignatius Press, 1988.

Heelan, Patrick. "Quantum Mechanics & Objectivity." *Problem of Scientific Realism*. Ed. E. MacKinnon. NY: Appleton-Century Crofts, 1972.

Heidegger, Martin. *An Introduction to Metaphysics*. Tr. by R. Manheim. NY: Doubleday Publishers, 1961.

Hesemann, M. and P. Mantle. *Beyond Roswell: The Alien Autopsy Film, Area 51, & the US Government Coverup of UFOs*. With a Foreword by Jesse Marcel, M.D. (NY: Marlowe & Co., 1997).

Hitchcock, Helen. Ed. *The Politics of Prayer: Feminist Language and the Worship of God*. San Francisco: Ignatius Press, 1992.

Holt, A. *Field Resonance Propulsion Concept*. Houston, TX: NASA, Lyndon B. Johnson Space Center, 1979.

Hoover, Richard B. and Rozanov, Alexei Yu. "Evidence for Biomarkers and Microfossils in Ancient Rocks and Meteorites," *The Search for Extraterrestrial Intelligence in the Optical Spectrum III – SPIE Proceedings of the 2nd Int. Conf. on Optical SETI*, Vol. 4273-35, 2001.

Hopkins, Bud. *Intruders: The Incredible Visitations at Copely Woods*. NY: Random House, 1987.

Horowitz, Dr. Irving Louis. *Sociology & the Common Culture*. Retirement dinner. Sponsored by the Dept of Sociology, Rutgers University, at the Eagleton Institute of Politics, May 14, 1998.

Hough, P. and Jenny Randles. *Looking for the Aliens*. NY: Barnes & Noble Books, 1991.

Howe, Linda. "1989: The Harvest Continues." *The UFO Report 1991*. Ed. by T. Good. London: Sidgwick and Jackson, 1990.

Howe Linda. *An Alien Harvest*. Huntingdon Valley, Pennsylvania: Linda Moulton Howe Productions, PO Box 538, 19006, 1989.

Hynek, J. Allen. *The UFO Experience: A Scientific Inquiry*. Chicago: Henry Regnery Co., 1972.

Jacobs, David. *Secret Life: Firsthand Documented Accounts of UFO Abductions*.

Foreword by John E. Mack, M.D. Professor of Psychiatry, Harvard Medical School. New York: Simon & Schuster, 1992.

Jahn, Robert G. (Professor Aerospace Science & Dean Emeritus of the Mech.– Aerospace Engineering Dept., Princeton University). "Introduction of the Dinsdale Award Lecture, William G. Roll," *Journal of Scientific Discovery*, Vol. 17, No. 1, 2003, article 4.

Jahn, Robert G. and Dunne, Brenda. "A Modular Model of Mind/ Matter Manifestations (M5)," *Journal of Scientific Discovery*, Vol. 15, No. 3, 2001, article 1.

Jahn, G., J. Mischo, D. Vaitl et al. "Mind/Machine Interaction Consortium: PortREG Replication Experiments," *Journal of Scientific Discovery*, Vol. 14, No. 4, 2000, article 1.

Jaki, Fr. Stanley L. (Ph.D. Physics, Ph.D. Th., Distinguished University Professor Princeton, N.J., Seton Hall University "The Last Word in Physics." *Philosophy in Science* V 1993, pp. 9-32.

Jean, Clinton. *Behind the Eurocentric Veils: The Search for African Realities.* Amherst, MA: The University of Massachusetts Press, 1992.

Kant, Immanuel. *Groundwork of the Metaphysics of Morals.* NY: Harper & Row, 1948.

Kenny, Anthony Ed. *The Oxford History of Western Philosophy.* Oxford: Oxford University Press, 1994.

Kheel, Marti "Ecofeminism and Deep Ecology." *Covenant for a New Creation.* Ed. by C.S. Robb & C.J. Casebolt. NY: Orbis Books, 1991.

King, Moray B. *Quest for Zero Point Energy Engineering Principles for Free Energy.* NY: Adventures Unlimited Press, 2002.

Kitchener, R. "Toward a Critical Philosophy of Science." *New Horizons in the Philosophy of Science.* Ed. by D. Lamb. London: Ashgate Publishing Co., Avebury, 1992.

Korner, Stephan. *Experience and Theory.* London: Routledge & Kegan Paul, 1969.

Kragh, Helge. *Quantum Generations: A History of Physics in the Twentieth Century.* NJ: Princeton University Press, 1999.

Kuhn, Thomas. "2nd Thoughts on Paradigms." *Structure of Scientific Theories.* Ed. F. Suppe. Chicago: University of Illinois Press, 1977.

Kuhn, Thomas *The Structure of Scientific Revolutions.* Chicago: The University of Chicago Press, 1969.

Kuznik, F. "The Must-Win Mission." Interview of Astronaut S. Musgrave, M.D. *USA Weekend.* November 26-28, 1993, p. 6.

Bibliography

LaFollette, H. and N. Shanks. "Chaos Theory: Analogical Reasoning in Biomedical Research." *Idealistic Studies: An Interdisciplinary Journal of Philosophy*, Vol. 24 (Fall 1994) 241-254.

Larkins, Lisette. *Calling on Extraterrestrials: 11 Steps to Inviting Your Own UFO Encounters*. NY: Hampton Roads Pub. Co., 2003.

Laudan, Larry. *Science and Relativism: Some Key Controversies in the Philosophy of Science*. Chicago: University of Chicago Press, 1990.

Lear, John (former CIA member of Lear-Jet family fame). "The UFO Cover-Up." *The International UFO Congress*. Tucson, Arizona, 1991.

Letter, David L. "The Pathology of Organized Skepticism," *Journal of Scientific Discovery*, Vol. 16, No. 1, 2002, article 7.

Lewis, C.S. (Theologian, University of Cambridge). *Christian Reflections*. Grand Rapids, MI: W. B. Eerdsmans Publishing Co., 1971.

Lewontin, R. *Biology As Ideology: The Doctrine of DNA*. NY: Harper Collins Publishers, 1998.

Mack, John E., M.D. (Professor of Psychiatry, Harvard Medical School). *Passport to the Cosmos: Human Transformation and Alien Encounters*. NY: Crown Publishing Group, 1999.

Mack, John E., M.D. *Abduction: Human Encounters with Aliens*. NY: Charles Scribner's Sons, 1994.

Malcolm, Norman (Dept. Philosophy, Cornell University). *Ludwig Wittgenstein: A Memoir*. NY: Oxford University Press, 1984.

McErlean, Jennifer, Ed.. *Philosophies of Science: From Foundations to Contemporary Issues*. Belmont, CA: Wadsworth/Thomson, 2000.

Menzel, Donald (Chair, Dept. of Astronomy at Harvard and suspected member of MJ-12). "UFOs: The Modern Myth." *UFOs – A Scientific Debate*. Ed. C. Sagan & T. Page. NY: Cornell University Press, 1975.

Merleau-Ponty, Merleau. *Phenomenology of Perception*. London: Routledge & Kegan Paul, 1978.

Millis, Marc G. (Space Propulsion Tech. Div. NASA Lewis Res.Ctr.). "Emerging Possibilities for Space Propulsion Breakthroughs," *Interstellar Propulsion Society Newsletter*, Vol. 1, No. 1, 1995.

Monk, Ray. *Ludwig Wittgenstein: The Duty of Genius*. London: Penguin Books, Ltd, 1990).

Monoaghan, P. "Encounters With Aliens: Before a Tough Audience, A Psychiatry Professor Defends His Research on 'Experiences'." *Chronicle of Higher Education*, July 6, 1994, A10, A16.

293

Moran, Mary, Ed. *What You Should Know About Women's Lib: A Christian Approach to Problems of Today*. Conn: Keats Pub. 1974.

Nakamura, Hiroshi. "Video Analysis of an Anomalous Image Filmed During Apollo 16," *Journal of Scientific Discovery*, Vol. 17, No. 3, 2003, article 2.

Newton-Smith, W.H. (Dept. Philosophy, Oxford University). *The Rationality of Science*. London: Routledge & Kegan Paul, 1981.

Nolt, John. *Logics*. Belmont, CA: Wadsworth/Thomson, 1997.

O'leary, B.T. "The 'Face' on Mars & Possible Intelligent Origin," *J. Brit. Interplan. Soc.* 32 (1990).

O'dell, Dale. *UFO's 2004 Calender: Official Photo-Documented Images*. NY: Willow Creek Press, 2004.

Olum, Ken (Inst. Cosmology, Dept. Physics & Astronomy, Tufts University). "Conflict Between Anthropic Reasoning and Observation," *Physics arXiv: gr-gc/0303070*, v2, 4 Feb 2004, pp. 1-7.

Paglia, Camille. "A Call for Lustiness: Just Say No to the Sex Police." *Time*, March 23, 1998, p. 54

Patai, Daphne and Koertge, Noretta. *Professing Feminism: Education and Indoctrination in Women's Studies*. NY: Lexington Books, 2003,

Petroski, H. "The Draper [Engineering] Prize." *American Scientist* Vol. 82, March-April 1994, pp. 114-117.

Pine, Ronald. *Science and the Human Prospect*. Belmont, CA: Wadsworth Publishing Co., 1989.

Plato. *Republic*. From *Plato: The Republic*, Tr. by Paul Shorey. Cambridge, Mass. and London: Loeb Classical Library, 1953 & 1956.

Pope, Nick (Air Staff Secretariat, British MoD). *Open Skies, Closed Minds: For the First Time a Government UFO Expert Speaks Out*. NY: Dell Publishing, 1998.

Popper, Sir Karl. *Realism & the Aim of Science*. Totowa, NJ: Roman & Littlefield Publishers, 1983.

Ragl, A. J. S. "Inside the Military UFO Underground." *Omni*. Vol. 16, April 1994, pp. 50-54.

Randle, Kevin and David Schmitt. *UFO Crash at Roswell*. NY: Avon Books, 1991.

Ratzinger, J., Cardinal Ed., *Catechism of the Catholic Church*. Liguori, MO: Liguori Publications, 1994.

Redfern, Nick. *Strange Secrets: Real Government Files on the Unknown*. NY: Pocket Books, 2003.

Rescher, Nicholas "Where Wise Men Fear to Tread." *American Philosophical*

Quarterly, Vol. 27, July 1990, p. 259.

Richards, Cara. "Problems Reporting Anomalous Observations in Anthropology," *Journal of Scientific Discovery*, Vol. 17, No. 1, 2003, article 1.

Saler, Benson. "Explanation, Conspiracy, and Extraterrestrials: The Roswell Incident." *93rd Annual Meeting of the American Anthropological Association*, Dec. 4, 1994, Atlanta, Georgia.

Schaaffs, Wilhelm. *Theology, Physics, and Miracles*. Tr. by R. L. Renfield. Washington DC: Canon Press, 1974.

Schick, T. and Vaughn, L. *How to Think About Weird Things*. Mountain View, CA: Mayfield Publishing Co., 1995.

Schuessler, J. "Cash-Landrum Case Closed." *MUFON UFO Journal*. No. 222, Oct. 1986, pp.12-17.

Sheffield, Derek. *UFO: A Deadly Concealment. The Official Cover-Up?* London: A Blanford Book, 1996.

Sheldrake, Rupert. *The Rebirth of Nature*. Rochester, VT: Inner Traditions International Lmt., 1994.

Shermer, Michael. *How We Believe: The Search for God in an Age of Science*. NY: WH Freeman & Co., 2000.

Shermer, Michael and Gould, Stephen Jay. *Why People Believe Weird Things.* 2nd ed. WH Freeman & Co., 2002.

Shor, R. E. "The Fundamental Problem Viewed From Historic Perspective." *Hypnosis: Research and Perspectives*. Ed. by E. Fromm and R. E. Shor. Chicago: Aldine-Atherton, 1972.

Sommers, Christina. *The War Against Boys*. NY: Simon & Schuster, 2000.

Sommers, Christina. "Argumentum Ad Feminam." *Journal of Social Philosophy*, Vol. 22, 1991, pp. 5-19.

Spencer, J. *The UFO Encyclopedia*. NY: Avon Books, 1991.

Stein, M. and Carlotto, "A Method for Searching for Artificial Objects on Planetary Surfaces," *J. Brit. Interplan. Soc.* 43 (1990).

Sturrock, Peter (Stanford University Professor of Space Science & Astrophysics in Applied Physics Dept.). *The UFO Enigma: A New Review of the Physical Evidence*. NY: Aspect, 2000.

Sturrock, Peter (Stanford University Professor of Space Science & Astrophysics in Applied Physics Dept.) Physical Evidence Related to UFO Reports," *The Proceedings of a Workshop Held at the Pocantico Conference center, Terrytown, New York*, 1997.

Szpir, Michael (Dept. Physics, University of Wales). "Spacetime Hyper-surfing?"

American Scientist – The Sigma Xi Scientific Research Society. Vol. 82, 1994, pp. 422-423.

Talbot, Michael. *The Holographic Universe*. NY: HarperTrade, 1992.

Tipler, Frank. *The Physics of Immortality: Modern Cosmology, God & the Resurrection of the Dead*. NY: Anchor Books, 1994.

Truman, President Harry S. *Memoirs by Harry S. Truman: Years of Decisions. Vol. I*. New York: Doubleday & Co., 1955.

Trundle, Robert and Filer, Maj. George (US Air Force Intelligence, Ret). *Illustrated News of the Unbelievable: Paradigm Reports of Alien Worlds for the 21st Century*. NC: Granite Publishing, Ltd., 2004.

Trundle, Robert and Vossmeyer, Michael, M.D. "Inferring Prescriptions from Medical Descriptions," *Aquinas*, Vol. 47, 2004, pp. 1–14.

Trundle, Robert. "A First Cause and the Causal Principle: How the Principle Binds Theology to Science." *Philosophy in Science X*, edited by William Stoeger, S.J., Ph.D. Astrophysics, Cambridge University (Kraków Papal Academy of Theology, Vatican Astronomical Observatory & University of Arizona), December 2003, 7-35.

Trundle, Robert. "Philosophy and Controversy Over UFOs," *Filosofia: International Journal of Philosophy*, Vol. 30, No. 1, 2001, pp. 39-57.

Trundle, Robert. *UFOs: Politics, God & Science – Philosophy on a Taboo Topic*, Forewords by John E. Mack, M.D., NASA's Dr. Richard Haines, & Maj. George Filer. Florence, IT: European Press, 2001.

Trundle, Robert. *From Physics to Politics: The Metaphysical Foundations of Modern Philosophy*. With a Foreword by Peter Redpath, St. John's University. 2nd Ed. Piscataway, NJ: Transaction Publishers – Rutgers, The State University, 2001.

Trundle, Robert. *Medieval Modal Logic & Science*. With a Foreword by Dr. David Lamb, Medical School, Univ. of Birmingham, England, and by J. Roland Ramirez, Ph.D., Institut Catholique de Paris. Lanham, MD: Rowman & Littlefield Publishing, 1999.

Trundle, Robert. "From Experience to Modal Scientific Reasoning." Invited by Professor of Physics Maria Falbo-Kenkel for the *The Scientific Research Society of Sigma Xi*, Oct. 8, 1997.

Trundle, Robert. "Has Global Ethnic Conflict Superseded Cold War Ideology?" *Studies in Conflict & Terrorism*, RAND Corp., Washington D.C., Vol. 19, No. 1, 1996, pp. 93-107.

Trundle, Robert. "Cold War Ideology: An Apologetics for Global Conflict." *Res Publica*, Leuven University, Institute of Political Science, Belgium, Vol. 37, 1996, pp. 61-84.

Trundle, Robert. "Thomas' 2nd Way: A Defense by Modal Scientific Reasoning." *Logique et Analyse: The Belgium National Center for Research in Logic.* Vol. 146, 1994, pp. 145-168.

Trundle, Robert. *Ancient Greek Philosophy: Its Relevance to Our Time.* Avebury Series in Philosophy. London: Ashgate Publishing, 1994.

Trundle, Robert. "Extraterrestrial Intelligence and UFOs." *Method & Science: Journal for Empirical Study of the Foundations of Science,* The Netherlands, Vol. 27, No. 2, 1994, 73-79.

Trundle, Robert. "Religious Belief and Scientific Weltanschauungen." *Laval Théologique et Philosophique.* Vol. 45, 1989, 405-422.

Trundle, Robert. "Philosophical Reflections at the American Bicentennial." *National Forum.* Vol. 57, 1975, pp. 3-9.

US Government. "Addonizio, H.J." *Biographical Directory Of The American Congress 1794-1961.* Wash. DC: US Gov Print. Off., 1961.

Vallée, Jaques (former Principle Investigator for the National Science Foundation Project for Computer Networking). *Confrontations: A Scientist's Search for Alien Contact.* NY: Ballantine Books, 1990.

Waller, B. N. *Critical Thinking.* NY: Prentice Hall, 1998.

Walters, Edward and Bruce Maccabee, Ph.D. *UFOs are Real: Here's the Proof.* With an Introduction by Maj. George Filer, US Air Force Intelligence Officer (Ret.). NY: Avon Books, 1997.

Washington Post. "Report Urges Study of UFO Sightings." From *The Cincinnati Enquirer,* June 29, 1998, p. A2.

Webb, Stephen. *If the Universe Is Teeming with Aliens... Where Is Everybody? Fifty Solutions to Fermi's Paradox and the Problem of Extraterrestrial Life.* NY: Copernicus Books, 2002.

Welch, J. RAL Director, University of California at Berkeley. "Water Discovered In Another Galaxy." *Assoc. Press,* June 3, 1994, p. A6.

Whipple, F. "Introduction." *The UFO Enigma,* by Dr. D. H. Menzel and Dr. E. H. Taves. NY: Doubleday & Co., 1977.

Wickramasinghe, Chandra, M. Wainright, and J. Narlikar. "SARS Virus Introduced by Cosmic Dust?," Letter, *The Lancet: International Journal of Medical Science & Practice,* Vol. 361 (May 2003), p. 1832.

Wickramasinghe, Chandra and Hoyle, Fred. "The Unity of Cosmic Life and the Inevitability of Evolved Life Forms," *The Search for Extraterrestrial Intelligence in the Optical Spectrum III – SPIE Proceedings of the 2nd Int. Conf. on Optical SETI,* Vol. 4273-01, 2001.

Wittgenstein, Ludwig *On Certainty*. Ed. by G. E. M. Anscombe and G. H. von Wright. Tr. by D. Paul and G. E. M. Anscombe. London: Harper & Row, 1969.

Wright, S. *UFO Headquarters: Investigations on Current Extraterrestrial Activity*. NY: St. Martin's Press, 1999.

Yavorsky, B. and Yu Seleznev. *Physics: A Refresher Course*. Tr. from Russian by G. Leib with contributions by N. Boguslavskaya. Moscow: MIR Publishers, 1979.

Youssef, Saul. "Is Quantum Physics An Exotic Probability Theory?," *Fundamental Problems in Quantum Theory*. A Conference Held in Honor of Professor John A. Wheeler. Ed. by Dr. D. Greenberger and Dr. A. Zeilinger. NY: Annals of the NY Academy of Sciences, 1995.

Zimmerman, Michael (Tulane University Professor of Philosophy). "The 'Alien Abduction' Phenomenon: Forbidden Knowledge of Hidden Events." *Philosophy Today* Vol. 41, 1997, pp. 235-253.

Index of Names and Subjects

Heavenly Lights

The Apparitions of Fátima and the UFO Phenomenon

Dr. Joaquim Fernandes & Fina d'Armada

Translated from Portuguese and Edited by
Andrew D. Basiago & Eva M. Thompson

Foreword by Dr. Jacques F. Vallée

A meticulous synthesis of history, anthropology, and science, **Heavenly Lights** establishes that the Fátima incident of 1917 involved not "Marian" apparitions, as is conventionally understood, but rather a series of close encounters with alien beings. It does this by subjecting all of the available pertinent facts about Fátima to a sweeping evidentiary analysis that is at once very thorough and highly readable. In periodic digressions entitled "Parallels in UFOlogy," the authors draw many relevant connections between the facts of the Fátima incident and similar events in UFO history that have occurred at different times and places around the world. The text, which is rich in historical detail, speculative insight, and paranormal allure, is illustrated with dozens of photographs and drawings that depict what actually occurred during the many events associated with the enigmatic Fátima incident.

ISBN 0-973534-3-3

www.eccenova.com

Religious & Scientific Non-Fiction

www.eccenova.com

Printed in the United States
47344LVS00003B/28

9 780973 534122